T0407946

# The Law of Nature: Spherical Quantity in Dynamic Calculus

# THE LAW OF NATURE: SPHERICAL QUANTITY IN DYNAMIC CALCULUS

KUNMING XU

**Nova Science Publishers, Inc.**
*New York*

For permission to use material from this book please contact us:
Telephone 631-231-7269; Fax 631-231-8175
Web Site: http://www.novapublishers.com



**LIBRARY OF CONGRESS CATALOGING-IN-PUBLICATION DATA**

Xu, Kunming.
The law of nature : spherical quantity in dynamic calculus / Kunming Xu.
p. cm.
Includes index.
ISBN 978-1-61668-345-0 (hardcover)
1. Dynamics. 2. Atomic structure--Mathematical models. I. Title.
QA851.X85 2009
530.15--dc22

2009054181

*Published by Nova Science Publishers, Inc. † New York*

# Contents

# PREFACE

Publish or perish! For the past fifteen years, I have been considering how to express mathematically the motion of electrons within an atom as well as the life of organisms on the earth. This book is the release of the long time effort. Basically, people tend to describe the material world by linear methods, such as relating three axes of Cartesian coordinates to three dimensions of space, and a one-dimensional ray to time flow, but entities from a single atom, a cell, an organism to the earth itself show otherwise in spherical shape. The nature of a sphere is rotation and cyclic oscillation, distinguished from linear motion. Starting from the only assumption of harmonic oscillation, this thesis characterizes the atomic structure, molecular configuration, and cellular model in spherical view that features spherical quantities in dynamic calculus, alternative to physical quantities in linear algebra.

By decomposing a spherical layer into points, lines, surfaces, and solids orthogonally, this new book establishes a four-dimensional space theory for explaining harmonic oscillation of electrons within inert atoms. In particular, electronic orbitals of 2s2p within a neon atom are defined in calculus, trigonometry, and geometry rigorously. Both 1s electrons are dimensionally encapsulated in neon shell by a mechanism reverse to differential chain rule and integral chain rule. The geometric shapes of 2s2p electrons account for the molecular configurations and reaction mechanisms of methane, ethylene, ethyne, benzene, alkyl halides, and mono-substituted benzenes satisfactorily, better than orbital hybridization. Anisotropic 2s2p orbitals give rise to carbon chirality. Due to the selective electron transfer property of chiral carbons in deoxyriboses, a DNA molecule is characterized as stepwise *LC* oscillatory circuits that transfer charges from one base pair to another, in agreement with experimental evidence. Charge transfer through DNA double helix follows harmonic oscillation.

All physical quantities are ordered into a periodic table according to their space and time dimensions. The natural law that governs the relationship between electrons within atoms and the relationship between cellular organelles alike boils down to dynamic calculus of spherical quantities. Dynamic calculus is a new variant of calculus, which can be expressed as a sine or cosine operation on a rotating radian angle. A spherical quantity is a quantity that assumes a fixed magnitude and a continuous varying dimension, complementary to physical quantities. It represents true physical existence that transforms sinusoidally relative to uniform Newtonian time. Spherical quantities in dynamic calculus comply with Pythagorean theorem and general Stokes's theorem, observe Maxwell's equations, and characterize the rhythms of life essentially.

So what has this book mainly accomplished? It has broken new ground in calculus, trigonometry, geometry, electromagnetism, relativity, organic chemistry, biology, and Chinese medicine, etc. Specifically, for the first time it has

1. Introduced the concept of spherical quantity in complement to physical quantity; and listed all physical quantities in a periodic table according to their dimensions.
2. Invented dynamic calculus that is conceptually different from infinitesimal calculus; and defined it as the operational rule of spherical quantity. Spherical quantity in dynamic calculus is regarded as the law of nature.
3. Discovered four-dimensional spacetime structure in neon shell; and characterized 2s2p electron octet by calculus, trigonometry, and geometry in a consistent manner. This is the core of this book. Anisotropic 2s2p orbitals in carbon atoms account for the properties of chemical bonds in many organic molecules better than orbital hybridization.
4. Discovered dimension encapsulation mechanism in reverse order to differential chain rule and integral chain rule.
5. Discovered the role of chiral carbons in selective electron transfer; and depicted a DNA molecule as stepwise *LC* oscillatory circuits.
6. Characterized life phenomena by definite wave functions and rope structure; and explained the probability density of biological variables by trigonometric function.
7. Proved in rigorous mathematics the core concepts of traditional Chinese medicine and philosophy, such as yin yang theory and five element theory; and hence established the inherent connections between Eastern and Western sciences.

This book is intended for professional researchers as well as the general public who are interested in fundamental science. Although the topics discussed are quite broad and conceptually rich, minimum requirements are demanded of readers for this book. Most of the major topics are developed from scratch. The background assumed is that usually obtained in the freshman to sophomore mathematics and science sequence. Most of the ideas are original and unconventional, so readers are encouraged to keep an open mind in studying the creation. However, one should not expect to completely understand the contents during the first round of skimming. Some theories are more complicated to grasp than they first seem and require careful reading with enormous patience. Space imagination and idea abstraction are exploited in many cases. Undoubtedly, creating a new science is a daunting task. I write this book not for profit, but under a strong feeling that it is my life mission to introduce the law of nature into the realm of science. Never can I give up this responsibility.

This book is divided into two parts. The first part develops the concepts and theory of spherical quantity in dynamic calculus in four-dimensional space through the description of the atomic structure. The second part is the application of these concepts and theory to molecules, cells, and organisms. The law of nature proves very powerful in elucidating molecular orders and biological patterns. In the appendices, I give some illuminative questions for each chapter, provide short explanations for key concepts, and recapitulate the law of nature in mathematical formulae. Indeed, until spherical quantity in dynamic calculus has been fully realized, the controlling mechanism underneath the life phenomena will remain a mystery.

Kunming Xu
Xiamen University, China
October, 2009

# Part I. The Atomic Structure

Scientific research in its various branches seeks not only to record particular occurrences in the world of our experience, but also to discover the general laws hidden behind the apparent phenomena. Although the laws are independent of the observers, people watching a specific event from various perspectives or in different contexts usually arrive at different depths of understanding and explanations. Some accounts are more fundamental than the others. Newton's perception of a falling apple under gravity was more fundamental than other people's casual observations. This treatise describes my thrilling discovery of spherical quantity in dynamic calculus, which is so wonderful and fascinating that I believe it reflects nature on its utmost fundamental level.

The theoretical basis of this research lies in the novel conception of time and space. In Euclidean space, it is difficult to understand a system with more than three dimensions of space. When describing the motion of an object, we tend to treat the object as a point particle and record its position in X, Y, and Z directions at every instant of time. Such a three-dimensional curve with time is known as the world-line of the point particle. This linear algebraic approach does solve a lot of classical mechanics problems, but it assumes uniformity of time and hence is not feasible to characterize the oscillation of electrons where multiple levels of sinusoidal time and space are involved. Instead of adopting physical quantity in Euclidean space, we introduce spherical quantity in quaternity spacetime. A spherical quantity assumes a fixed magnitude and a continuous varying dimension, in complement to a physical quantity. A spherical quantity characterizes electronic motion by dimension shift instead of position movement. This nonlinear approach turns out to be extremely powerful in describing electronic orbitals within atoms.

The first part of this book explains the basic rationale of spherical quantity in dynamic calculus through the description of electronic orbitals within inert atoms. The first chapter introduces harmonic oscillation and space and time symmetry in the two-dimensional sphere of a helium atom. Time and space are dynamically represented by two electrons. The second chapter provides a geometrical model of four-dimensional spacetime with electronic octet in neon shell as a material instance. The theory of quaternity regards space and time as inherent physical properties that cannot be divorced from physical objects themselves. In the atomic sphere, three orientations of X, Y, and Z do not mean three dimensions automatically, but rather each dimension is associated with a real electron. So the description of electrons naturally constitutes an illustration of canonical spacetime properties. As add-on material, the third chapter characterizes spherical quantities in two-dimensional helium and four-dimensional neon in terms of vector calculus. It is helpful to understand electronic motion by dynamic calculus. The fourth chapter unifies quantum mechanics and special relativity under quaternity perspective, and hence demonstrates the great capacity of the new science. By the principle of rotatory operation, the fifth chapter formulates a wave function for every electron in various inert atoms and arranges all chemical elements into a quaternity periodic table. The table provides sharp insight into the complicated relationship between various electrons within a large atom. These five chapters establish quaternity spacetime step by step in rigorous mathematical logic. They outline the atomic structure in an elegant and consistent manner.

Modern physicists and chemists are too embarrassed to confess that they don't know the structure of a helium atom, the simplest molecule in existence, not to mention the organization of the world in general. Something fundamental must have been missing from the traditional sciences, which prevents them from making a breakthrough in understanding the world order. Here we shall start by plumbing the basis of the most successful theory in physics, namely classical mechanics, to find out what has been missing. Curiously, every concept is developed at the expense of losing generality unconsciously. New knowledge of alternative perception or complementary view will help us understand the natural pattern better. Open to us is the jungly frontier of fundamental science. So please fasten your seat belt and enjoy an exotic view as we drive through a vast, unbroken stretch of wild territory.

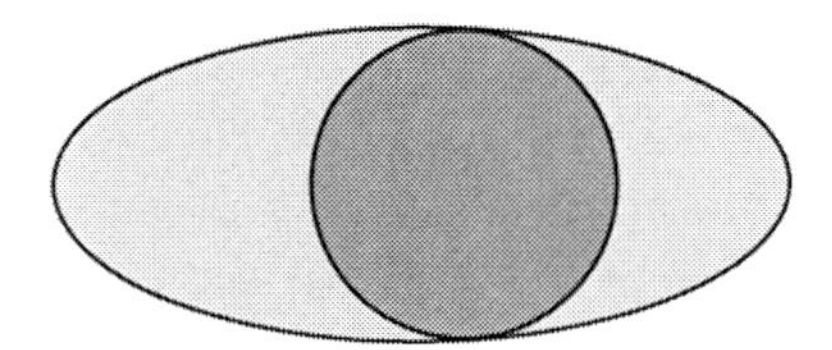

Motif 1. Circle, sphere, convolution, pupil, portal, and perception.

*Chapter 1*

# SPHERICAL QUANTITY IN TWO-DIMENSIONAL SPACETIME

We begin by discussing some of the concepts fundamental to classical mechanics. The first concept on which to base all physics is physical quantity. Classical mechanics is the study of physical quantities and is framed by the conception of them. The second fundamental concept is three-dimensional linear space with one-dimensional uniform time. Linear space is normally modeled by Cartesian coordinate system, and Cartesian space that satisfies the properties of scalar product operation on pairs of vectors is termed Euclidean space. Physical quantities and Euclidean space are the basic assumptions, either explicit or implicit, of classical mechanics. They impart it both strengths and weaknesses in describing the natural phenomena. The strengths are well known, but the weaknesses of the assumptions are hardly recognized. The purpose of this chapter is to dig out the roots of the weaknesses and introduce the concept of spherical quantity in sinusoidal form to complement physical quantity in linear space. From a fresh perspective, we view both electrons within a helium atom as space and time dimensions of a two-dimensional system and characterize their harmonic oscillations by dynamic calculus. We shall establish an atomic spacetime theory from scratch.

## 1.1. INTRODUCTION TO SPHERICAL QUANTITY

A physical quantity has a measurable magnitude and a definite dimension, the magnitude being expressed by a certain number and the dimension by certain units. If we adopt one variable to account for the measurable number and another variable to account for the dimension units, we soon discover that while the number variable could be either a continuous variable or a discrete one depending on the physical quantity of concern, the dimension variable is a discrete variable. A continuous variable can assume an infinite number of values between any two fixed points at least theoretically. For example, the water level in a dam may rise from 1.15m to 1.16m high continuously in an hour. A discrete variable has only certain fixed values, with no intermediate values possible in between. For example, the units of velocity and acceleration are m/s and $m/s^2$, respectively, but no physical quantity has a unit of $m/s^{1.16}$. For all physical quantities, the dimension variables take some discrete units composed of seven base units, say meter, kilogram, second, ampere, kelvin, mole, and candela. For a specific physical quantity, the dimension variable is assumed to be unchangeable or static, in a fixed unit.

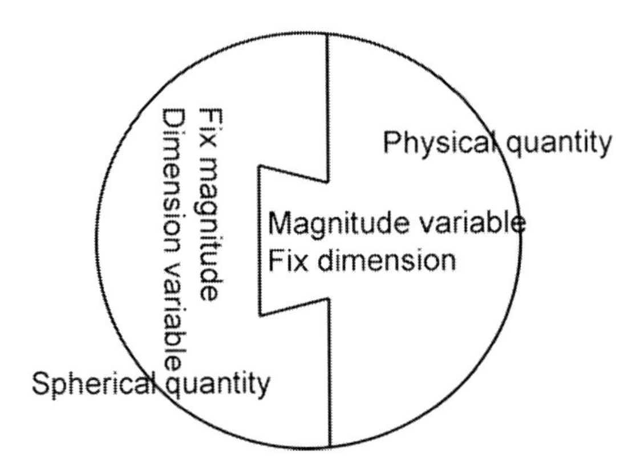


Figure 1.1. Complementary properties of spherical quantity and physical quantity.

The basic assumption of variable magnitude and invariable dimension underlying a physical quantity, which is often taken for granted in physics, has shaped our mind in the direction of linear algebra that we can model any events by a certain combination of some physical quantities. Yet these physical quantities are discrete quantities in terms of their dimensions. Any combinations of finite discrete quantities cannot describe the evolution of an event in a continuous manner, inevitably leaving holes in spacetime without being covered properly (please see review question 1.5). Based on the recognition of this limitation, we propose the concept of spherical quantity as an alternative to physical quantity. Switching the properties of both variables for a physical quantity the other way around, we get a spherical quantity, one quantity that assumes a fixed magnitude and a continuous varying dimension (Figure 1.1).

The concept of spherical quantity relies on the atomic spacetime that will be established shortly. Only when two physical quantities are inter-converted smoothly in certain situations is it rational to introduce a spherical quantity with a continuous varying dimension between the dimensions of both physical quantities. This is the case in electromagnetic waves where a changing electric field may produce a changing magnetic field, and vice versa. The dimensions of electric field strength and magnetic field strength differ by a velocity factor, which may be ascribed to the interval between space and time dimensions. We shall define these two basic dimensions within helium shell in rigorous mathematics. But before that, it is helpful for us to examine the conventional space and time concepts, their implicit assumptions, and limitations associated with the assumptions.

## 1.2. Linear Spacetime and Limitations

Since antiquity, human beings have believed that they know about space and time because of their direct experience, but scientific conception of space and time turns out to be elusive. While their meanings seem intuitively clear, attempts to define them encounter remarkable difficulty. This is not because space and time are so complex but they are so fundamental that there is not any preceding rule for reference. Like the stage of an act, the theme of a song, or the context of a paragraph, space and time are the foundation and background of all sciences. Their importance can never be overestimated. How we define space and time determines our perspective and standpoint in making observations of the world around us. It goes without saying that any successful theories must handle the basic concepts properly. Significant examples are Newtonian mechanics and Einstein's relativity.

The ancient Egyptians learnt to measure lands and constructed pyramids thousands of years ago. By Newton's time, Cartesian coordinate system had been set up to account for the three-dimensional space and people had been pretty good at elementary Euclidean geometry. But what is about time? Is it one-dimensional? Isaac Newton thought about this and defined that "Absolute, true and mathematical time, of itself, and from its own nature, flows equably, without relation to anything external." Given such a postulate, he developed three laws of motion that form the core of classical mechanics. Simple as it was, yet it has been the most practical and fruitful interpretation of time ever since.

Space and time had customarily been considered to be separate physical quantities until the 1900s when Minkowski proposed a four-axis spacetime continuum, in which a time dimension is coupled together with three space dimensions through events. Minkowski's four-axis coordinates became the framework for Einstein's special theory of relativity, which explains that space is relative, and time is relative, too. They are relative in the sense that when an object travels at a high speed, its time dilates and length contracts compared with those of rest objects. Such an unusual conclusion is demonstrated by Lorentz's transformations connecting an inertial reference frame and another frame moving at a constant velocity relative to it. In general theory of relativity, Albert Einstein further proposed that time curves in the space around stars and other massive objects and came up with an equation relating the curvature tensor of the distance function to the distribution of matter and energy in spacetime.

In spite of the progress, modern conception of space and time is framed by Euclidean geometry and Newtonian time. Since Euclidean geometry does not incorporate time axis into three-dimensional Cartesian space, time remains isolated from space. Einstein's relativity does not surpass this framework either. In both special and general relativities, timelike and spacelike dimensions are clearly distinguishable. Time is assumed to be one-dimensional and hence is not considered to be the counterpart of or in symmetry to space. Moreover, the three-dimensional space abstracted by Cartesian coordinate system is a linear vector space, dissociated from real entities that are spherical in nature; and the recording of uniform time by the ticking of a mechanical clock is an idealized counter, dissociated from circadian organisms, the natural subjects of time sensing and recording. The detachment of space and time and the mechanical abstractions of them must have limited their applicability in reality. For example, the earth is a solid sphere, positioning a rectangular coordinate system on it is cumbersome. And the rotation and revolution of the earth are more important in determining four seasons than any accurate mechanical clock. Although spherical polar coordinate system is available, we have not really mastered its essence yet.

The applicability of classical mechanics based on Euclidean geometry and Newtonian time is a good tester of conventional space and time concept. Classical mechanics characterizes the motion of objects on the ground to a certain precision, but it breaks down when applied to a large astronomical distance. For example, objects are no longer traveling in a straight line as predicted by Newton's first law of motion; instead geodesics represent the paths of freely falling objects in a given cosmic space. Relativity takes over in explaining the discrepancy. This indicates that the universe is curved or spherical that defies linear characterization.

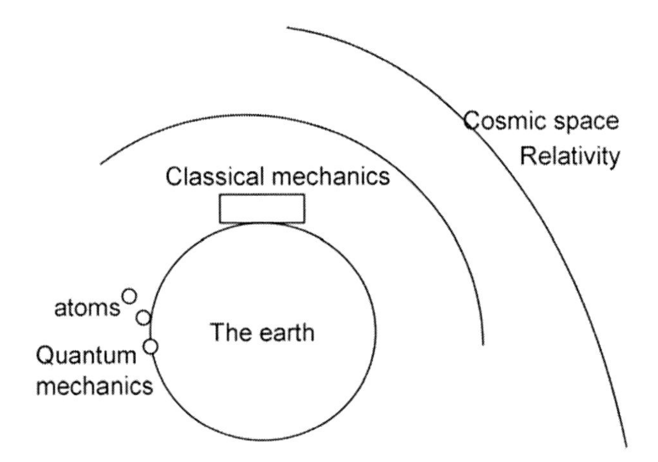


Figure 1.2. Between the scales of atoms and cosmic space, classical mechanics is only applicable in the narrow crevice.

Turning our focus to the microscopic world, the principle of uncertainty precludes the application of Newton's laws either. Quantum mechanics has to be called for to study the behavior of electrons. Schrödinger's equation, the basis of quantum mechanics, could not have been solved without transforming the wave function from Cartesian coordinates into spherical polar coordinates. Why? It is because hydrogen atom is spherical in spacetime that defies linear characterization. If so, why don't we start with a wave equation formulated by spherical polar coordinates instead of one by Cartesian coordinates? This is exactly what we shall consider in this treatise. From macrocosm to microcosm, the result of the validity test indicates that classical mechanics barely survives in a narrow crevice between both ends (Figure 1.2).

Using rectangles to measure the lands on the earth is only an approximation to the curved surface of the globe. The approach fails when the scale of the rectangles is larger than millions of miles or smaller than millimeters.

By the same manner, physical quantities cannot describe the spherical nature of atoms and the universe as a whole, to which quantum mechanics and relativity should be applied, respectively.

The limitation of linear sciences in the spherical world is conceivable. Unfortunately, modern physics does not relate the cosmic spacetime curvature to the atomic spacetime on a different scale. While general relativity furnishes tensors to account for spacetime curvature, quantum mechanics adopts statistical probability to describe electronic orbitals.

Departing from classical mechanics in the contrary directions, both fields remain detached and to be unified. Since space and time are the most fundamental physical quantities, the limitations of classical mechanics and the wanting unification of relativity and quantum mechanics prompt us to search for new space and time concepts on a more profound level and in a wider scope.

## 1.3. TWO-DIMENSIONAL SPACETIME

If in the atomic microcosm or in the universe, space and time on certain scales are nonlinear, given the validity test of classical mechanics has indicated so, then we need to be cautious on any presumptions that we have inadvertently introduced after Euclidean geometry or Newtonian time. Let's discard every antecedent belief and premise except saying that space and time are a pair of fundamental spherical quantities. Our common sense is inclined to support two first-degree spherical quantities, which may refer to up and down, positive and negative, yin and yang, male and female, on and off, yes and no, etc. Many entities, such as a plant leaf, a cashew nut, a human body, and a planet, exhibit two symmetric sides or two contrary poles. Technically, to describe a certain property, one must establish a reference first, e.g., up position is relative to down, north is opposite to south, and male is in contrast to female. Hence at least two quantities are needed as fundamental spherical quantities to explain an event. Using one quantity as a reference, the other as a ruler, both spherical quantities can sufficiently measure everything in the world. We have seen that binary numbers can actually express everything that a computer or a robot does, the capabilities of which appear unlimited. Given space and time quantities, any third quantity could be derived or computed from them through certain combinations or transformations. Hence it is reasonable to assume that there are exactly two fundamental or basic spherical quantities, namely space and time. From this reasoning, the definitions of space and time are relative, i.e., if the first one is designated as space, then the second one as time, or the other way around, even though we tend to associate space with volume and time with flow. Volume and flow are coupled intimately since reservoir volumes at different altitudes lead to a river flow between them and a flow naturally accumulates into a reservoir volume.

To expatiate on spherical quantities of space and time in a helium atom with exactly two electrons, we may associate space with one electron, time with the other, or in a more general manner as will be introduced. Helium atom is such a conservative system that both electrons should best represent two basic dimensions. It is by this approach that we explore the property of space and time through the description of the atomic structure as follows.

### 1.3.1. Harmonic Oscillation

Consider the motion of electrons in a helium atom as harmonic oscillation. An electron living in two-dimensional spacetime possesses space and time components. As shown in Figure 1.3, an electron is confined to a spatial sphere as well as to a temporal sphere, both being inseparable. Space and time components are inherently coupled together making the electron. They cannot be divorced from the electron as if two sides of a coin cannot be divorced from the coin itself. Because a helium atom is inert, both 1s electrons must be undergoing certain kind of constant oscillations without damping; otherwise, helium shell would not be a stable system. So the basic assumption of the atomic system is simple harmonic oscillation:

$$\frac{d^2\Psi}{dt^2} = -\omega^2\Psi, \tag{1.1}$$

where $\Psi$ is time component of 1s electron and $\omega$ denotes angular velocity. One is tempted to compare this equation with Schrödinger's equation, which is much more complicated in expression. But bear in mind that we are dealing with spherical quantities in which space and time have non-classical meanings, so do their related parameters. We shall discuss Schrödinger's equation in section 1.3.4 and suffice it to say here that the complexity of Schrödinger's equation is unnecessary in spherical quantities. The simple harmonic oscillation equation provides much useful information on electronic orbitals with its typical solution as

$$\Psi = C_1(\cos\alpha - i\sin\alpha), \tag{1.2}$$

where $C_1$ is a dimension constant and $\alpha$ is a radian angle related to time component of the electron, which satisfies the relation of

$$-\frac{d\alpha}{dt} = \omega, \tag{1.3}$$

where the negative sign implies contrary aligning direction of $\alpha$ displacement relative to time dimension orientation. The solution can be verified by carrying out differential operations twice upon $\Psi$ with respect to time $t$.

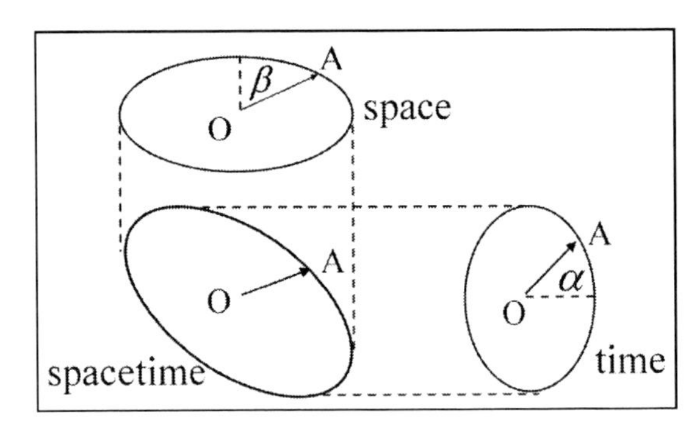


Figure 1.3. Harmonic oscillation of an electron within a helium atom where its time factor is determined by the rotation of $\alpha$ angle and space factor by the rotation of $\beta$ angle.

Similarly, since space and time are relative and symmetric, the spherical quantity describing space component of the electron satisfies:

$$\frac{d^2\psi}{dl^2} = -\frac{1}{r^2}\psi, \tag{1.4}$$

$$\psi = C_2(\cos\beta + j\sin\beta), \tag{1.5}$$

$$\frac{1}{r} = \frac{d\beta}{dl}, \tag{1.6}$$

$$v = \omega r, \tag{1.7}$$

where $C_2$ is a dimension constant, $r$ is orbital radius, and $v$ is velocity, and $j$ is a complex number notation like $i$, but it describes imaginary space instead of imaginary time component.

Since an electron is composed of time and space components that are orthogonal at any moment, we may express electronic orbitals by their product as

$$\Omega = \Psi \cdot \psi \,, \tag{1.8}$$

whence

$$\Omega = C_1 C_2 (\cos\alpha \cos\beta - i \sin\alpha \cos\beta - ij \sin\alpha \sin\beta + j \cos\alpha \sin\beta)\,. \tag{1.9}$$

Here the dot operation means coordination or association of two orthogonal spherical quantities making the electrons. However, to understand the significance of the spherical quantities, we must decipher the meanings of $i$ and $j$ notations in them. (In principle, a complex number is introduced when real numbers cannot express a two-dimensional vector. On Cartesian X-Y plane, the identifier $i$ is an operator casting a real number in X-axis into an imaginary component in Y-axis. The imaginary and the real parts of a complex number denote two perpendicular space orientations to express a vector on the plane.) How to express various dimensions in two-dimensional helium shell? If we start with a dimensionless spherical quantity, change it in the direction of reducing a time dimension and increasing a space dimension, and then reverse the dimension changes the other way around to complete a cycle, then we get four distinct types of dimensions. They are (1, $\omega$, $v$, $r$) with SI units of (1, 1/s, m/s, m) respectively. These four dimensions represent possible combinations of one-dimensional space and one-dimensional time. Thus, the significance of complex notations in $\Omega$ is interpreted dimensionally as

$$\begin{pmatrix} 1 & i \\ ij & j \end{pmatrix} = \begin{pmatrix} 1 & \omega \\ v & r \end{pmatrix}. \tag{1.10}$$

Since $\Psi$ and $\psi$ represent two orthogonal time and space components, a partial differentiation on $\Omega$ with respect to time or space dimension only affects its time or space factor.

$$\frac{\partial^2 \Omega}{\partial t^2} = \psi \frac{d^2 \Psi}{dt^2}\,, \tag{1.11}$$

$$\frac{\partial^2 \Omega}{\partial l^2} = \Psi \frac{d^2 \psi}{dl^2}\,. \tag{1.12}$$

Combining equations (1.1), (1.4), (1.7), (1.11), and (1.12) yields an oscillation equation for electrons within helium shell:

$$\frac{\partial^2 \Omega}{\partial t^2} = v^2 \frac{\partial^2 \Omega}{\partial l^2}, \tag{1.13}$$

which we shall call duality equation. Conversely, upon separation of space and time variables, this partial differential equation would revert to two ordinary differential equations (1.1) and (1.4). There are four characteristic roots to duality equation, representing four spherical quantities:

$$\begin{pmatrix} \Omega_0 \\ \Omega_1 \\ \Omega_2 \\ \Omega_3 \end{pmatrix} = C_1 C_2 \begin{pmatrix} \cos\alpha\cos\beta \\ -\omega\sin\alpha\cos\beta \\ -v\sin\alpha\sin\beta \\ r\cos\alpha\sin\beta \end{pmatrix}. \tag{1.14}$$

However, since there are only two electrons within a helium atom, each must comprise two adjacent roots. For example, one electron may take the form of ( $\Omega_0 + \Omega_1$ ) while the other ( $\Omega_2 + \Omega_3$ ). In this way, we have defined both electrons in a helium atom by four spherical quantities that observe duality equations. Disregarding constant dimensions of $C_1C_2$ , four spherical quantities are in space and/or time dimensions only.

### 1.3.2. Duality in Helium

Among four roots to duality equation, there are strict differential or integral relationships:

$$-\frac{\partial \Omega_0}{\partial t} = \Omega_1, \tag{1.15}$$

$$\int \Omega_1 dl = \Omega_2, \tag{1.16}$$

$$-\int \Omega_2 dt = \Omega_3, \tag{1.17}$$

$$\frac{\partial \Omega_3}{\partial l} = \Omega_0. \tag{1.18}$$

How to interpret these relationships? Each mathematical equation must have its corresponding physical process. To specify, when an electron is at the state of ( $\Omega_0 + \Omega_1$ ), its $\Omega_0$ component is transforming into $\Omega_1$ component according to equation (1.15). As this process completes, the integral process of equation (1.16) starts off. Or in terms of steady state, both differential and integral processes are undergoing simultaneously. In short, one electron is shifting its state from $\Omega_0$ to $\Omega_2$ losing a time dimension and gaining a space dimension while the other electron evolves from state $\Omega_2$ to state $\Omega_0$ increasing a time dimension and reducing a space dimension. Hence both electrons in helium shell are not static, but are switching states continuously and periodically.

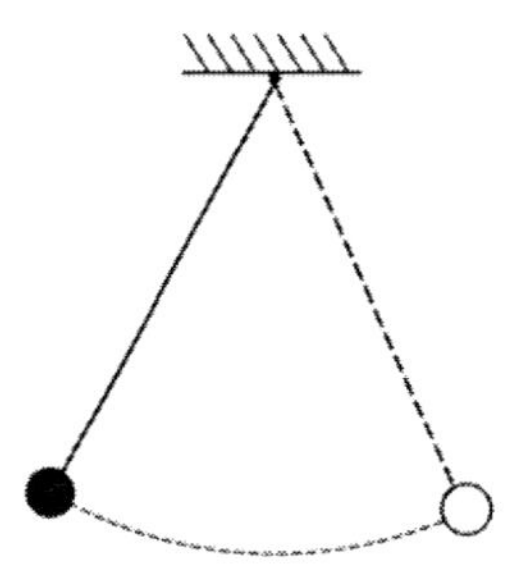

Figure 1.4. An eternal pendulum of which harmonic oscillation is the theme.

By simple interpretation, the oscillation of the electrons is somewhat similar to a pendulum where a body suspended from a fixed support swings freely back and forth under the influence of gravity (Figure 1.4). One significant difference is that the period of the electronic cycle is very short. If each electron were orbiting around the nucleus like planets around the sun kinematically as was suggested by Niels Bohr, then it would emit energy due to its high frequency. As a result, the oscillation would be damped quickly and the system could not maintain conservative for long. To overcome this, the electrons must oscillate through changing states so that both electrons exchange energy internally, i.e., each electron receives the momentum and energy emitted by the other. In this manner, the system will not lose energy to the outer environment so that the oscillatory cycle proceeds forever. Here the electron is revolving in the sense that it changes physical state continuously and periodically as point, $A$, orbits around the origin O (Figure 1.3). The circular track of point $A$ represents the course of electronic transformation or induction rather than kinematical movement.

Combining equations (1.15) and (1.16) together, and (1.17) and (1.18) together yields:

$$-\frac{\partial \Omega_0}{\partial t} = \frac{\partial \Omega_2}{\partial l}, \tag{1.19}$$

$$-\int \Omega_0 dl = \int \Omega_2 dt, \tag{1.20}$$

which mean that the changing rate of one electron in time is equivalent to the varying rate of the other electron in space. The derivative form conforms to Faraday's law of induction,

$$\nabla \times E = -\frac{\partial B}{\partial t}, \tag{1.21}$$

where the *curl* operator, $\nabla \times$, declines to $\partial/\partial l$ in two-dimensional spacetime. If we treat $\Omega_0$ as magnetic field strength and $\Omega_2$ as electric field strength, then equation (1.19) is indeed an expression of Faraday's law. This indicates that electronic oscillation within the atom is an electromagnetic phenomenon. Because $\Omega_0$ and $\Omega_2$ are not necessarily magnetic field and electric field strengths, we may put a more general statement that both electrons are in certain states obeying the calculus relationship. The relationship defines both electrons at a velocity dimension interval in the atomic spacetime. For example, if we treat $\Omega_0$ as a probability density function and $\Omega_2$ as probability current, then equation (1.19) indicates that a change in probability density in region $l$ is compensated by a net change in flux into that region. This agrees with quantum mechanics on probability, too.

Considering $\omega^2 = -1$ and $r^2 = -1$ dimensionally from their complex notation property, we may rewrite calculus relationships of (1.17) and (1.18) as:

$$-\frac{\partial \Omega_2}{\partial t} = \Omega_3 , \tag{1.22}$$

$$\int \Omega_3 dl = \Omega_0 . \tag{1.23}$$

The reason for their equivalence is that in two-dimensional world, reducing a time dimension from a spherical quantity for the second times is equivalent to increasing it while increasing a space dimension from a spherical quantity for the second times is equivalent to reducing it. This is somewhat analogous to the situation of a man who walks along a circle. As he walks forward a distance of half circumference, he arrives at the other end of a diameter; but as he continues to go forward for another half circumference, he returns to the original place, forming a circle. In the same manner, equations (1.15), (1.16), (1.22), and (1.23) form a closed loop.

During the electronic transformation, the process of increasing a space dimension is always accompanied by the process of decreasing a time dimension, or vice versa. In other words, expansion of space is undergoing with release of time wrinkles whereas contraction of space results in condensation of time. When space fully unfolds, it loses all density and wraps back spontaneously according to the stipulated cycle, so does time. Space and time components are coupled together in such an intimate way that the atomic spacetime is a finite and yet unbounded continuum. This spacetime view agrees well with the theory of relativity that spells out time dilation and length contraction.

From the perspective of waves, if we regard each root to duality equation, $\Omega_i$, as a waveform, then each electronic orbital comprises two adjacent roots, and therefore has two waveforms. Because every pair of adjacent roots are exactly one dimension apart, separated by either a time or a space dimension, both waveforms are orthogonal to each other. In other words, an electron manifests as two perpendicular waveforms. For instance, an electron may exist as a pair of interwoven electric wave and magnetic wave. An electronic wave propagates from one waveform to another following the differential and integral rule.

How to explain electronic duality? We believe electrons are real particles, but they may not exist as solid particles in the atom all the time. No one has ever captured a single electron in its static particle form as a biologist captures a bacterium under a microscope. If we associate time with a condensed particle and space with an expanded volatile cloud, then electronic motion can be interpreted as oscillation between point particle and cloud medium (see section 1.5.2). An electron demonstrates wave and particle behavior because of its varying state between space and time. Time represents density, inertia, or propensity whereas space represents volume, medium, or amount at large. Thus wave and particle duality of an electron is a manifestation of space and time oscillation.

It should be realized that when we regard helium shell as a conservative system, it does not mean energy conservation or momentum conservation for each electron because each electron is transforming between states so that any particular physical quantity is not constantly maintained within a single cycle, but rather that both electrons are inter-converted in such a way that keeps spherical quantity of each electron the same value at the same phase of the next cycle. Energy and momentum conservations apply to both electrons as a whole.

### 1.3.3. Duality in an *LC* Circuit

One will not be satisfied with the foregoing abstract description of harmonic oscillation for electrons. Neither quantum mechanics nor the standard model of particles and forces provides clearer explanation of electronic motion in the inert atom. We therefore investigate the property of an *LC* oscillator in this section with the hope that readers will gain better insight into the behavior of electrons and the significance of spherical quantities.

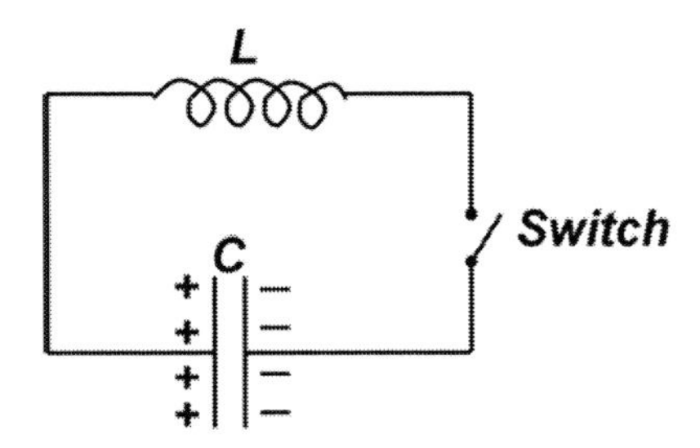


Figure 1.5. An *LC* circuit.

Consider an idealized circuit shown in Figure 1.5 containing only a switch, a capacitor $C$, and an inductor $L$ with $N$ turns of a coil (and then treating $N$=1 for brevity), ignoring the resistance in the wire. Suppose the capacitor is initially charged so that one plate has positive charge $Q$ and the other plate has negative charge of the same amount. As soon as the switch turns on, the capacitor discharges. The inductor initially opposes the growth of the current in the circuit by creating a change in magnetic flux through itself. The change in the magnetic flux induces an electromotive force in the circuit. After the capacitor discharges completely, the electromotive force drives a current along the circuit in the opposite direction, charging the capacitor in the reverse polarity. The cycle then repeats itself in the opposite direction. In the absence of any electric and magnetic energy loss, the oscillation will continue back and forth indefinitely, which may be characterized by

$$\frac{d^2q}{dt^2} = -\frac{1}{LC}q, \tag{1.24}$$

where $q$ indicates charge in the capacitor varying in sinusoidal form:

$$q = Q\cos\omega t, \tag{1.25}$$

$$\omega = \sqrt{\frac{1}{LC}} . \tag{1.26}$$

In such an electromagnetic resonance, electric current *I*, electromotive force *V*, and magnetic flux $\Phi$ have the following basic relationships:

$$I = \frac{dq}{dt}, \tag{1.27}$$

$$V = -L\frac{dI}{dt}, \tag{1.28}$$

$$-N\frac{d\Phi}{dt} = V , \tag{1.29}$$

whence

$$I = -Q\omega \sin \omega t , \tag{1.30}$$

$$\Phi = -\frac{LQ\omega}{N} \sin \omega t , \tag{1.31}$$

$$V = LQ\omega^2 \cos \omega t . \tag{1.32}$$

We may describe the harmonic oscillation by spherical quantities as well as by physical quantities. Instead of electric charge, the following two complex functions satisfy the oscillation equation in general:

$$\begin{aligned} q_1 &= q + I \\ &= Q(\cos \omega t - \omega \sin \omega t) , \end{aligned} \tag{1.33}$$

$$\begin{aligned} q_2 &= \Phi + V \\ &= QL\omega(-\sin \omega t + \omega \cos \omega t) . \end{aligned} \tag{1.34}$$

Here an electron may take the form of a static particle charge, flow as electric current, transform into magnetic flux, and build an electromotive force. We also realize that the charge is stored in the capacitor while the magnetic flux exists inside the inductor, both electric elements being orthogonal in that a capacitor allows alternating current to pass but cuts off direct current whereas an inductor allows direct current to pass but impedes alternating current. Taking the hardware of the circuit into consideration, we may associate $q_1$ with $\cos \beta$ for a typical capacitor while associate $q_2$ with $\sin \beta$ for a typical inductor. As an electron undergoes electromagnetic oscillation, it sets off from the capacitor, takes a path along the circuit, enters the coiled wire of the inductor, transforms into magnetic flux through the inductor, backs into the circuit as electromotive force, and returns to the capacitor. Besides temporal sinusoidal waves, spherical quantities $q_1$ and $q_2$ have spatial sinusoidal waves in terms of hardware association as electrons travel along the wire. Since the temporal wave depends on the spatial transition of the electron along the circuit, both temporal and spatial signals are synchronized automatically. After considering electron and hardware association in the circuit system, we get

$$q_1 = Q(\cos \omega t - \omega \sin \omega t)\cos \beta, \tag{1.35}$$

$$q_2 = QL\omega(-\sin \omega t + \omega \cos \omega t)\sin \beta . \tag{1.36}$$

Here charge and current constitute a complex function associated with the capacitor while magnetic flux and electromotive force form another complex function associated with the inductor. Careful readers surely find that both functions correspond to both electrons of ( $\Omega_0 + \Omega_1$ ) and ( $\Omega_2 + \Omega_3$ ) within a helium atom pending

$$\frac{1}{L} = \frac{\partial \beta}{\partial l} \tag{1.37}$$

until Chapter 10. We do not imply that both electrons in helium shell are in the same physical states as those in the idealized *LC* oscillator, but it is certain that they are undergoing analogous harmonic oscillation. Man can design an *LC* oscillator with trivial resistance in the circuit, but the idealized *LC* oscillator without any damping as was described can only be created by nature, in the form of an inert atom. How wonderful the nature is! How much have we known about its secret?

### 1.3.4. Duality Equation vs. Schrödinger's Equation

Our description of electronic orbitals is not only self-consistent but also compatible with the well-established quantum mechanics. Schrödinger's equations for the motion of a particle are the starting point for the development of quantum mechanics. For an electron within a hydrogen atom, the one-dimensional Schrödinger's equation is as follows:

$$i\hbar \frac{\partial \xi}{\partial t} = -\frac{\hbar^2}{2m}\frac{\partial^2 \xi}{\partial x^2}, \tag{1.38}$$

where $\xi$ is the wave function for the electron. In order to compare this equation with duality equation, we write down the following basic physical quantity relationships:

$$r = \frac{\lambda}{2\pi}, \omega = 2\pi f , \tag{1.39}$$

$$E = hf \ , \ p = \frac{h}{\lambda}, \tag{1.40}$$

$$\hbar = \frac{h}{2\pi}, \tag{1.41}$$

$$E = \frac{p^2}{2m}, \tag{1.42}$$

where $h$, $\lambda$, $f$, $E$, and $p$ refer to Planck's constant, wavelength, frequency, energy, and momentum, respectively. For characterizing the kinematics of an oscillating entity, parameters $\omega$ and $r$ are more descriptive and pertinent than $\hbar$ and $m$. They are closely related to energy and momentum through the rationalized Planck's constant:

$$\omega = \frac{E}{\hbar}, \ \frac{1}{r} = \frac{p}{\hbar}. \tag{1.43}$$

After converting parameters $\hbar$ and $m$ into $\omega$ and $r$, equation (1.38) becomes

$$-\omega \frac{i\partial\xi}{\partial t} = \omega^2 r^2 \frac{\partial^2 \xi}{\partial x^2}. \tag{1.44}$$

Treating time and space dimensions in this equation as spherical quantities yields

$$\omega \frac{i\partial\xi}{\partial t} = v^2 \frac{\partial^2 \xi}{\partial l^2}. \tag{1.45}$$

We cancel out the negative sign because the first derivative and the second derivative of $\xi$ are anti-parallel with respect to dimension *l*. ( $\partial l^2 = -\partial x^2$ ). Moreover, from wave function $\Psi$ in equation (1.2), we obtain

$$i\Psi = C_1(i\cos\alpha + \sin\alpha) = -\frac{d\Psi}{d\alpha}, \tag{1.46}$$

which indicates that factor $i$ may be regarded as a differential operator when placed before a spherical quantity directly. The differential operator transforms its operand to the dimension orthogonal to it. This is different from the interpretation of factor $i$ as angular velocity $\omega$ when placed before a trigonometric function. Based on equations (1.46) and (1.3), we have spherical quantity

$$\begin{aligned}\omega i\xi &= -\omega\frac{\partial\xi}{\partial\alpha} \\ &= \frac{\partial\xi}{\partial\alpha}\cdot\frac{\partial\alpha}{\partial t} \\ &= \frac{\partial\xi}{\partial t}.\end{aligned} \tag{1.47}$$

Substituting $\omega i\xi$ value of this equation into the left-hand side of equation (1.45) yields a duality equation. Thus the one-dimensional Schrödinger's equation is indeed another expression of duality equation. However, the same wave equation has different significance in the concepts of spherical quantity and of physical quantity. In physical quantity, equation (1.38) is a one-dimensional wave function for an electron. In spherical quantity, since the second derivative contains the process of the first derivative, duality equation has two dimensions. Duality equation is a two-dimensional harmonic oscillation equation in the atomic spacetime.

## 1.4. Dynamic Calculus

The foregoing perception of two-dimensional spacetime enables us to make a rigorous definition on the property of spherical quantity. Although simple harmonic oscillation equations in spherical quantities and in physical quantities have a similar form, they have different meanings. Spherical quantities have a unique set of rules of operations such as plus, multiplication, and calculus. They also have special dimensional and trigonometric correspondences that deserve consideration in details. After discarding any premises along with physical quantities, we need to establish new ground upon which to base spherical quantities firmly.

### 1.4.1. Infinitesimal Calculus Review

Before introducing dynamic calculus, it is helpful for us to review the basic concept of traditional infinitesimal calculus. Calculus includes integral calculus and differential calculus, the former originating from the attempt to determine the area under a curve while the later to measure the steepness of a curve at a certain point.

It is believed that Newton and Leibniz, quite independently of one another, were responsible for developing the ideas of calculus that is hitherto so powerful to solve a variety of problems in science and engineering.

The central idea of differential calculus is the notion of derivative. The derivative of function $f$ with respect to $x$ is the function $f'$ defined by

$$f'(x) = \lim_{\Delta x\to 0}\frac{f(x+\Delta x)-f(x)}{\Delta x} \tag{1.48}$$

provided the limit exists. It represents the slope of the tangent line to the graph of *f* at any point (*x*, *f(x)*), or the rate of change of *f* at *x*. The process of obtaining the derivative is called differential operation, or differentiation. Geometrically, differentiation is the approximation of a curve or surface by a tangent line or plane for calculation in linear algebra.

The integral of a continuous function *f(x)* on the closed interval [*a*, *b*] is defined by

$$\int_a^b f(x)dx = \lim_{\Delta x \to 0} \sum_{i=1}^{n} f(x_i)\Delta x \tag{1.49}$$

provided the limit exists. The interval [*a*, *b*] is divided into *n* equal subintervals of width $\Delta x$, where $\Delta x$ is $(b - a)/n$, and $x_i$ is the rightmost point in the *i*th interval. According to the fundamental theorem of calculus, the definite calculus can be calculated by

$$\int_a^b f(x)dx = F(b) - F(a), \tag{1.50}$$

where *F* is any antiderivative of *f*. It is equal to the signed area bounded by lines $x = a$, $x = b$, *x*-axis and the curve defined by function *f*.

If $f(x) = \cos x$, then according to the above definitions of infinitesimal calculus we obtain

$$f'(x) = -\sin x; \int_0^x f(x)dx = \sin x. \tag{1.51}$$

This indicates that the slope of $\cos x$ at any point *x* within the first quadrant is equal to the negative number of the area under the curve from the origin to *x* (Figure 1.6). We may also write

$$f''(x) = -f(x), \tag{1.52}$$

or

$$\frac{d^2 f}{dx^2} = -f, \tag{1.53}$$

which conforms to the equation of simple harmonic oscillation. Performing differentiation twice upon a cosine function returns it to the original cosine with a minus sign, i.e., the trigonometric function wraps back upon repeated calculus operations. The study of cosine and sine conversion in calculus leads to the invention of dynamic calculus in stark contrast to infinitesimal calculus.

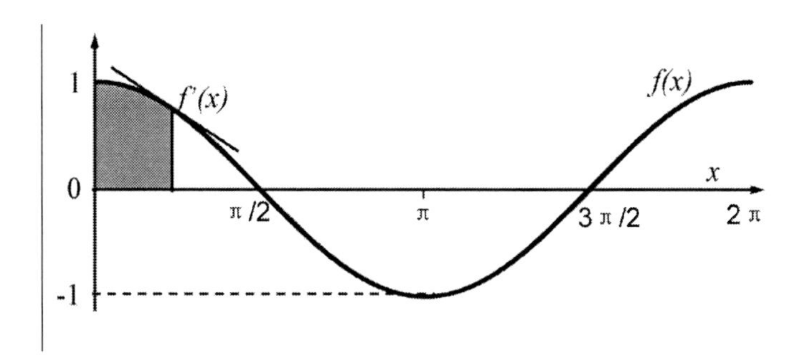


Figure 1.6. The slope of cosine function is equal to the negative area of its corresponding shaded region in the first quadrant.

### 1.4.2. Dynamic Calculus

Because in the atomic spacetime, electronic motion follows differential and integral operations as were demonstrated by equations (1.15) to (1.18) and equations (1.22) and (1.23), here we examine the significance of calculus on spherical quantities. A differential operation upon a trigonometric function with respect to a time dimension, such as that dictated by equation (1.15), does not physically happen in a flash, but it is carried out gradually and smoothly. For example, when $\cos\alpha$ receives the differentiation command, the angle $\alpha$ is then rotating gradually up to ($\pi/2 + \alpha$), at which point the differential operation completes so that $\cos\alpha$ transforms into $-\sin\alpha$ accordingly. Differential and integral operations upon trigonometric functions can be interpreted as follows:

$$\frac{d}{d\alpha}\cos\alpha = \cos(\alpha + \frac{\pi}{2})\,; \tag{1.54}$$

$$\frac{d}{d\alpha}\sin\alpha = \sin(\alpha + \frac{\pi}{2})\,; \tag{1.55}$$

$$\int \cos\beta\; d\beta = \cos(\beta - \frac{\pi}{2})\,; \tag{1.56}$$

$$\int \sin\beta\; d\beta = \sin(\beta - \frac{\pi}{2})\,. \tag{1.57}$$

The correctness of these expressions can be easily verified by infinitesimal calculus. But here we regard the plus sign in the parentheses as continuous radian angle rotation up to $\pi/2$ displacement and the minus sign as continuous radian angle rotation down to $-\pi/2$ displacement. In light of the special relationships between calculus and radian angle rotation, here we introduce the concept of dynamic calculus. A typical expression of differential operation upon a trigonometric function $\Psi_0 = C_1\cos\alpha$ is

$$-\frac{d\Psi_0}{dt} = -C_1 \cos\alpha, where\ \ (\alpha \mapsto \frac{\pi}{2} + \alpha)\ , \tag{1.58}$$

where the arrow $\mapsto$ indicates the process of radian angle rotation from $\alpha$ to ($\pi/2 + \alpha$ ). A differential operation is realized through the smooth increment of $\alpha$ angle from $\alpha$ to ($\pi/2 + \alpha$ ), which of course results in the change of $\cos\alpha$ term into $-\sin\alpha$ term, followed by differential chain rule of $d\alpha/dt$ in the end. Disregarding the differential chain rule for a moment, we may express a differential operation in terms of radian angle rotation. Since electronic motion is a dynamic process, by treating $\alpha$ as a continuously changing variable, we may omit the subordinate clause in equation (1.58) and express a differential operation by a simple trigonometric function no matter whether the differential process is carried out completely or not. If it is completely done, then $\alpha$ increases a displacement of $\pi/2$ so that $\cos\alpha$ transforms into $-\sin\alpha$ and $-d\Psi_0/dt = -C_1\,\omega\sin\alpha$ by differential chain rule; otherwise, the differential process is still at a state represented by the magnitude of the radian angle. Because $\alpha$ is a dynamic variable, the difference between both trigonometric terms ( $\cos\alpha$ and $-\sin\alpha$ ) is a matter of $\alpha$ value difference in cosine function. Thus, both $\cos\alpha$ and $-\sin\alpha$ can be used to express the differentiation on the left-hand side of equation (1.58), even though we traditionally associate the final result of the differentiation with $-\sin\alpha$ , and the initial condition with $\cos\alpha$ . Thus, we define dynamic calculus as follows.

*Definition 1.* Given a spherical quantity $f(\alpha) = \cos\alpha$ or $f(\alpha) = \sin\alpha$ , a differential operation on $f(\alpha)$ with respect to $\alpha$ means increasing $\alpha$ variable a displacement of $\pi/2$ in the function.

*Definition 2.* Given a spherical quantity $g(\beta) = \cos\beta$ or $g(\beta) = \sin\beta$ , an integral operation on $g(\beta)$ over $\beta$ means decreasing $\beta$ variable a displacement of $\pi/2$ in the function.

Wave functions are spherical quantities in sine or cosine functions. Dynamic calculus on spherical quantities can be explained in terms of translational movements of a curve. If we plot $f(\alpha) = \cos\alpha$ on Cartesian plane, then increasing $\alpha$ variable a displacement of $\pi/2$ in the function is a translational movement of $f(\alpha)$ along $\alpha$ axis $\pi/2$ so that $\cos\alpha$ transforms into $-\sin\alpha$ in the graph. Likewise, if we plot $g(\beta) = \cos\beta$ on Cartesian plane, then reducing $\beta$ variable a displacement of $\pi/2$ in the function is a translation of $g(\beta)$ along $\alpha$ axis $-\pi/2$ so that $\cos\beta$ becomes $\sin\beta$ in the graph. Thus a calculus operation upon a wave function may be illustrated by a translational movement of its curve on Cartesian plane.

The difference between infinitesimal calculus and dynamic calculus is that the latter is a process covering $\pi/2$ range of radian angle rotation in a trigonometric function whereas the former only captures the terminal state of the trigonometric function at a specific radian value. A rotatory radian angle of a trigonometric function is the underlying implementation of dynamic calculus whereas an infinitesimal limit of a function is the underlying implementation of infinitesimal calculus. Dynamic calculus stands for a physical process rather than a mathematical limit.

Taking differential chain rule into consideration beforehand, we may implements equations (1.15) and (1.16) by:

$$-\frac{d}{dt}\cos\alpha = \omega\cos(\alpha + \frac{\pi}{2}) ; \tag{1.59}$$

$$\int \cos\beta \; dl = r\cos(\beta - \frac{\pi}{2}) . \tag{1.60}$$

As shown in Figure 1.7(a), transformation of the time component of $\Omega_0$ from $C_1 \cos\alpha$ to $-C_1\omega\sin\alpha$ can be expressed by dynamic motion of point C along semicircle ACB. At any specific point C, chord BC denotes $C_1 \cos\alpha$ while chord AC denotes $-C_1\omega\sin\alpha$ where factor $\omega$ is a complex notation indicating mutually perpendicular relationship between BC and AC. As radian angle $\alpha$ rotates from 0 to π/2, chord BC disappears while chord AC increases up to the maximum of diameter AB. The time component of the electron transforms from $C_1 \cos\alpha$ into $-C_1\omega\sin\alpha$ . This is a trigonometric interpretation of differential operation $-\partial\Omega_0/\partial t$ where the minus sign indicates time component decrement.

Likewise, Figure 1.7(b) illustrates the course of integral operation of equation (1.60). As radian angle $\beta$ decreases from π/2 to 0, point C traces along semicircle ACB from A to B. At any specific point C on the course, chord AC denotes $C_2 \cos\beta$ while chord BC denotes $C_2 r\sin\beta$ , both being perpendicular at any time as noted by factor $r$ as complex number identifier. Orthogonality means AC⊥BC at any moment even though both spherical quantities are transforming between each other dynamically. Thus the integral operation of equation (1.16) is implemented by continuous $\beta$ angle decrement in $\Omega_1$ . The space component of the electron transforms from $C_2 \cos\beta$ into $C_2 r\sin\beta$ . Because $\alpha$ and $\beta$ angle are complementary and synchronized at any moment, equation (1.19) holds describing the relationship between $\Omega_0$ and $\Omega_2$ .

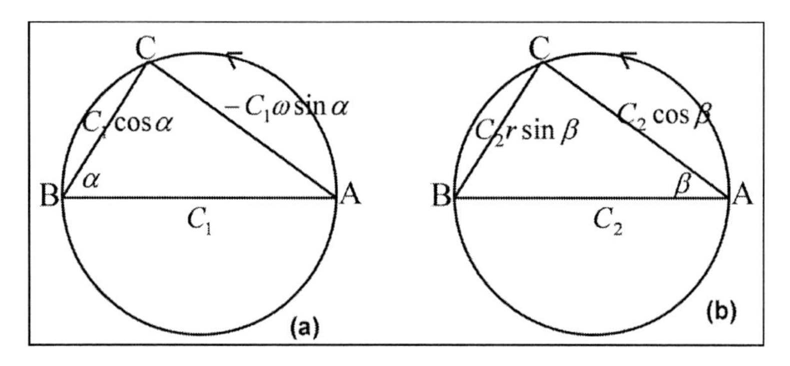


Figure 1.7. Dimension diagram of dynamic calculus of spherical quantity for the motion of an electron with synchronized (a) time and (b) space components.

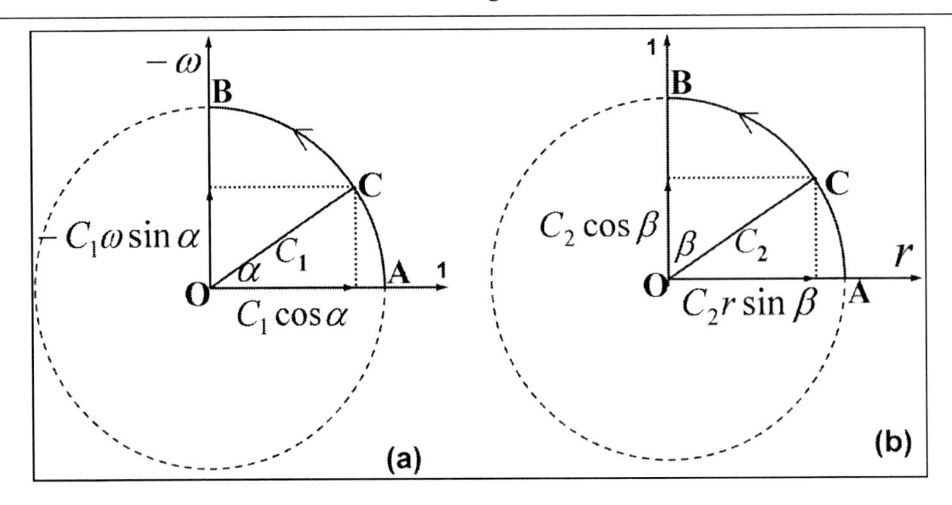


Figure 1.8. Schematic diagram of dynamic calculus of spherical quantity for the motion of an electron in time and space.

In this way, we have explained electronic motion by dynamic calculus and the implementation of dynamic calculus by trigonometry in dimension diagram.

The electronic transformation can be illustrated by another diagram. As shown in Figure 1.8a, as the electron travels from point A towards point B along quarter arc ACB, it traverses a time dimension from a dimensionless quantity along X-axis to negative angular velocity dimension along Y-axis. At any specific moment of point C on the course, the time components of both electrons are shown as both vectors $C_1\cos\alpha$ and $-C_1\omega\sin\alpha$ in the graph, respectively. Differential operation $-\partial\Omega_0/\partial t$ transforms $C_1\cos\alpha$ into its orthogonal dimension of $-C_1\omega\sin\alpha$ with complex notation $\omega$ as its dimension identifier. Angular velocity $\omega$ from the differential chain rule is consistent with the usage of $\omega$ as a complex notation that transforms its operand into its orthogonal dimension.

Likewise, as shown in Figure 1.8(b), the quarter circle from A to B represents electronic transformation of $\int\Omega_1 dl$ implemented by the continuous decrement of radian angle $\beta$ from $\pi/2$ to zero in the process. At any specific moment of point C on the course, radian angles $\alpha$ and $\beta$ are complementary. Integral operation $\int\Omega_1 dl$ transforms $C_2\cos\beta$ into its orthogonal dimension $C_2 r\sin\beta$ with complex notation $r$ as its dimension identifier. From points A to B, the electron changes dimension from spherical quantities $\Omega_0$ to $\Omega_2$ eventually.

### 1.4.3. Spacetime Continuity

Because a spherical quantity changes by differential and integral operations, there are interesting relationships between a spherical quantity and its first and second derivatives. As shown in Figure 1.9(a) for equation (1.1) in two-dimensional spacetime, dimension $d^2\Psi/dt^2$ must be equivalent to dimension $\Psi$, i.e., performing differential operations twice upon a spherical quantity returns it to the original quantity provided $\omega^2=-1$.

Moreover, since operator $d^2/dt^2$ casting upon $\Psi$ constitutes a complete loop in two-dimensional spacetime, dimensions $\Psi$ and $-d\Psi/dt$ must be symmetric within the cycle. It follows that if $\Psi_0$ is a valid solution to equation (1.1), then $-d\Psi_0/dt$ must be a valid solution as well for it is an intermediate step for spherical quantity $\Psi_0$ to evolve towards $d^2\Psi_0/dt^2$. For example, if a solution to equation (1.1) is

$$\Psi_0 = C_1 \cos\alpha\ , \qquad (1.61)$$

then there is another solution in the derivative term $-d\Psi_0/dt$:

$$\Psi_1 = -C_1\omega\sin\alpha\ . \qquad (1.62)$$

Likewise, as shown in Figure 1.9(b) for equation (1.4) in two-dimensional spacetime, the solutions to it can be compactly written in a complex function of equation (1.5).

Dynamic calculus of a spherical quantity is a continuous process. As shown in Figure 1.10(a), equations (1.15) (1.16) and (1.22)(1.23) constitute a loop. And if we regard the differential and integral operations as a simultaneous event, then functions $\Omega_1$ and $\Omega_3$ may be deleted from the loop so that $\Omega_0$ and $\Omega_2$ represent both electrons in a helium atom. As shown in Figure 1.10(b), both spherical quantities have an intervening spacetime distance of reducing a time dimension while increasing a space dimension. Performing $\int(-\partial/\partial t)dl$ operation upon $\Omega_0$ produces $\Omega_2$ and performing the same operation upon the latter yields the former, both operations forming a cycle. Each spherical quantity is the symmetric counterpart of the other in the cycle.

The implementation of dynamic calculus by smooth angle rotation bridges discrete physical quantities into a continuous spherical quantity. Spherical quantities observe the rule of dynamic calculus instead of linear algebra.

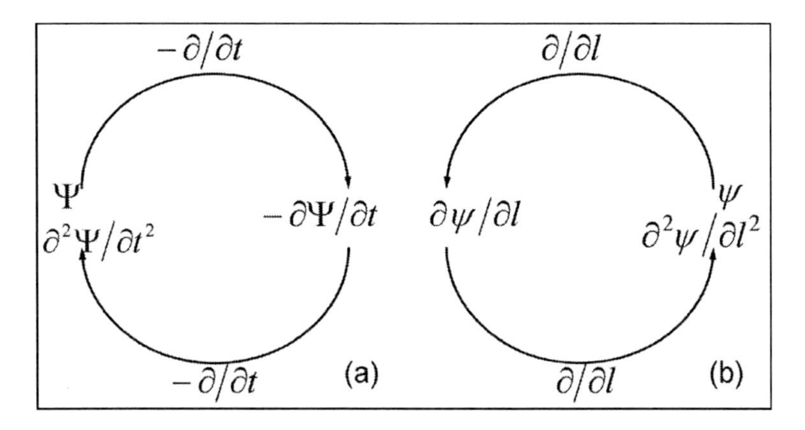


Figure 1.9. The equivalence between a spherical quantity and its second derivative and the symmetry of a spherical quantity and its first derivative for (a) time component and (b) space component of the electrons within a helium atom.

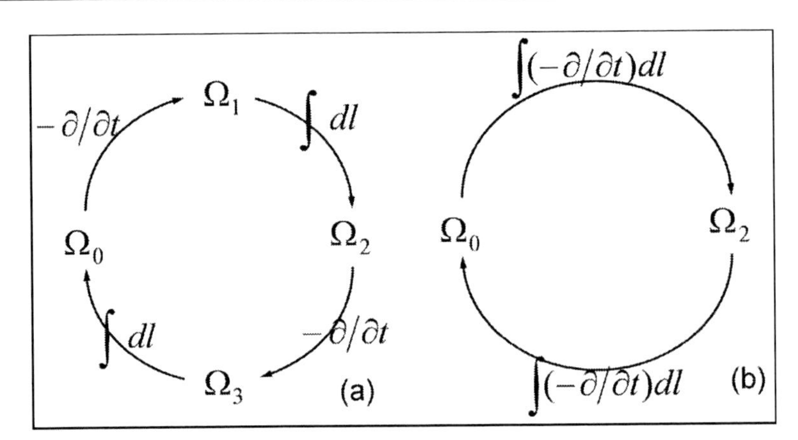


Figure 1.10. Calculus relationships (a) between four characteristic roots to duality equation and (b) between two representative roots that form a rope structure.

During the course of a dynamic calculus operation on a spherical quantity, there is not any break or void or abrupt jump in the spherical quantity but smooth transition in the physical dimension. The atomic spacetime is continuous in that space and time components are transforming smoothly and continuously from one state to another. Here spacetime continuity has non-classical meaning. To give an analogy, when we say that a father and a son are spacetime continuous, we do not mean that they sit on the same bench and in close touch, but that the son is developed from the seed of the father. Space refers to the biological bodies, not the location or volume of air that they empty off. By the same way, the atomic spacetime continuity does not mean that both electrons occupies two positions close enough, but that both electrons are transforming from each other smoothly without any interruption even though both electrons do occupy the atomic sphere without a clear physical distinction between them. Spacetime continuity refers to smooth calculus transformation implemented by trigonometry. This concept is more fundamental and stricter than conventional function continuity at a specific position under Cartesian coordinates.

After introducing dynamic calculus and spacetime continuity concepts, we shall make a modification concerning the general solution to duality equation. While four characteristic roots to duality equation set the dimensional framework for electrons, every point of C along the semicircle (Figure 1.7) or the quarter circle (Figure 1.8) constitutes a valid solution. Because the course of point C is a continuous arc, given a specific pair of complementary $\alpha$ and $\beta$ values, there is one solution to duality equation corresponding to it. In other words, there are countless solutions to duality equation and the arc constitutes the set of all solutions. Among others, two boundary points deserve further scrutiny in mathematics.

Both electrons in a helium atom transform one another smoothly and continuously by the rule of dynamic calculus. At a specific moment, both electrons constitute two basic dimensions, time and space, separated by a phase of $\pi/2$ in their waveforms. For example, an electron ( $\Omega_0+\Omega_1$ ) may represent time as $\alpha$ equals 0 while the other electron ( $\Omega_2+\Omega_3$ ) represents space. At that specific moment, both $\Omega_1$ and $\Omega_3$ vanish so that one electron, $\Omega_0=C_1$, indicates a full time dimension while the other, $\Omega_2=C_2$, denotes a full space dimension. In these calculations, the expression of spherical quantity $\Omega_0=C_1C_2\cos\alpha\cos\beta$ as a wave function should not be construed as conventional multiplication of orthogonal time and space components, instead it is only an expression of time and space coordination. Coordination does not share the common property of conventional multiplication. For example, at the point of $\alpha = 0$ and $\beta = \pi/2$, space component $C_2 \cos\beta = 0$, the spherical quantity is not equal to zero, but equal to time component $\Omega_0 = C_1 \cos\alpha = C_1$ with space component disappearing away. Dynamic calculus is a physical process. When $0< \alpha <\pi/2$, both electrons contain a mix of space and time components. Since both electrons in helium shell at a certain phase ( $\alpha = 0$ and $\beta = \pi/2$ ) represent space and time dimensions, they are symmetric and orthogonal as if space and time dimensions are symmetric and orthogonal.

### 1.4.4. Properties of Spherical Quantity

The foregoing discussion has demonstrated three ways to express a differential operation on a spherical quantity. The first is by a derivative form such as $-d\Psi_0/dt$ ; the second is by a trigonometric function such as $-C_1 \cos\alpha$ ; and the third is by a shorthand notation such as $\dot{\Psi}_0$, similar to $\omega$ or $r$ expression as a dimension identifier. These three usages agree with each other in expressing the dynamic transformation of a spherical quantity.

By expressing $-d\Psi_0/dt$ in terms of $-C_1 \cos\alpha$, we deliver at least three messages. First, the radian angle $\alpha$ is a dynamic variable rotating from 0 to $\pi/2$ smoothly. At any specific moment, the value of the spherical quantity is represented by $-C_1 \cos\alpha$ (Figure 1.11). Second, spherical quantity $\Psi_0$ changes cosinusoidally corresponding to the uniform rotation of angle $\alpha$ from 0 to $\pi/2$, i.e., $-d\Psi_0/dt$ is a special quotient expression with a constant denominator $t$ and a cosinusoidal nominator. Third, operator $-d/dt$ casting on spherical quantity $\Psi_0$ is equivalent to cosine operator casting on uniform rotatory angle $\alpha$ in expressing the same differentiation. These three meanings agree with each other.

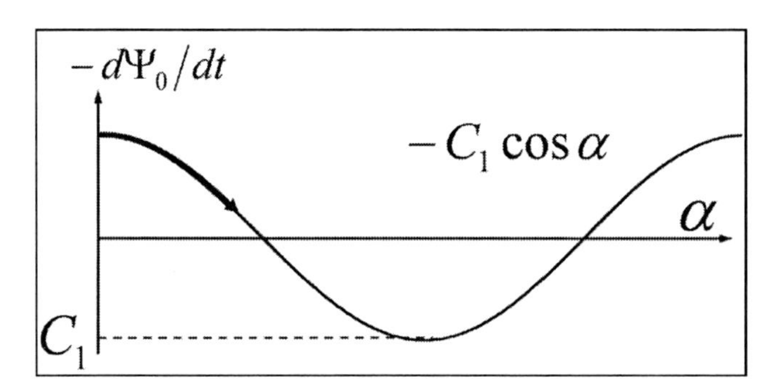


Figure 1.11. The value of a spherical quantity in dynamic differentiation. As $\alpha$ rotates from 0 to $\pi/2$ smoothly, the spherical quantity traces the cosinusoidal curve.

We have encountered many unconventional consequences mathematically in spherical quantities. Firstly, addition has different physical meanings from conventional one. Since electronic motion follows dynamic calculus transforming from one state to another, the plus sign in complex function ( $\Omega_0 + \Omega_1$ ) indicates a spherical quantity at a state shifting from $\Omega_0$ to $\Omega_1$ as was expressed by equation (1.15). The plus sign indicates a calculus operation. By this sense, the addition of spherical quantities ( $\Omega_0 + \Omega_1$ ) has different physical meaning from ( $\Omega_1 + \Omega_0$ ). Hence addition of two spherical quantities does not observe commutative law. The atomic space is not a linear vector space. Spherical quantities are governed by the rule of dynamic calculus.

Secondly, symmetry has different meanings for spherical quantities. There are orbital symmetry and dimension symmetry in the atomic spacetime. We consider electron orbitals $\Omega_0$ and $\Omega_2$ to be symmetric because performing a rotatory operation upon $\Omega_0$ results in $\Omega_2$ and performing a rotatory operation upon $\Omega_2$ results in $\Omega_0$ . This symmetry was illustrated by Figure 1.10(b). We consider space and time to be symmetric because they are two complementary dimensions, each being the counterpart of the other. This symmetry was illustrated by Figure 1.9 and by equations (1.1) and (1.4). The derivative of a spherical quantity with respect to time and the derivative of the spherical quantity with respect to space are equal. For example,

$$-\frac{\partial \Omega_0}{\partial t} = v\frac{\partial \Omega_0}{\partial l}, \tag{1.63}$$

where $v$ is a velocity to compensate the dimensional difference between both sides and the minus sign indicates anti-parallel alignment of time and space dimensions. This equation may be derived from the following calculations:

$$-\frac{\partial \Omega_0}{\partial t} = -C_1 C_2 \omega \sin\alpha \cos\beta; \tag{1.64}$$

$$\frac{\partial \Omega_0}{\partial l} = -C_1 C_2 \frac{1}{r} \cos\alpha \sin\beta . \tag{1.65}$$

Assigning $\alpha = \beta$ for both differential processes, we get

$$\sin\alpha \cos\beta = \cos\alpha \sin\beta , \tag{1.66}$$

which leads to the symmetric equation (1.63).

Thirdly, orthogonality has fresh mathematical significance in the atomic spacetime. Besides symmetry between $\Omega_0$ and $\Omega_1$, we regard both as orthogonal because they are exactly one dimension apart as was shown in equation (1.15). Orthogonality is defined as an exact differential and/or integral relationship between two spherical quantities. By this definition, we may deduce that four characteristic roots to duality equation are all mutually orthogonal from equations (1.15) to (1.18). Orthogonal quantities are coordinative, which permits dot operation between them. As was shown in equation (1.8), dot operation is applied to space and time components of an electron to indicate coordination. Orthogonal quantities are interdependent and perpendicular. As was shown in Figure 1.7, chords AC and BC are perpendicular. They form two sides of a right triangle in dimension diagram. In trigonometric functions, two orthogonal spherical quantities have a $\pi/2$ radian phase difference, but they are transforming from one spherical quantity to another according to dynamic calculus. This is radically different from orthogonality in Euclidean space where orthogonal physical quantities are supposed to be mutually independent and inexchangeable.

## 1.5. Linear Interpretation of Spherical Quantity

We have described the motion of 1s electrons by spherical quantities in harmonic oscillation equation, trigonometry, and dynamic calculus. Readers naturally wish to know how to relate spherical quantities in the atomic spacetime to variables in Euclidean space that we are familiar with. The answer is not so straightforward as the common translation between Cartesian coordinates and spherical polar coordinates because it is about relating two systems that are rooted in fundamentally different concepts. Nonetheless, the translation between both systems is critical to the success of the atomic spacetime theory because only when it is properly integrated with conventional knowledge can it be recognized and gain its place in science. Spherical quantities must correspond to physical quantities in certain way; and the atomic structure in spherical quantities must have its geometric correspondence in Euclidean space. This section tries to establish such correspondences.

### 1.5.1 Spherical Quantity by Rotatory Vector

A spherical quantity is a wave function for harmonic oscillation. Harmonic oscillation can be treated as circular motion or the projection of circular motion onto a certain profile. Figure 1.12 shows the vibration of a ball under the action of springs. The displacement of the ball from the equilibrium position is governed by both Newton's second law and Hooke's law

$$m\frac{d^2x}{dt^2} = -kx\,, \tag{1.67}$$

where $k$ is spring constant and $m$ is the mass of the ball. This is equivalent to equation (1.1) in noting

$$\omega^2 = \frac{k}{m}\,. \tag{1.68}$$

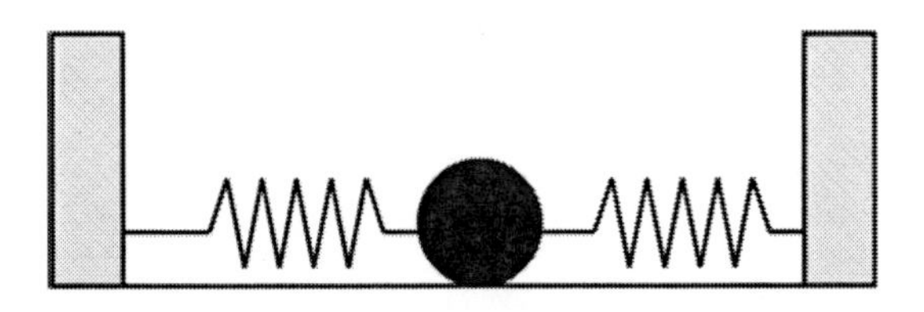

Figure 1.12. Harmonic oscillation of a ball under the influence of springs after shifting the ball a displacement from the central equilibrium position and letting it go without friction.

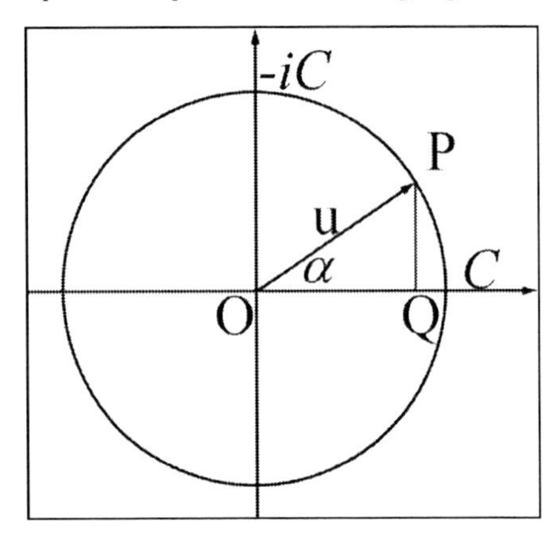


Figure 1.13. Schematic illustration of a rotatory vector to denote a spherical quantity in harmonic oscillation.

Harmonic oscillation represents a large class of natural phenomena whose far-reaching implications have not been recognized yet. Defining harmonic oscillation as circular motion, we may adopt a rotatory vector u around a central point to describe the state of the oscillating entity. As shown in Figure 1.13, a circle is situated at a Cartesian coordinate system whose abscissa axis is of dimension $C$ and ordinate axis of its orthogonal dimension $-iC$. Any point P located at the circle can be expressed by complex function ( $C\cos\alpha - iC\sin\alpha$ ). Vector OP, denoted by u, rotates from dimensions $C\cos\alpha$ to $-iC\sin\alpha$ as $\alpha$ increases from 0 to $\pi/2$, i.e., the rotation transforms the vector along the abscissa to that along the ordinate. Here we are concerning about the direction change rather than quantity change of the vector. The length of the vector is a fixed one while the direction of the vector varies continuously. The direction shift represents dimension change in the wave function. Thus the rotatory vector represents a spherical quantity. We define the rotatory process by dynamic differentiation upon the spherical quantity with respect to time dimension:

$$-\frac{dC\cos\alpha}{dt} = -iC\sin\alpha \tag{1.69}$$

with the assignment of $i = \omega$. This dynamic process rotates the vector from the abscissa axis to the ordinate axis on Cartesian plane. It is implemented by the continuous counterclockwise rotation of radian angle $\alpha$ from 0 to $\pi/2$.

Defining differential calculus of a spherical quantity as a rotatory vector, we may express harmonic oscillation with a rotatory vector. As the vector, u, rotates counterclockwise, at $\alpha = 0$, vector u is at its initial state corresponding to spherical quantity $\Psi = C\cos\alpha$ ; at $\alpha = \pi/2$, the rotatory vector reaches the ordinate axis position corresponding to the first derivative of the spherical quantity $-d\Psi/dt = -Ci\sin\alpha$ ; at $\alpha = \pi$ , the rotatory vector returns to the abscissa axis corresponding to the second derivative $d^2\Psi/dt^2 = C\cos\alpha$ , which is equal to the original function in terms of dimensions; and so on (Figure 1.14). In the range of $0 < \alpha < \pi/2$, the spherical quantity is transforming from $\Psi$ towards $-d\Psi/dt$ with the tip of vector u sweeping along the arc in the first quadrant; between $\pi/2 < \alpha < \pi$, the spherical quantity is transforming from $-d\Psi/dt$ towards $d^2\Psi/dt^2$ with the tip of vector u sweeping along the arc in the second quadrant as was shown in Figure 1.13. From the trigonometric perspective, the differentiation is a gradual course or a smooth process rather than a single step operation. Thus, although $-d\Psi/dt$ is conventionally associated with the final calculation result of the differentiation, it should be understood here as the process of a position vector along the curve, bridging the gap between two dimensions smoothly.

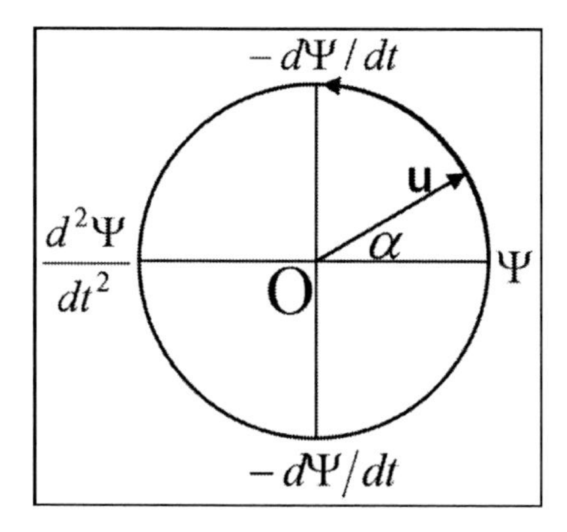


Figure 1.14. A rotatory vector u at various positions corresponding to a spherical quantity and its derivatives.

Since rotatory vector u indicates the state of a spherical quantity during the harmonic oscillation, the physical significance of the spherical quantity can be interpreted by the vector. If physical quantity $x_0$ denotes the initial maximum displacement of the ball from its equilibrium position (Figure 1.12), then the magnitude of $x_0$ is the norm of the vector ( $|\mathbf{u}| = x_0$ in Figure 1.13) and displacement is the initial dimension of the spherical quantity ( $\Psi = C = meter$ in SI unit). As the ball moves to a certain position with displacement $x$, vector u rotates to a position that forms an angle $\alpha = \arccos(x/x_0)$ with the abscissa axis, and the spherical quantity becomes $\Psi = (C\cos\alpha - iC\sin\alpha)$ . As the ball reaches the equilibrium position ($x$=0), vector u reaches the ordinate axis, and the spherical quantity becomes $-d\Psi/dt = -C\omega\sin\alpha$ with $\alpha = \pi/2$. In terms of SI unit, the spherical quantity is $\Psi = C\omega = meter/second$ . At this moment, the spherical quantity denotes velocity dimension instead of displacement. When $\alpha$ rotates from 0 to $\pi/2$ , displacement is decreasing while velocity increases (see Figure 1.7). A spherical quantity in dynamic calculus, represented by rotatory vector u, describes both trends. In SI unit, the spherical quantity is $\Psi = meter \cdot second^{-\sin\alpha}$ when $0 < \alpha < \pi/2$ with the minus sign indicating anti-parallel alignment of space and time dimensions. Dimension transformation of a spherical quantity from the abscissa axis to the ordinate axis accounts for the transformation between physical quantities of displacement and velocity, or the conversion between potential energy and kinetic energy of the ball under the influence of springs.

So is a spherical quantity measurable? Yes. The state of a spherical quantity is always expressed as a sine or cosine function, which is the ratio of two physical quantities. For example, in the above case, the space component of the spherical quantity is $\cos\alpha = x / x_0$ where $x_0$ is the maximum displacement off the equilibrium position and $x$ is the current displacement; and the time component of the spherical quantity is $\sin\alpha = s / s_0$ where $s_0$ is the maximum speed at the equilibrium position and $s$ is the current speed. Spherical quantities can be derived from physical quantities.

**Table 1.1. A property comparison of physical quantity and spherical quantity**

| Quantity | Physical quantity | Spherical quantity |
|---|---|---|
| Variable | Quantity changes, dimension fixed | Dimension transforms, Quantity fixed |
| Calculus | Infinitesimal calculus | Dynamic calculus |
| Complex notation | Imaginary part identifier $i$ | Dimension factor identifier $\omega$ |
| Favorable spacetime | Euclidean space Newtonian time | The atomic spacetime |
| Applicable laws | The laws of physics | The law of nature |
| Quantification | Measurable | As a ratio of two physical quantities |

Factor $\omega$ as a physical quantity of angular velocity is also a dimension factor in the spherical quantity. As expressed in equation (1.68), it describes the interaction between the ball and the spring, the struggle of the guided and the guider, and the unity of Newton's second law and Hooke's law that govern the system. As a dimension factor, $\omega^2 = -1$ indicates to and fro of an entity in the harmonic oscillation, the restoration of the initial dimension after two consecutive differentiations. The differences between physical quantities and spherical quantities are summarized in Table 1.1.

### 1.5.2. Spherical Quantity in Euclidean Space

As shown in Figure 1.15, if we interpret the oscillations of 1s electrons as spherical expansion and contraction in Euclidean geometry, then as electron $\Omega_0$ is undergoing sinusoidal expansion, the other electron $\Omega_2$ experiences harmonic contraction in the meanwhile. At any specific moment, the spherical radius of $\Omega_0$ depends on OB as point B orbits along a semicircle OBA, and the spherical radius of $\Omega_2$ depends on OC as point C orbits along a semicircle ACO, both being synchronized by complementary radian angles $\beta$ in Figure 1.15(a) and $\alpha$ in Figure 1.15(b). When point B reaches position A, electron $\Omega_0$ attains maximum volume and begins to wrap back following the case of $\Omega_2$ in Figure 1.15(b). On the other hand, when $\Omega_2$ diminishes to minimum space at point O, it starts to expand following the case of $\Omega_0$ in Figure 1.15(a). Both electrons inflate and deflate their spatial spheres harmonically with their center at the origin and their radii stretching out or drawing back via the dashed circle, like two cranks driving the dashed wheel. However, due to their dimensional difference, both electrons are orthogonal and not in the same physical state during the oscillations. They may exist as mutually inductive electric field and magnetic field (Figure 1.15c and 1.15d).

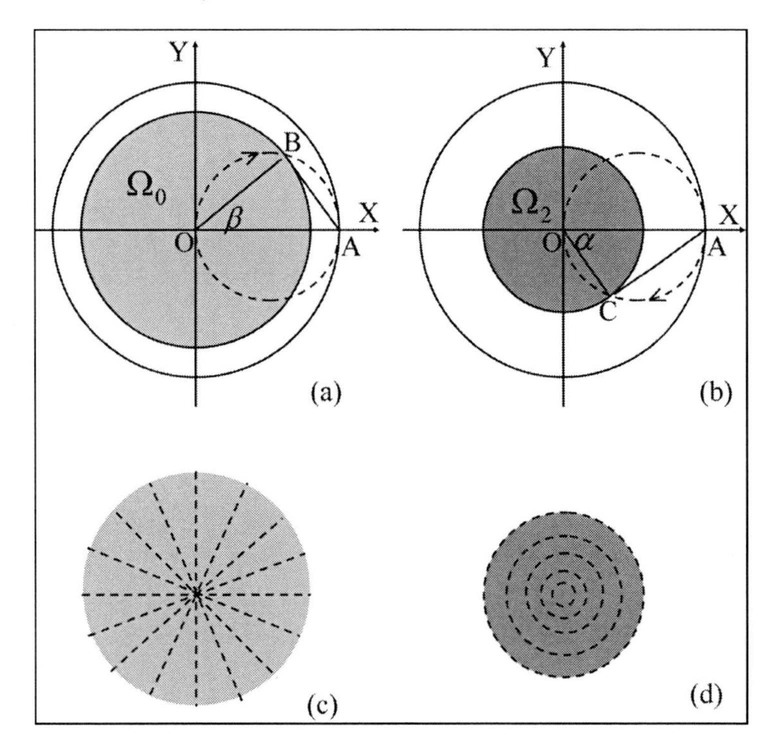


Figure 1.15. Interpretation of both 1s electrons under Euclidean space where shaded volumes were electron clouds.

Given the foregoing scenario, we may calculate the probability density of electron $\Omega_0$ within the spatial sphere in the case of Figure 1.15(a). At any specific moment, the spatial volume of the electron is:

$$V = \frac{4}{3}\pi R^3, \tag{1.70}$$

where the radius $R$ corresponds to chord OB along the dashed circle. If we designate the maximum radius OA as $R_0$, then

$$R = R_0 \cos\beta, \tag{1.71}$$

where radian angle $\beta$ is rotating from $\pi/2$ to 0 uniformly. Suppose the electron cloud distributes homogenously within the occupied sphere, then the electronic density must vary with the increment of the spherical volume as follows:

$$dP = -D\frac{4\pi R^2 dR}{V}, \tag{1.72}$$

where $dP$ mean the differential of electronic density, and $D$ is a constant. Integrating both sides of the equation yields

$$P = \int -D\frac{4\pi R^2}{V}dR. \tag{1.73}$$

Substituting equation (1.70) into (1.73) gives the probability density of the electron cloud at a specific R value as

$$P = -3D\ln(R), \tag{1.74}$$

which, upon normalizing $D$ value and ordering $R_0$=1, transforms into

$$P = -\ln(\cos\beta), \tag{1.75}$$

which indirectly involves two transcendental numbers: $\pi$ in trigonometry and $e$ in natural logarithm.

From an alternative perspective of two-dimensional spacetime, when the space component of the electron is $R_0 \cos\beta$, its time component is $R_0 \sin\beta$ so that

$$dP = -\frac{R_0 \sin\beta}{R_0 \cos\beta}d\beta, \tag{1.76}$$

which also leads to equation (1.75) upon integration. Figure 1.16 shows the logarithmic function with radius ratio $R/R_0$ in the range of (0,1). The graph indicates that electronic density is zero at the outer boundary of $R=R_0$ so that the electron never actually reaches radius $R_0$.

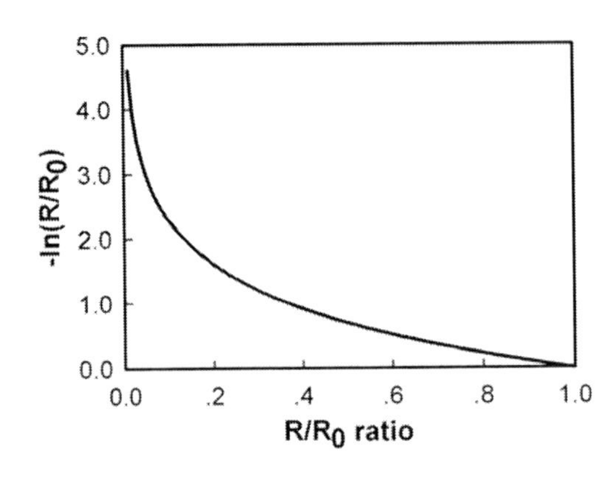


Figure 1.16. Natural logarithmic function $-\ln(R/R_0)$ as the probability density of an electron at R distance from the nucleus where $R_0$ denotes the maximum radius.

On the other hand, as $R$ approaches 0 at the origin, the electronic density becomes infinitely large, but $R$ cannot be assigned to 0 because it is a fraction denominator in equations (1.72) and (1.73). We assume that the electron is neutralized or annihilated at the nuclear center. Thus the sphere of the electron is an open set in Euclidean space. By statistics, this probability density is the chance of finding the electron at $R$ distance from the nucleus upon certain reliable physical conversion from the continuous medium to discrete counts. Although the mathematical result looks simple, exact and elegant, this tentative deduction should be further verified theoretically and experimentally in the future.

### 1.5.3. Introduction to Spherical View

Based on the mere assumption that electrons observe harmonic oscillation equations, we have given an orbital interpretation of space and time orthogonality in two-dimensional world. Time and space are inherently associated with physical entities that instantiate them. They are symmetric and relative quantities referring to two modes or two dimensions in spacetime as were represented by two electrons in helium. In the atomic spacetime, space is no more a conventional three-dimensional volume, and time is no more a unidirectional flow. The meanings of space and time are different from their usual meanings. Space still has X, Y, and Z orientations in Euclidean space, so does time, but it is more proper to adopt sine and cosine functions in various dimensions ($1, \omega, v, r$) to characterize their intricacy than to use isotropic and linear X, Y, and Z coordinates independent of real objects. Spherical quantity time is true physical existence that transforms sinusoidally in a certain dimension with relation to space; and spherical quantity space is true physical existence that transforms sinusoidally in a certain dimension with relation to time. The atomic spacetime is continuous via differential and integral operations. This new outlook of spacetime is termed spherical view, beyond our usual mental concept. Spherical view is the scientific approach of adopting spherical quantities to characterize natural phenomena.

We have defined electronic motion by harmonic oscillation, formulated harmonic oscillation by spherical quantity in duality equation, explained the equation by dynamic calculus, implemented dynamic calculus by trigonometry, and interpreted the trigonometry in terms of density probability in Euclidean space. The atomic spacetime is about spherical quantity in dynamic calculus; it is not a linear vector space with a uniform time flow, neither Euclidean space nor Hilbert space.

As infinitesimal calculus and complex number drive mathematics towards idealism farther and farther, our new definition of dynamic calculus and dimensional interpretation of complex number bring it back to reality. However, granting the new interpretation of space and time does not mean to overthrow existing successful theories. Just as Einstein's relativity does not invalidate Newtonian mechanics, but sees objects in a wider scope and effectively extends it, our example here supplements physical quantities with spherical quantities for the description of electronic orbitals in the atomic spacetime where classical mechanics ceases to be effective and the power of quantum mechanics is so limited.

Because of the great success of Newtonian mechanics, the biosphere of the earth is customarily assumed to be three-dimensional space with one-dimensional time from the perspective of human beings. But it might not be the case when observed from the perspective of an animal, say a worm under the ground, or from the perspective of the planetary god or extraterrestrial being, if exists at all. We never know how a virus sees its living environment, or how a plant seed feels about itself in the air and in the soil. In order to achieve real objectivity in the quest for the truth of nature, one must try to perceive the world from an alternative subject, at a different angle, and by a new standard. Such original perceptions and tastes are this book all about.

By viewing space and time in a helium atom differently from conventional three-dimensional space with one-dimensional time medium, we are actually examining the world from a new perspective and by a fresh mindset. Just as things may take various shapes from different angles, it is not surprising that the results we get might be quite different from those of quantum mechanics. For instance, a Chinese sees an apple in red color whereas an Englishman weights it half a pound. In this case, both are right and compatible. It is easy to understand that different perspectives and approaches can be taken to attack a problem. In analytic geometry, an equation in polar coordinates is normally different from that in Cartesian coordinates. The question is not which one is correct but which one is more informative and elegant to tackle the problem under consideration. Indeed, our theory could be established without challenging any successful theories, but adding fresh insight to them instead.

Although this introductory chapter looks deceptively naïve, it is merely the beginning of a series of discoveries with two-dimensional helium as the simplest case of the atomic spacetime. A unique perspective by a single mind is required to develop a complex theory coherently. The fundamental law of nature will not be easily explained in a couple of pages because it actually relates to every corner of science. We shall unfold this great beauty step by step leading to its full feather, but we still cannot exhaust every feature of it, nor can we integrate it with established sciences in all aspects. Our solutions and logic may seem radical or even perverse on occasion, which may attract many criticisms from conservative physicists. Readers should grasp the heart and soul of spherical view for integrity instead of probing the details impatiently even though every specific issue is believed to be important. We therefore urge you to put up with the rough until the end of the book.

*Chapter 2*

# THE THEORY OF QUATERNITY

Electronic motion does not like mechanical movement in Euclidean space with Newtonian time where linear algebra is applied to physical quantities. Electronic motion obeys electromagnetism in the atomic spacetime where dynamic calculus of spherical quantities is the rule. To extend two-dimensional spacetime model in helium to four-dimensional in neon shell, here we put forward a novel spacetime theory, defining four space dimensions in geometry, trigonometry, and calculus consistently in the context of harmonic oscillations. Four-dimensional spacetime is explained and instantiated by 2s2p electron octet, which are built mathematically upon the wave functions of both 1s electrons in helium shell. Ten electrons within a neon atom constitute a multi-dimensional spacetime continuum where wave transmission from 1s electrons towards 2s2p electrons is realized by dimension encapsulation mechanism, and wave reflection inwards from neon shell to helium shell follows calculus chain rule.

## 2.1. FOUR SPACE DIMENSIONS

In Chapter 1, we introduced spherical quantities for describing electronic orbitals within a helium atom. Filled by both 1s electrons, helium shell is a perfect two-dimensional spacetime continuum. However, the most important spacetime is of four dimensions, as will be found in neon shell with 2s2p electron octet. We think that the best way to delineate four-dimensional space is to define four geometric elements corresponding to four dimensions. This approach is markedly different from modern geometry, which tends to give pure abstraction (such as $n$-dimensional hyperspace) instead of intelligible pictures. When mathematical formulation dissociates from reality, it is losing the ground and value and sailing to nowhere. Therefore, we adhere to laying out quaternity spacetime on a solid basis through geometric and physical instantiations. Frankly speaking, accurately identifying four orthogonal geometric elements was a milestone in the early development of the atomic spacetime theory. There are definitely four orthogonal geometries or elements in Euclidean space corresponding to four dimensions, but they are not so obvious at first glance or on first thought. This is because we are not taught in traditional geometry at schools in that way. So please keep an open mind in searching the basic elements of geometry in the following sections.

### 2.1.1. Four Geometric Elements in a Sphere

How to divide a spatial sphere into four dimensions logically is the subject of this section. Euclidean geometry is the de facto foundation of all knowledge of space, so we shall begin our discussion from it. Nowadays, every middle school student makes acquaintance with the basic elements of geometry: points, lines, planes, and solids. Two points determine a one-dimensional straight line and two parallel lines determine a flat plane. Besides straight lines and flat planes, there are circular arc and spherical surface. The relationship between a circle and straight lines deserves further scrutiny, so does that between a spherical surface and flat planes.

Because the center of a circle is the most unique and pivotal point of it, we shall study a circle from that standpoint. As shown in Figure 2.1, a circle corresponds to two lines and two points. For every point $a$ in the circle, we can always find a point $A$ in the lines corresponding to it except at points $P_1$ and $P_2$. Conversely, for every point $B$ in line $L_1$ or $L_2$, we can always find a point $b$ in the circle corresponding to it. There are projective correspondences between the circle and two lines and two points. Points $P_1$ and $P_2$ and lines $L_1$ and $L_2$ are two basic elements of the circle. If we regard the two points as a geometric dimension and the two lines as another, then the circle contains two dimensions. Both dimensions are, of course, not the same. Like in number 88, the first digit referring to eighty and the second digit to eight only, a circle has two dimensions of different geometric structures, two points and two straight lines. Thus a circle has two dimensions as if we would say that number 88 has two digits.

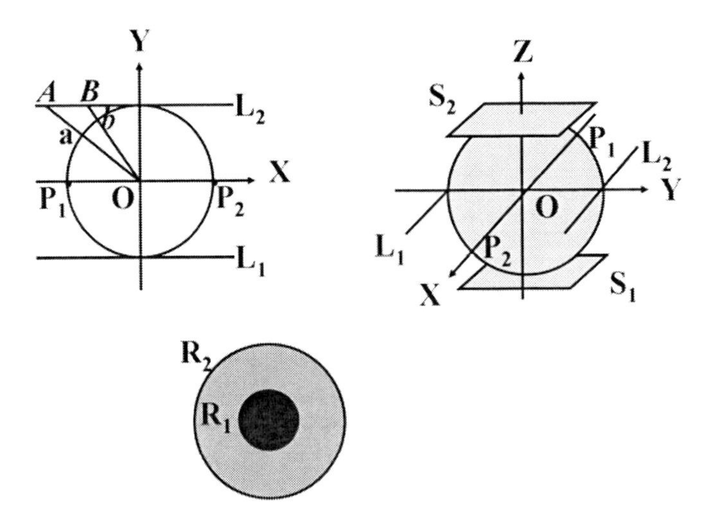


Figure 2.1. Geometric analyses of dimensions in a spherical layer. A circle is projectively decomposed into two lines and two points; a spherical surface into two planes $S_1$ and $S_2$, two lines $L_1$ and $L_2$, and two points $P_1$ and $P_2$; and a spherical layer is an outer solid sphere $R_2$ minus an inner solid sphere $R_1$.

By the same way, a spherical surface is composed of two flat planes, two straight lines, and two points. There is a one to one point projective correspondence between a spherical surface and three mutually exclusive elements: $P_1$ and $P_2$, $L_1$ and $L_2$, and $S_1$ and $S_2$. In other words, these three geometric elements are a distorted expression of the spherical surface.

Thus a spherical surface possesses three dimensions of various geometric structures, the third one being planes $S_1$ and $S_2$.

Because two points, $P_1$ and $P_2$, may determine a straight line, we regard these two points as one-dimensional in shape; because two parallel lines, $L_1$ and $L_2$, may define a flat plane, we treat both parallel lines as two-dimensional; and likewise, we regard parallel planes $S_1$ and $S_2$ combined together as a three-dimensional geometric element. Thus a spherical surface has three dimensions in the sense that it is composed of three structural elements: one-dimensional dual points, two-dimensional dual lines, and three-dimensional dual planes. This is analogous to say that number 888 has three digits: ones, tens, and hundreds.

A solid sphere has one additional dimension in the radial direction compared with a spherical surface. So we regard it as a four-dimensional shape. A radius represents a dimension, the addition of which to a spherical surface yields a solid sphere. A spherical layer is an outer sphere minus a concentric inner sphere or the medium between two concentric spherical surfaces. As we study a spherical layer, the outer sphere is treated as a four-dimensional space whereas the inner sphere as a zero-dimensional point. In traditional geometry, a point is assumed to be infinitesimal in size, but here a point has the size of the inner sphere. Since the inner sphere may be treated as a point, the spherical layer may be treated as a solid sphere as well, ignoring the existence of the inner sphere. Throughout this book, a sphere always means a solid spherical object, not a spherical surface.

Four space elements (spheres, planes, lines, and points) that we have defined are of various spatial significance. This is somewhat analogous to number expression. For example, in number 8888, the first 8 means eight thousand, the second means eight hundred, and so forth. Four digits actually denote different magnitudes. In fact, it is such a kind of structural grading that makes it feasible to express a large number. Quaternity space follows the same logic. Four space elements form four rungs in a ladder (i.e., four, three, two, one, and zero) in terms of dimensions, contrary to Cartesian coordinates where dimensional properties of X, Y, and Z orientations are assumed to be the same. Due to their grading, four various dimensions are able to express properly a spherical layer. The logic of combining various dimensions to account for a spherical layer is similar to that of adopting various digits to express a large number (Table 2.1). We focus on the geometry of a spherical layer because it represents a large class of atomic shells, such as 2s2p electron octet within a neon atom.

**Table 2.1. Various geometric elements within a spherical layer analogous to various digits in a decimal number**

| Geometries | | Number expression | |
|---|---|---|---|
| An inner sphere | Zero-dimensional | 0 | Starting point |
| Two points | One-dimensional | 8 | First digit |
| Two lines | Two-dimensional | 80 | Second digit |
| A Circle | Has two dimensions | 88 | A two digit number |
| Two planes | Three-dimensional | 800 | Third digit |
| A spherical surface | Has three dimensions | 888 | A three digit number |
| A radius (of outer sphere) | Four-dimensional | 8000 | Fourth digit |
| An outer sphere | Has four dimensions | 8888 | A four digit number |

### 2.1.2. Orthogonality and Scalability of Four Dimensions

We have characterized a spherical layer by four dimensions: two spheres, two planes, two lines, and two points. As was shown in Figure 2.1, points $P_1$ and $P_2$ are along X-axis; the perpendiculars to lines $L_1$ and $L_2$ are along Y-axis; and the normal lines to planes $S_1$ and $S_2$ are along Z-axis. These three geometric elements are mutually exclusive in positions and mutually perpendicular in orientations. Together they form a spherical surface. As the fourth dimension, the radius of the sphere is perpendicular to the spherical surface. Thus, four space dimensions are mutually orthogonal in Euclidean geometry.

In the atomic spacetime, orthogonality is defined as exact dimension apart in spacetime. As was mentioned previously, points $P_1$ and $P_2$ determine a one-dimensional line; lines $L_1$ and $L_2$ determine a two-dimensional plane; planes $S_1$ and $S_2$ determine a three-dimensional surface; and spheres $R_1$ and $R_2$ determine a four-dimensional solid. Thus these four geometric elements are orthogonal because of whole dimension interval between them (Figure 2.2).

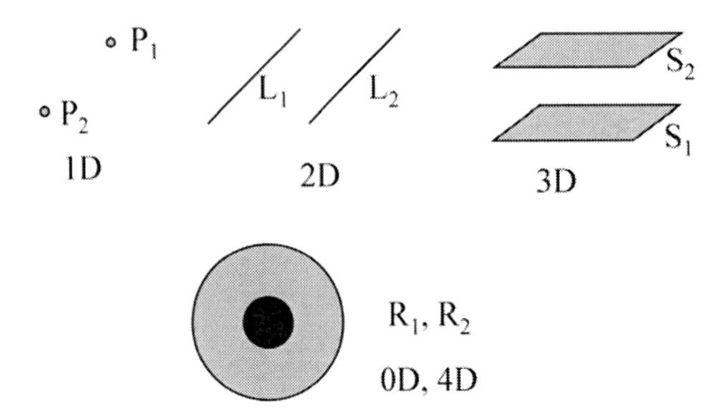


Figure 2.2. Four orthogonal elements of geometry for characterizing a spherical layer with reference to Figure 2.1.

There are interesting differential and/or integral relationships between various spherical structures. As shown in Figure 2.3, if we start with a four-dimensional outer sphere $E_0$ and reduce it a dimension, then we get a spherical surface $E_1$. The dimension reduced is the highest dimension of the geometric structure, a space dimension oriented in radial direction. Such an action is like taking away 8000 from number 8888 because a radius is four-dimensional. We may write this process in a differential way as:

$$\frac{dE_0}{dl_0} = E_1, \tag{2.1}$$

where $l_0$ denotes a space dimension in radial direction. If we further deduct a dimension from the spherical surface, we get a circle $E_2$. The immediate space dimension taken away is represented by both parallel planes $S_1$ and $S_2$, which has three-dimensional space property. The action is analogous to subtracting 800 from number 888. Likewise, reducing a subsequent dimension from the remaining circle leads to two symmetric points $E_3$ about the center; and reducing the last dimension from $E_3$ exposes the central point $E_4$, a zero-dimensional inner sphere. These geometric transformations may be expressed by differentiations of

$$\frac{dE_1}{dl_1} = E_2; \ \frac{dE_2}{dl_2} = E_3; \ \frac{dE_3}{dl_3} = E_4 \qquad (2.2)$$

where $l_1$ is a three-dimensional geometric element represented by two parallel planes, $l_2$ is a two-dimensional geometric element represented by two parallel lines, and $l_3$ is a one-dimensional geometric element represented by two points dislocated from the spherical center. A differentiation upon a geometric structure is always operated with respect to the highest outstanding dimension, analogous to taking away the leftmost digit from a number.

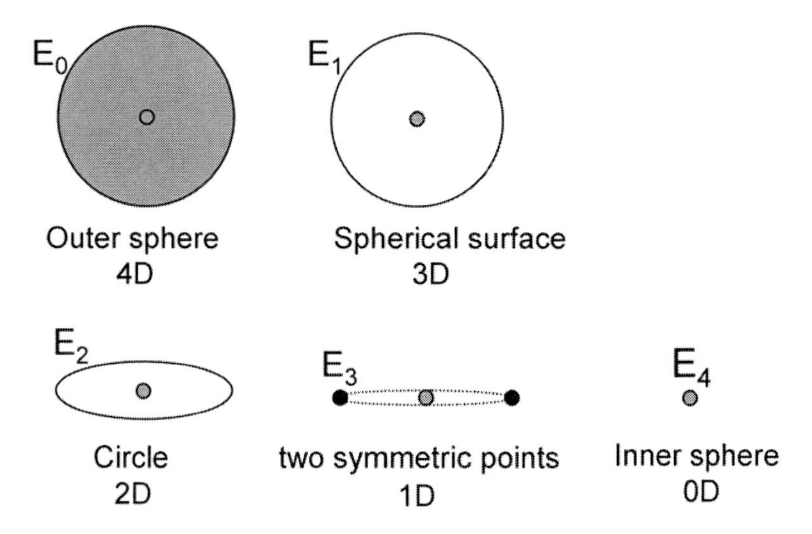


Figure 2.3. Calculus relationships between spherical structures where the inner sphere are hidden and located at the geometric centers until the last one.

It is interesting to find that a sphere reduced four dimensions consecutively produces the same shape of different size. The four space dimensions reduced are a radius, two planes, two lines, and two points in sequence, and the final result is a smaller sphere represented by an inner radius. As we study a well-organized structure with multiple spherical layers, the dimension of a specific sphere depends on whether it is the outer sphere or the inner sphere within the spherical layer of concern. If it is the outer sphere, then it must be regarded as containing four dimensions in space; and if it is the inner sphere, then it is a zero-dimensional point encapsulating all complexity of itself. That performing differential operations upon a sphere four times consecutively produces another smaller sphere gives scalability for the four dimensions, which can be peeled off in sequence repeatedly from a large and well-structured sphere, such as a krypton atom. As an exception, a two-dimensional sphere, after being reduced one dimension, results in a point immediately. This case has been explained in Chapter 1.

## 2.2. QUATERNITY SPACETIME

We have analyzed the dimensional structure of a sphere in space. We shall further explain the dimensional structure of a sphere as time component is added into the scene. The complexity of four-dimensional spacetime is remarkably greater than that of two-dimensional spacetime as if a neon atom is compared with a helium atom. The fundamental law of the atomic spacetime is spherical quantity in dynamic calculus. The appropriate initial concept on which to base the theory of quatenity is that of four-dimensional structure in a sphere. The most difficult part of quaternity is perhaps dimension encapsulation. If dynamic calculus and four-dimensional structure are two cornerstones of the theory of quaternity, then dimension encapsulation is the keystone of the mathematical architecture. So readers are encouraged to peruse the following material for grasping the delicate structure of a neon atom.

### 2.2.1. Symmetry and Complementarity of Time and Space

Geometric analysis indicates that a spherical layer possesses four orthogonal space dimensions. Because we view time and space as two relative and symmetric spherical quantities, a spherical layer must contain four time dimensions in a similar structure to space. Traditionally we take it for granted that time is one-dimensional and space is three–dimensional. After giving up this presumption and treating it as a special case in the atomic spacetime, we get four kinds of space and time combinations in a four-dimensional sphere (Table 2.2).

Space and time dimensions are complementary. When the total number of dimensions is set for a given entity, reducing a dimension in space inevitably increases a dimension in time accordingly, and vice versa. If we use a real number 1 to indicate a spherical quantity of four-dimensional time with zero-dimensional space, then by reducing a time dimension while increasing a space dimension we obtain a velocity dimension $u$ to indicate a spherical quantity of three-dimensional time with one-dimensional space. Dimension $u$ refers to velocity dimension in four-dimensional spacetime whereas $v$ refers to velocity dimension in two-dimensional spacetime, both belonging to different spherical layers.

**Table 2.2. Quaternity spacetime in a four-dimensional sphere based on four orthogonal dimensions**

| Quaternity axis | Space dimension | Time dimension | Projective geometries | Spherical layer correspondences |
|---|---|---|---|---|
| 1 | 0 or 4 | 4 or 0 | Two radii | Inner and outer spheres |
| $u$ | 1 | 3 | Two points | Two symmetric poles |
| $u^2$ | 2 | 2 | Two lines | Two semicircular arcs |
| $u^3$ | 3 | 1 | Two planes | Two hemispherical surfaces |

By the same manner, we use $u^2$ to represent a spherical quantity of two-dimensional time with two-dimensional space and $u^3$ to represent a spherical quantity of one-dimensional time with three-dimensional space. These four distinct combinations of time and space dimensions form four orthogonal axes in quaternity coordinates (Figure 2.4). The home axis indicates the current spacetime dimension of a spherical quantity. The purpose of quaternity coordinates is to illustrate the intricate dimensional relationships between various spherical quantities in dynamic calculus instead of facilitating metric measurements between positions in Cartesian coordinates.

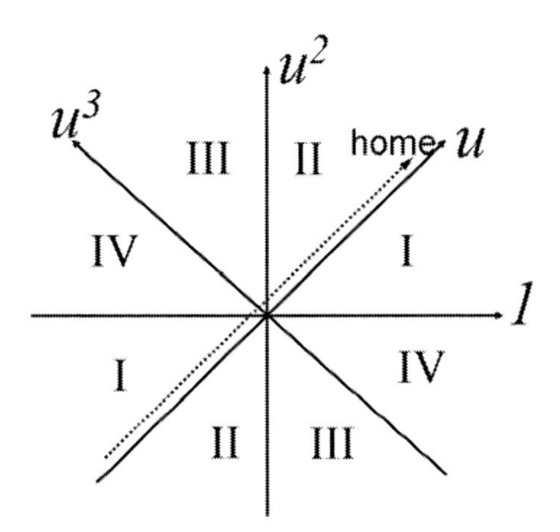


Figure 2.4. Four quaternity axes dividing a spacetime sphere into four spherical quadrants.

As shown in Figure 2.4, four quaternity axes divide a four-dimensional spacetime continuum into four spherical quadrants. The spacetime dimension between *1* and $u$ axes is the first spherical quadrant; the spacetime dimension between $u$ and $u^2$ axes is the second spherical quadrant; the spacetime dimension between $u^2$ and $u^3$ is the third spherical quadrant; and the spacetime dimension between $u^3$ and *1* axes is the fourth spherical quadrant. Spherical quantities in these four spherical quadrants are termed variously: a polor is a spherical quantity of the first spherical quadrant, a metor of the second, a vitor of the third, and a scalor of the fourth. Quaternity means four spherical quadrants in one sphere. For the literal meanings of these four spherical quantities, please refer to section 11.3.

Time and space are not necessarily symmetry in each spherical quadrant, but when we consider four spherical quadrants together, time and space are indeed symmetry. For example, the polor is in symmetry to the vitor with regard to space and time switch. Four spherical quadrants reflect the symmetry and complementarity of space and time in the whole sphere. We are not accustomed to multiple time dimensions. One way to grasp it is to ponder over physical quantity of acceleration with a unit of $m/s^2$ that contains two dimensions of time, getting the feel of propensity. In spherical view, multiple time dimensions define harmonic oscillations in multiple orientations consecutively.

The proposition of quaternity spacetime carries a further step beyond Minkowski's four-dimensional events where time is one-dimensional. Quaternity is not framed by that traditional worldview. To illustrate, suppose that we are observing the far away heavenly bodies from the earth, which is at the center of the imaginary celestial sphere, receiving the light from the stars. The distance between the earth and a star is best characterized by the time interval that light travels between them. The unit of light-year is normally used in this sense. So at the center of the celestial sphere, we are watching the stars located on a spherical surface, the radial direction being the coming light. The light showering on us indicates the flow of time. If the universe is a medium of one-dimensional time with three-dimensional space, then a spherical surface must be three-dimensional in space because it is exactly one dimension less along the radial direction than the whole celestial sphere, the dimension along the radial direction being the time flow. However, since the time flow can be expressed by a distance, the celestial sphere is indeed four-dimensional in space. Thus a celestial sphere may be characterized by one-dimensional time with three-dimensional space or by four-dimensional space only. If a dimension in time and in space may be traded according to certain perceptual interpretation, then the complementarity of time and space as was listed in Table 2.2 are understandable.

### 2.2.2. Rotatory Operation and Quaternity Equation

When the total number of spacetime dimensions is fixed for an electron, the electron decreasing a dimension in time inevitably increases a dimension in space accordingly, and vice versa. Because electronic motion undergoes concurrent differential and integral operations, i.e., a differentiation upon a spherical quantity with respect to a time dimension must be accompanied by an integral operation over a space dimension. This can be expressed in quaternity coordinates by a counterclockwise rotation of the home axis that reduces a time dimension while increasing a space dimension for a spherical quantity. If $\Phi$ is a spherical quantity, then its counterclockwise rotation $\Phi^0$ is defined as

$$\Phi^0 \equiv \int \left( -\frac{\partial \Phi}{\partial t} \right) dl \,, \qquad (2.3)$$

or

$$\frac{\partial \Phi^0}{\partial l} = -\frac{\partial \Phi}{\partial t} \,, \qquad (2.4)$$

where $l$ represents the highest space dimension within $\Phi^0$, $t$ represents the highest time dimension within $\Phi$, and the minus sign indicates that space and time dimensions are anti-parallel, aligning in opposite directions. The principle of rotatory operation applies to spherical quantities in various spherical quadrants. For example, if the current home axis of $\Phi$ is at the location of three-dimensional time with one-dimensional space, then a rotatory operation brings it to the location of two-dimensional time with two-dimensional space as $\Phi^0$ . The rotation of a polor results in a metor.

It is important to realize that the principle of rotatory operation is not invented out of pure imagination, but deeply rooted in Faraday's law, one of Maxwell's equations, as was demonstrated in Chapter 1 (see equation 1.21). We may explain the principle of rotatory operation in a simply way. For example, a boy grows into a man over years; and a seed develops into a tall tree over a certain time period. In both natural phenomena, time elapsing is in return for size gaining of the organisms. A spherical quantity losing a dimension in time is always accompanied by gaining a new dimension in space.

It can be seen from Figure 2.4 that rotating the home axis eight times consecutively brings it to the same axis position; and rotating the home axis four times consecutively brings it to the same axis but in the opposite direction. Because neon shell can be regarded as a four-dimensional spherical layer, a proper spherical quantity describing electronic oscillations within it must satisfy the following equations:

$$\Phi^{08} = \Phi , \tag{2.5}$$

$$\Phi^{04} = b\Phi , \tag{2.6}$$

where $\Phi^{08}$ and $\Phi^{04}$ denote rotating counterclockwise eight times and four times from spherical quantity $\Phi$ respectively, and $b$ is a dimension parameter. Since the distance between $\Phi^{04}$ and $\Phi$ dimensions is four axes apart, $b$ actually bears a spacetime factor of $u^4$ so that equation (2.6) can also be expressed in partial derivative form as

$$\frac{\partial^4 \Phi}{\partial t^4} = u^4 \frac{\partial^4 \Phi}{\partial l^4} , \tag{2.7}$$

which we shall call quaternity equation.

The physical significance of quaternity equation can be understood in a simple way. As we compared duality equation with Schrödinger's equation in Chapter 1, we noted that quantum mechanics extends one-dimensional Schrödinger's equation into three-dimensional by introducing Laplacian operator, which sees three dimensions of space in X, Y, and Z orientations so that linear algebra works. In contrast to this, quaternity views the space represented by X, Y, and Z axes as possessing different levels of dimensions. As shown in Figure 2.5, supposing that we are trying to locate a particular seat in a great auditorium, we would first locate the floor number as is represented by Z direction, and then within that proper floor search the tier number as is represented by Y axis, and finally look for X position in that tier. Here, three-dimensional Z direction refers to the arrangement of all the floors, each of which contains multiple tiers; two-dimensional Y direction refers to the sequence of the tiers, each of which contains multiple seats; and one-dimensional X direction refers to the order of seats. This simple logic is not beyond our everyday living experience, but mathematicians have not distinguished the dimensional meanings between these three axes. The wave functions in three orientations must represent different meanings and constitute three orthogonal dimensions. The natural shifting of the dimensions from floor order, to tier order, to seat order in locating a particular seat in the auditorium reflects the rule of dynamic differentiation, which is always operated with respect to the most immediate or highest dimension. In formulating multi-dimensional wave equations, quaternity climbs up to higher rank of derivatives than the second and sees four space dimensions in $\partial / \partial l$ , $\partial^2 / \partial l^2$ , $\partial^3 / \partial l^3$ , and $\partial^4 / \partial l^4$ forms instead of adopting Laplacian operator, $\frac{\partial^2}{\partial x^2} + \frac{\partial^2}{\partial y^2} + \frac{\partial^2}{\partial z^2}$. Thus a spherical quantity obeys dynamics calculus instead of linear algebra.

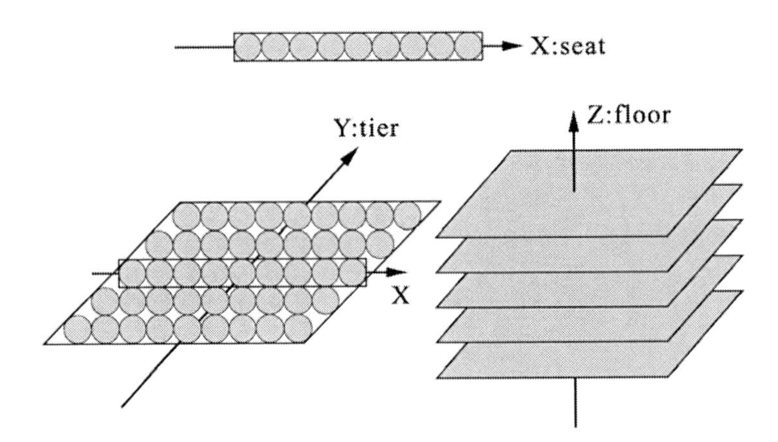


Figure 2.5. Even in a theater ticket, the meanings of X, Y, and Z axes in Euclidean geometry denote different levels of space dimensions.

Because the fourth derivative $\partial^4\Phi/\partial l^4$ contains the operations of the third, second, and first derivatives, quaternity equation governs electronic oscillations in four various dimensions. Quaternity equation holds true no matter where the home axis of a spherical quantity is. For example, if the home axis is at the four-dimensional time with zero-dimensional space position, then the equation governs the scalor. Likewise, if the home axis is at the three-dimensional time with one-dimensional space location, then the equation is for that specific polor as well. Spherical quantities in four spherical quadrants have four different kinds of space and time combinations. They satisfy quaternity equation simultaneously and individually in their respective dimensions. In other words, quaternity equation is an invariant form for four spherical quantities in four spherical quadrants. It is not a one-dimensional linear differential equation, but a four-dimensional harmonic oscillation equation, the solution to which will give spherical quantities for eight electrons in neon shell. Within neon shell, four electrons colonize four spherical quadrants in the positive directions of quaternity axes whereas another quartet colonize spherical quadrants in the negative directions of the same axes.

### 2.2.3. Electronic Orbitals in Neon Shell

Since electrons in neon shell occupy a spherical layer outside helium shell, spherical quantities for describing 2s2p are closely related to those for $1s^2$ electrons. We return to the first chapter where time and space components of 1s electrons are represented by

$$\Psi = C_1(\cos\alpha - i\sin\alpha); \tag{2.8}$$

$$\psi = C_2(\cos\beta + j\sin\beta). \tag{2.9}$$

As shown in Figure 2.6, semicircle ACB represents the course of dynamic calculus upon spherical quantities $\Psi$ and $\psi$. Both $C_1\cos\alpha$ and $-C_1 i\sin\alpha$ terms together determine a time state point C on semicircle ACB; and both $C_2 \cos\beta$ and $C_2 j \sin\beta$ terms together determine a space state point C on the same semicircle. As point C travels from A towards B along the semicircle, a 1s electron transforms into another by rotatory operation:

$$\Omega_0^{\ 0} = \Omega_2 , \tag{2.10}$$

which indicates orthogonality between both 1s electrons. When treated as a spherical quantity, $\Psi$ is measured by chord BC= $C_1 \cos\alpha$ and chord AC= $-C_1 i \sin\alpha$ ; but when treated as a radian angle, $\Psi$ is measured by arc AC, which increases from 0 to $\pi/2$ during the differentiation. As radian angle, the whole semicircle ACB corresponds to circumferential angle $\pi/2$ in dimension diagram. Arc $\Psi$ equals radian angle $\alpha$ provided that the differential course is a strict semicircle. But they measure radian angles in different scopes. Radian angle $\alpha$ belongs to helium shell while radian angle $\Psi$ belongs to neon shell. As point C travels from A towards B, arc BC is retreating while arc AC is stretching. If in the scope of neon shell functions $\Psi$ and $\psi$ represent both arcs AC and BC respectively, then they are complementary in terms of radian angles:

$$\Psi + \psi = \pi/2 . \tag{2.11}$$

Electronic orbitals are in ever-changing dynamic state. At any moment, state point C is codetermined by four segments: $C_1 \cos\alpha$ , $-C_1 i \sin\alpha$ , $C_2 \cos\beta$, and $C_2 j \sin\beta$ in right triangle ACB. This redundancy check on electronic state reinforces synchronization of time and space components and complementarity of radian angles $\Psi$ and $\psi$ during the rotatory operation.

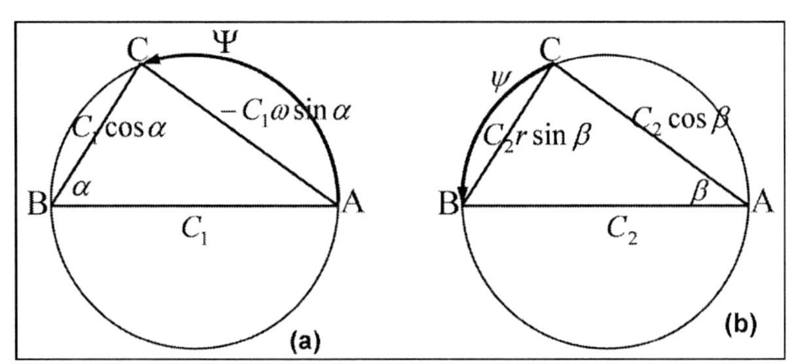


Figure 2.6. Synchronized spherical quantities $\Psi$ and $\psi$ as complementary radian angles in bold arcs in (a) time and (b) space dimension diagram.

With $\Psi$ and $\psi$ representing time and space radian angles respectively, we shall construct eight spherical quantities to represent $2s^2 2p^6$ electrons in neon shell. Given an initial function of

$$\Phi_{00} = A_1 A_2 \cos\Psi \cos\psi \tag{2.12}$$

where $A_1$ and $A_2$ are time and space dimension constants, we may perform a rotatory operation on it as follows:

$$\begin{aligned}\Phi_{00}^{0} &= -\frac{dA_1 \cos\Psi}{dt} \int A_2 \cos\psi \, dl \\ &= (A_1 \frac{d\Psi}{dt} \sin\Psi) \cdot (-A_2 \frac{1}{\psi'} \int \cos\psi \, d\psi) \\ &= -A_1 A_2 (-\frac{1}{\psi'} \dot{\Psi}) \sin\Psi \sin\psi \\ &= -A_1 A_2 u \sin\Psi \sin\psi,\end{aligned} \tag{2.13}$$

where

$$\dot{\Psi} = -\frac{d\Psi}{dt}, \tag{2.14}$$

$$\psi' = -\frac{d\psi}{dl}, \tag{2.15}$$

$$\begin{aligned}u &= -\frac{1}{\psi'} \cdot \dot{\Psi} = \frac{dl}{d\psi} \cdot (-\frac{d\Psi}{dt}) = \Omega_0^0 \\ &= \int C_2 \cos\beta dl \cdot (-\frac{dC_1 \cos\alpha}{dt}) \\ &= (r \int C_2 \cos\beta d\beta) \cdot (-C_1 \omega \sin\alpha) \\ &= -r\omega C_1 C_2 \sin\alpha \sin\beta \\ &= -v C_1 C_2 \sin\alpha \sin\beta \\ &= \Omega_2,\end{aligned} \tag{2.16}$$

where $dl/d\psi$ expands as an integral operation. After the execution of $\frac{1}{\psi'} \int \cos\psi \, d\psi = \frac{1}{\psi'} \sin\psi$ in equation (2.13), $dl/d\psi$ follows integral chain rule. Integral chain rule parallels differential chain rule in dynamic calculus. We made use of equations (1.3), (1.6), and (1.7) in the above calculation. Through consecutive rotatory operations, we obtain four spherical quantities for $2s^1 2p_x^1 2p_y^1 2p_z^1$ in the positive directions of quaternity axes:

$$\begin{pmatrix} \Phi_0 \\ \Phi_1 \\ \Phi_2 \\ \Phi_3 \end{pmatrix} = \begin{pmatrix} \Phi_{00} - \frac{\partial \Phi_{00}}{\partial t} \\ \Phi_{00}^{0} - \frac{\partial \Phi_{00}^{0}}{\partial t} \\ \Phi_{00}^{02} - \frac{\partial \Phi_{00}^{02}}{\partial t} \\ \Phi_{00}^{03} - \frac{\partial \Phi_{00}^{03}}{\partial t} \end{pmatrix}, \tag{2.17}$$

where $\Phi_{00}^{02}$ and $\Phi_{00}^{03}$ indicate performing counterclockwise rotations upon function $\Phi_{00}$ twice and thrice respectively. Substituting the initial value of $\Phi_{00}$ into the matrix produces

$$\begin{pmatrix} \Phi_0 \\ \Phi_1 \\ \Phi_2 \\ \Phi_3 \end{pmatrix} = A_1 A_2 \begin{pmatrix} \cos\Psi\cos\psi - \dot{\Psi}\sin\Psi\cos\psi \\ -u(\sin\Psi\sin\psi + \dot{\Psi}\cos\Psi\sin\psi) \\ u^2(\cos\Psi\cos\psi - \dot{\Psi}\sin\Psi\cos\psi) \\ -u^3(\sin\Psi\sin\psi + \dot{\Psi}\cos\Psi\sin\psi) \end{pmatrix}. \tag{2.18}$$

These four spherical quantities belong to four different spherical quadrants. They are the polor, the metor, the vitor, and the scalor in quaternity spacetime. After carrying out four consecutive rotatory operations upon $\Phi_{00}$, we arrive at function

$$\Phi_{40} = A_1 A_2 u^4 \cos\Psi\cos\psi\ . \tag{2.19}$$

At this moment, time and space radian angles switch their roles, i.e., $\Psi$ represents space radian angle while $\psi$ denotes time radian angles for further rotatory operations. For example,

$$\begin{aligned}\Phi_{40}^{0} &= -u^4 \frac{dA_2 \cos\psi}{dt} \int A_1 \cos\Psi dl \\ &= A_1 A_2 u^4 \frac{d\psi}{dt} \sin\psi \cdot \left(-\frac{1}{\Psi'} \int \cos\Psi d\Psi\right) \\ &= -A_1 A_2 u^4 \left(-\frac{1}{\Psi'}\dot{\psi}\right) \sin\psi \sin\Psi \\ &= -A_1 A_2 u^5 \sin\psi \sin\Psi,\end{aligned} \tag{2.20}$$

where

$$\dot{\psi} = -\frac{d\psi}{dt}, \tag{2.21}$$

$$\Psi' = -\frac{d\Psi}{dl}, \tag{2.22}$$

$$\begin{aligned}
-\frac{1}{\Psi'}\dot{\psi} &= \frac{dl}{d\Psi}(-\frac{d\psi}{dt}) \\
&= (\int C_1 \cos\alpha dt \frac{dl}{dt})(-\frac{dC_2 \cos\beta}{dl}\frac{dl}{dt}) \\
&= (-v\frac{1}{\omega}\int C_1 \cos\alpha d\alpha)(C_2 v \sin\beta \frac{d\beta}{dl}) \\
&= -C_1 C_2 \frac{v^2}{\omega r}\sin\alpha \sin\beta \\
&= u.
\end{aligned} \tag{2.23}$$

By the same manner, we may derive spherical quantities for another $2s^1 2p_x^1 2p_y^1 2p_z^1$ from

$$\begin{pmatrix} \Phi_4 \\ \Phi_5 \\ \Phi_6 \\ \Phi_7 \end{pmatrix} = \begin{pmatrix} \Phi_{40} - \dfrac{\partial \Phi_{40}}{\partial t} \\ \Phi_{40}^{0} - \dfrac{\partial \Phi_{40}^{0}}{\partial t} \\ \Phi_{40}^{02} - \dfrac{\partial \Phi_{40}^{02}}{\partial t} \\ \Phi_{40}^{03} - \dfrac{\partial \Phi_{40}^{03}}{\partial t} \end{pmatrix}, \tag{2.24}$$

which expands into trigonometric expression as

$$\begin{pmatrix} \Phi_4 \\ \Phi_5 \\ \Phi_6 \\ \Phi_7 \end{pmatrix} = A_1 A_2 u^4 \begin{pmatrix} \cos\psi\cos\Psi - \dot{\psi}\sin\psi\cos\Psi \\ -u(\sin\psi\sin\Psi + \dot{\psi}\cos\psi\sin\Psi) \\ u^2(\cos\psi\cos\Psi - \dot{\psi}\sin\psi\cos\Psi) \\ -u^3(\sin\psi\sin\Psi + \dot{\psi}\cos\psi\sin\Psi) \end{pmatrix}. \tag{2.25}$$

This electron quartet $2s^1 2p_x^1 2p_y^1 2p_z^1$ is in the negative direction of quaternity axes. The quaternity axes *1* refers to a four-dimensional time with zero-dimensional space quantity whereas factor $u^4$ denotes a four-dimensional space with zero-dimensional time sphere, both at the same quaternity axis but in the opposite directions. It has been stated that the home axis rotated four times consecutively returns to the same axis but in the opposite direction; and the home axis rotated eight times consecutively returns to its original position in quaternity coordinates so that

$$u^8 = 1. \tag{2.26}$$

Physically, being in the opposite direction of a quaternity axis has an important effect. The radian angles $\Psi$ and $\psi$ switch their time and space roles so that a counterclockwise rotation has a reverse meaning, i.e., reducing a time dimension from a spherical quantity actually means increasing a time dimension from it while increasing a space dimension from a spherical quantity means reducing a space dimension from it. This seemingly contradictory arithmetic can be understood in a simple way. For example, as you travel towards south, after passing through the Antarctic pole, you are actually heading towards north if you still maintain the original traveling direction of "south". Similarly, a spherical quantity exceeding the extremum of $u^4$ begins to wrap back. Spherical quantities in equations (2.18) and (2.25) form four pairs of electrons with opposite spins, each pair being located at the same quaternity axis but in the opposite directions.

Every spherical quantity contains two trigonometric terms at a time dimension interval. This indicates that each electron has two mutually orthogonal waveforms or two interrelated properties, which make it rigid in shape. Since a rotatory operation is in essence a Maxwell's equation in quaternity spacetime, an electron may, for example, manifest as electric field strength and magnetic field strength simultaneously and transform in a certain direction.

Between eight electrons, there are invariantly rotation relationships between adjacent ones, e.g., between $\Phi_0$ and $\Phi_1$, $\Phi_1$ and $\Phi_2$, $\Phi_7$ and $\Phi_0$, and so on. For instance, if $\Phi_{00}$ and $\Phi_{01}$ represent the first and second terms of spherical quantity $\Phi_0$, then there is a partial differential relationship between them:

$$-\frac{\partial \Phi_{00}}{\partial t} = \Phi_{01}, \tag{2.27}$$

and $\Phi_{01}$ has an integral relation with $\Phi_{10}$, the first term of spherical quantity $\Phi_1$.

$$\int \Phi_{01} dl = \Phi_{10} \tag{2.28}$$

An electron may shift its state from one waveform to another and further migrate to the second electronic state according to rotation principle. Equations (2.27) and (2.28) represent the pathway or mechanism of electronic transformation from state $\Phi_0$ to state $\Phi_1$. Four spherical quantities in equation (2.18) follow the similar counterclockwise rotation pathway whereas the other four in equation (2.25) switch time and space roles of radian angles $\Psi$ and $\psi$ in differential and integral operations, and lastly the rotation of $\Phi_7$ results in $\Phi_0$. At this moment the time and space roles of $\Psi$ and $\psi$ switch back to the default.

Within neon shell, eight electrons, $2s^2 2p^6$, circulate according to the principle of rotatory operation. An electron of $\Phi_0$ is traveling in quaternity spacetime and evolving into $\Phi_1$ while $\Phi_1$ is changing into $\Phi_2$, and $\Phi_2$ into $\Phi_3$, and so on. Each electron is gradually shifting its state in the following link direction:

$$\Phi_0 \mapsto \Phi_1 \mapsto \Phi_2 \mapsto \Phi_3 \mapsto \Phi_4 \mapsto \Phi_5 \mapsto \Phi_6 \mapsto \Phi_7 \mapsto \Phi_0, \tag{2.29}$$

where arrow $\mapsto$ indicates rotation transformation from one spherical quantity to another. Wave function $\Phi_0$ is initially a 2s orbital with zero-dimensional space with four-dimensional time. It goes through three 2p-oribtals and then returns to a 2s orbital $\Phi_4$ step by step. The number of space dimensions is increasing while that of time dimensions is decreasing, i.e., electronic space is expanding while time wrinkles are releasing during the rotations. Spherical quantity at $u^4$ is a medium of four-dimensional space with zero-dimensional time. By this dimension, space reaches maximum and begins to dwindle while time condenses. The 2s-orbital $\Phi_4$ with a $u^4$ factor means that radian angles $\Psi$ and $\psi$ must switch time and space roles so that a rotatory operation will transform time and space in the opposite direction. After going through eight electronic states, the electron returns to its original waveform, and another cycle begins. Within neon shell, a factor of $u^8$ is equivalent to a dimensionless number *1*. The electronic circulation constitutes harmonic oscillations prettily expressed by quaternity equation. Eight electrons fulfill a complete cycle in rotation transformations, which explains satisfactorily why a full octet configuration of neon shell is a stable system.

Neon shell is a spacetime continuum filled by eight electrons in an oscillatory cycle. From the standpoint of a specific electron, quaternity spacetime includes not only the current electron itself but also the other seven electrons in the downstream and upstream of rotatory operation pathway, or in the future and the past of the current electron. The future and the past are interconnected forming the whole neon shell continuum. Eight electrons represent eight spacetime dimensions, which graduate the scale of the spacetime continuum. Time is the flow of electronic link in equation (2.29), each electron having a unique sinusoidal time flight in the river of rotatory operations. For example, the current time of electron $\Phi_2$ is the future of electron $\Phi_1$. This cyclic spacetime worldview is more fundamental than linear Euclidean space of infinitely long X, Y, and Z axes with uniform time flow, or absolute time for all.

### 2.2.4. Dimension Encapsulation

As we derived spherical quantities for electrons in neon shell, we simply treated spherical quantities $\Psi$ and $\psi$ as two radian angles. What is the basis for this usage? Of course, to treat a quantity as an angle, it is necessary to use a sine or cosine function to cast on it. What is the physical significance of trigonometric functions that implement the conversions of $\Psi$ and $\psi$ from spherical quantities to radian angles? This section tries to answer these critical questions unambiguously.

It has been established that dynamic calculus is implemented by trigonometry. Upon the time component of 1s electron, we may perform a differential operation

$$-\frac{d\Psi}{dt} = -C_1 \cos\alpha \,, \tag{2.30}$$

where the left-hand side may be interpreted as a dynamic fraction whose denominator is the magnitude of a full time dimension $t$ and numerator is the current time quantity $\Psi$. In Figure 2.6, this fraction is BC/AB where C moves from A towards B along the semicircle. The differential notation indicates cosinusoidal reduction process of quantity $\Psi$ from a full dimension to vanishing, which corresponds to $\alpha$ angle rotation from 0 to $\pi/2$. Thus, when expressing a spacetime component changing from a full dimension to vanishing cosinusoidally, we may logically cast cosine operator upon a proper radian angle. This is the fraction interpretation of trigonometric function and hence of differential operation. Recognizing that electrons are in harmonic oscillations so that their spherical quantities are transforming in sinusoidal pattern, we set forth two theorems with regard to dynamic calculus.

*Theorem 1.* For a spherical quantity $f$, the derivative form $-\partial f/\partial t$ indicates cosinusoidal change of its time component from a full dimension *t* to vanishing such as

$$-\frac{\partial f}{\partial t}=-B_1\cos\alpha\,, \tag{2.31}$$

where $B_1$ is a dimension constant orthogonal to $\cos\alpha$, $t$ is a fix time dimension, and the cosinusoidal decrease of time component in the range of $f(t, 0)$ corresponds to the rotation of radian angle in the range $\alpha$ (0, $\pi/2$).

*Theorem 2.* For a spherical quantity $g$, the integral form $\int gdl$ indicates cosinusoidal change of its space component from naught to a full dimension *l* such as

$$\int gdl = B_2\cos\beta\,, \tag{2.32}$$

where $B_2$ is a dimension constant orthogonal to $\cos\beta$, $l$ is a fix space dimension, and the cosinusoidal increase of space component in the range of $g(0, l)$ corresponds to the rotation of radian angle in the range $\beta$ ($\pi/2$, 0).

By interpreting calculus in terms of a trigonometric function, and the trigonometric function in terms of a varying fraction or numerator, we have provided fresh physical significance for calculus. Trigonometrically, the cosinusoidal value of differential process $-\partial f/\partial t$ can be explained by a chord in dimension diagram. As shown in Figure 2.7(a), in a circle with diameter PR of dimension *t*, as Q moves from P towards R along the semicircle, radian angle $\alpha$ rotates from 0 to $\pi/2$ and chord QR reduces from a full dimension *t* to vanishing following cosine function. The minus sign indicates the reducing tendency of the spherical quantity. On the other hand, the integral process $\int gdl$ is represented by a corresponding dimension diagram of Figure 2.7(b). The integral is neither indefinite integral nor definite integral, but dynamic integral implemented by trigonometry. As Q moves from P towards R along the semicircle, radian angle $\beta$ rotates from $\pi/2$ to 0 so that chord PQ increases from naught to a full diameter dimension *l* following cosine function. As an electron Q traces along arc PQR, its time and space components are measured by chords QR and PQ, respectively.

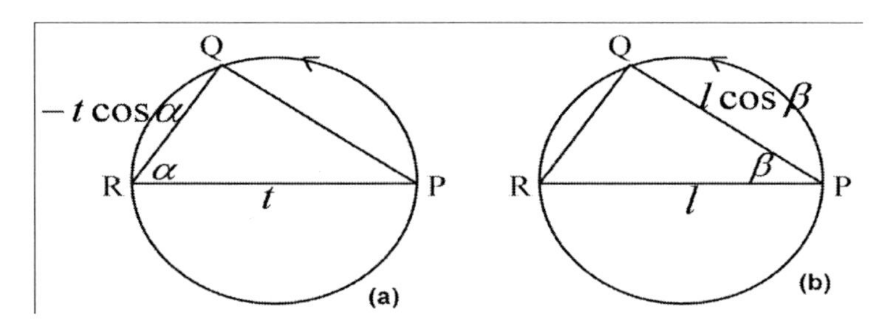


Figure 2.7. Trigonometric interpretation of dynamic calculus by cosinusoidal change of chords QR and PQ with point Q orbiting along the semicircle.

To express cosinusoidal change of time component within 1s electron $\Omega_0$, we had equation (2.30); and to express cosinusoidal change of time component within 2s electron $\Phi_0$, we may write by the same manner

$$-\frac{\partial \Phi_0}{\partial t} = -A_1 \cos \Psi, \tag{2.33}$$

where space component is omitted for brevity. The connection between both equations may be explained by rotatory vectors. As shown in Figure 2.8, vector $C_1$ rotates around origin O with its tip moving from A to B along arc APB, transforming spherical quantity $C_1 \cos \alpha$ into $-C_1 \omega \sin \alpha$ as radian angle $\alpha$ increases from 0 to π/2. Arc APB records the course of electronic transformation from $\Omega_0$ towards $\Omega_2$ in time component represented by $-C_1 \cos \alpha$ in equation (2.30). The tangential vector $A_1$ is another time dimension orthogonal to vector $C_1$. Shifting the perspective scope from helium shell to neon shell, we examine the motion of vector $A_1$ from the standpoint of P. Even though P is a moving point at Figure 2.8(a), it is treated as a stationary point at Figure 2.8(b), where vector $A_1$ rotates around P with its tip traveling from E towards F along arc EGF. Radian angle $\angle EPG$ (or arc AP) is $\Psi$, which has a phase $\pi/2$ ahead of $\alpha$. In other words, waveform $A_1 \cos \Psi$ is the future of waveform $C_1 \cos \alpha$ upon rotatory operation. As time component of a 1s electron in helium shell, $\Psi$ is represented by rotatory vector $C_1$; but under neon shell, it is treated as a radian angle, which equals $\alpha$ as was shown in Figure 2.6 disregarding $\pi/2$ phase difference between radian angles in both shells. So the cast of cosine operator on $\Psi$ in neon scope is justified mathematically.

Spherical quantity $\Psi$ is the solution to simple harmonic oscillation equation of 1s electron in time; and spherical quantity $\psi$ is the solution to simple harmonic oscillation equation of 1s electron in space. Simple harmonic oscillation is circular motion. So $\Psi$ and $\psi$ define a circle in time and a circle in space respectively (Figure 1.3). They represent a time dimension and a space dimension of helium shell respectively. After casting cosine operators upon $\Psi$ and $\psi$, we treat them as radian angles in the range between 0 and π/2. Trigonometric functions $\cos \Psi$ and $\cos \psi$ encapsulate both dimensions of helium shell.

As $A_1$ rotates from PE to PF, the time component of spherical quantity $\Phi_0$ transforms from $A_1 \cos \Psi$ to $-A_1 \dot{\Psi} \sin \Psi$ as $\Psi$ rotates from 0 to π/2. The rotation of $A_1$ from PE to PF corresponds to the movement of point P tracing arc APB, surrounding the boundary of 1s electrons. So both 1s electrons are enclosed by arc APB. As was shown in Figure 2.6(a), diameter $C_1$ is wrapped by arc ACB; and as was shown in Figure 2.8(a), vector $C_1$ is wrapped by arc APB. Allowing the radian angles to rotate a displacement of π/2, we are actually encapsulating helium shell in the scope of neon shell.

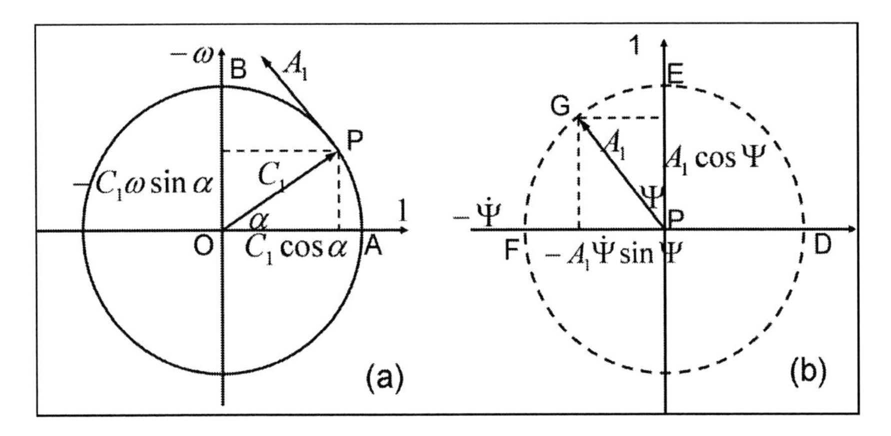


Figure 2.8. Rotatory vector interpretation of time component $\Psi$ as (a) spherical quantity $C_1(\cos\alpha - \omega\sin\alpha)$ and as (b) a radian angle encapsulating 1s electrons upon perspective shift.

Geometrically, function $\Psi$ in helium shell and $\cos\Psi$ in neon shell together determine a time dimension of 1s electrons. As was shown in Figure 2.6, a given length of chord $-C_1\omega\sin\alpha$ with subtending arc $\Psi$ (implemented by $\cos\Psi$ ) in a circle reflects the curvature of the circle and hence is an indicator of circular diameter $C_1$. To specify, when the arc subtending a fixed-length chord is in the radian range of (0, π/2), the more severely it curves, the smaller is its circular diameter, the smaller is $\cos\Psi$ value, and vice versa. The magnitudes of chord and its subtending arc codetermine the diameter of a circle. In this case, chord $-C_1\omega\sin\alpha$ and arc $\Psi$ codetermine the time dimension of helium shell.

Since from the perceptive of neon shell $\cos\Psi$ and $\cos\psi$ are two orthogonal diametrical indicators of a coiled time and space structure, the product of them encapsulates the spacetime block of both 1s electrons. As we built spherical quantity $\Phi_{00}$ based on $A_1A_2\cos\Psi\cos\psi$, we were actually dealing with a larger spatial sphere, i.e., using helium shell as the pattern to construct neon shell by dynamic calculus. In other words, we treated $A_1A_2\cos\Psi\cos\psi$ as zero-dimensional space with four-dimensional time in the scope of neon shell and derived spherical quantities for 2s2p electrons according to rotatory operation principle beginning from that spherical point. Such a trigonometric transformation reveals the mechanism of dimension curling up.

Since Theodor Kaluza proposed his fifth dimension curling up hypothesis in 1919, dimension encapsulation mechanism has not been demonstrated mathematically despite the popularity of trigonometric functions. Our wave functions construction has confirmed the phenomena of dimension curling up and instantiated the mechanism for hiding a sphere or two dimensions of helium shell from neon shell.

As a matter of fact, dimension encapsulation phenomena need not be esoteric. The instances of hidden dimensions are ubiquitous in our daily life. For example, as we are counting the members of the United Nations, the United States is one of them. We disregard the fact that there are 50 states within the United States because we treat the nation as a unit. The details of 50 states are hidden from direct visualization from the standpoint of the United Nations. Nonetheless, as we turn our focus towards the United States, we are interested in the 50 states that it comprises, but we still ignore the fact that the state of Delaware has three counties, namely Sussex, Kent, and New Castle, which are too detailed and irrelevant to the subject of study.

Thus dimension-hiding phenomena are closely associated with the scope of concern. For example, when we are talking about John's family, we usually refer to John and his wife, his daughters and sons. The details about John's daughter, her husband, and her kids are not of direct concern, which can be considered to be curled up dimensions in the scope of John's family. It is by the same logic that solid state physicists don't care about quarks.

### 2.2.5. Differential Chain Rule

The role conversion of $\Psi$ and $\psi$ from spherical quantities of 1s electrons to radian angles for 2s electrons is implemented by dimension encapsulation mechanism. Conversely, the role conversion of $\Psi$ and $\psi$ from radian angles in neon shell to spherical quantities in helium shell reflects the chain rule of dynamic calculus. For example, considering time component only, we have

$$\begin{aligned} -\frac{\partial \Phi_0}{\partial t} &= -A_1 \cos\Psi, \Psi \mapsto \Psi + \frac{\pi}{2} \\ &= -A_1 \dot{\Psi} \sin\Psi \\ &= -A_1 \sin\Psi \cdot (-C_1 \cos\alpha). \end{aligned} \tag{2.34}$$

where arrows $\mapsto$ indicate smooth rotation of the radian angles up to a displacement of $\pi/2$, and both differential operations are implemented by cosine functions stepwise. The first step of the differentiation $-\partial\Phi_0/\partial t$ is implemented by $-A_1\cos\Psi$ for radian angle $\Psi$ to increase $\pi/2$, which results in $-A_1\dot{\Psi}\sin\Psi$. This process is illustrated in Figure 2.8(b) through the rotation of vector $A_1$ around point P. Function $-A_1\cos\Psi$ of neon shell encapsulates the time dimension of helium shell. The second step of differentiation $-d\Psi/dt$ is hidden in neon shell by shorthand expression $\dot{\Psi}$. After the dimension of $-A_1\cos\Psi$ transforms into $-A_1\dot{\Psi}\sin\Psi$, $\dot{\Psi}$ is implemented by $-C_1\cos\alpha$ for radian angle $\alpha$ to increase $\pi/2$, which results in $-C_1\omega\sin\alpha$ finally. This second process is illustrated in Figure 2.8(a) through the rotation of vector $C_1$ from 0 to $\pi/2$. Both differential steps are coupled together with radian angle $\alpha$ lagging behind $\Psi$ a phase of $\pi/2$. In the first step, $\Psi$ is a radian angle; but in the second step, it is treated as a spherical quantity. The role changes due to perspective scope shift. Both differential steps are coupled together through differential chain rule.

It is worthwhile to note that $\dot{\Psi}$ is different from $\omega$ because $\dot{\Psi}$ contains a hidden dimension of $-C_1\cos\alpha$ to spread out by differential chain rule whereas $\omega$ is only a dimension constant. Angular velocity $\dot{\Psi}$ controls the pace of $-A_1\cos\Psi$ oscillation while radian angle $\Psi$ is indirectly determined by its corresponding chord $-C_1\omega\sin\alpha$ (Figure

2.6). Dimension constant $\omega$ is directly measured by the angular velocity with a value of $\pi/2\tau$ where $\tau$ is a quarter period of $-C_1 \cos\alpha$ oscillation. This constant ultimately sets the frequency for the resonances of all electrons. In other words, the electrons within the inner shell transmit motion to electrons in the outer shell.

As shown in equation (2.13), performing a rotatory operation upon $\Phi_{00}$ leads to spherical function $\Phi_{10} = -A_1 A_2 u \sin\Psi \sin\psi$ in neon shell where the dimension factor $u$ is implemented by $\Omega_0^0$ in helium shell (see equation 2.16). Rotatory operations in both shells are coupled together as if two cogwheels were geared to transmit circular motion (Figure 2.9). Helium shell is inside neon shell and their rotatory directions are all counterclockwise. As rotatory operations move on, we have $\Phi_{10}^0 = A_1 A_2 u^2 \cos\Psi \cos\psi$ in neon shell where $u^2$ is implemented by $\Omega_0^0 = \Omega_2$ followed by $\Omega_2^0 = \Omega_0$ in helium shell. The second $u$ factor starts a rotatory operation in helium shell from the ending point of the first rotation. Thus, both 1s electrons complete a cycle inside helium shell corresponding to two consecutive rotations $\Phi_{00}^0 = \Phi_{10}$ and $\Phi_{10}^0 = \Phi_{20}$ in neon shell. When 2s2p octet complete a cycle, both 1s electrons have looped four cycles because radian angles $\Psi$ and $\alpha$ are synchronized, so are radian angles $\psi$ and $\beta$. In other words, the oscillatory rhythm of 2s2p electrons is controlled by that of 1s electrons.

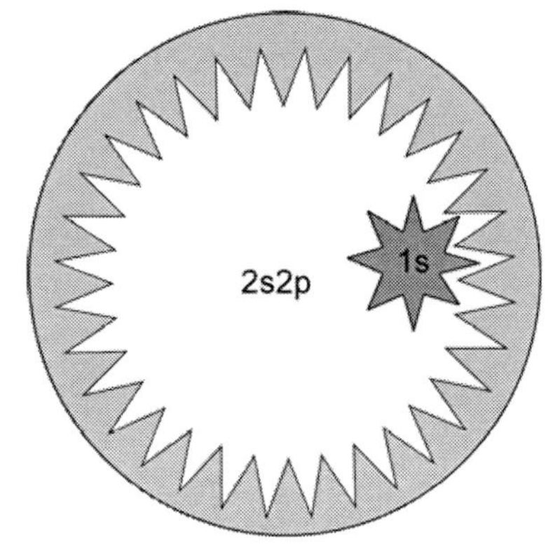


Figure 2.9. Conceptual illustration of differential chain rule between 2s2p and 1s electrons as cogwheels in mesh.

Bridging both differentiation steps, $\dot{\Psi}$ actually acts as a gateway between helium and neon shells. The copulation of both shells observes differential chain rule mathematically. Such an elegant relationship between both 1s and 2s spherical layers is a demonstration of delicate atomic structure. Strictly speaking, although $\dot{\Psi}$ is expressed by equation (2.14), there is subtle difference between $\dot{\Psi}$ and $-d\Psi/dt$ under the context of dynamic differentiation. The former represents the initial state of the differential operation with respect to a time dimension while the latter term is traditionally associated with the final state of the differentiation. The former hides a time dimension of helium shell in neon scope while the latter is an unfolding time dimension of helium under focus. The observer's perspective shift accompanies differential chain rule and the observer's standpoint determines the physical meaning of a function. In neon shell, $\dot{\Psi}$ is only a dimension indicator like a complex number identifier as was shown in Figure 2.8(b) because differential process of $-d\Psi/dt$ does not unfold yet. The outspread of the curled up dimension manifests only in helium shell, where $\dot{\Psi}$ propagates through equations (1.15) to (1.18), describing both 1s electrons.

We call $\dot{\Psi}$ a gateway function for it connects both 1s and 2s electrons into a spacetime continuum within a neon atom. It really serves as a portal between the outer waves (2s2p electrons at this case) and the inner waves (1s electrons). The gateway function is also a complex number identifier like $i$. This becomes evident when we compare spherical quantity $\Phi_0$ with that of a 1s electron having the form of

$$\Omega_0 + \Omega_1 = C_1 C_2 (\cos\alpha\cos\beta - \omega\sin\alpha\cos\beta), \tag{2.35}$$

where $\omega$ is another expression of $i$ as was indicated by equation (1.10). By the same way, we call $\omega$ a gateway function too because it communicates between 1s electrons and the nucleus. Thus, the derivative term $\dot{\Psi}$ is indeed a dimension indicator, a complex number notation, and a gateway function. Spherical quantities of 1s and 2s2p electrons are spacetime continuous. They are in a uniform that features sine and cosine functions, complex numbers, and gateway communications that synchronize waves between both spherical layers. In these ways, we have plotted a unique version of multi-dimensional hyperspace where ten dimensions represented by 1s2s2p orbitals are continuous via consecutive rotatory operations. The coupling of 1s and 2s2p electron waves is by dimension encapsulation mechanism in outwards direction and by calculus chain rule inwards.

By the way, we have discussed differential chain rule of time components in dynamic differentiations. For space components in dynamic integrations, there is integral chain rule parallel to differential chain rule. Integral chain rule is a unique feature of dynamic calculus even though it does not exist in infinitesimal calculus.

## 2.3. Four-Dimensional Harmonic Oscillations

We have characterized electronic orbitals in neon shell by quaternity equation and provided solutions to the equation in calculus and trigonometry. We shall further examine electron transformation from one state to another in geometry and establish the connection between spherical quantities and four geometric elements in a spherical layer as was discussed in section 2.1. Because the electron octet as were listed in equations (2.18) and (2.25) comprise four types of orbitals, $2s2p_x2p_y2p_z$, the transformation of spherical quantities includes four distinct geometric paths. This section illustrates the geometry of four electronic transformations $\Phi_0 \mapsto \Phi_1 \mapsto \Phi_2 \mapsto \Phi_3 \mapsto \Phi_4$ in details. On the negative directions of quaternity axes, another four rotation steps $\Phi_4 \mapsto \Phi_5 \mapsto \Phi_6 \mapsto \Phi_7 \mapsto \Phi_0$ follow similar geometric paths but with space and time roles reversal (Figure 2.10).

Every two adjacent electrons have rotatory operation relationship. For brevity, here we only consider the relationship of the first terms of spherical quantities $\Phi_0$ and $\Phi_1$ because differential and integral processes are undergoing concurrently so that their second terms may be dismissed. Given

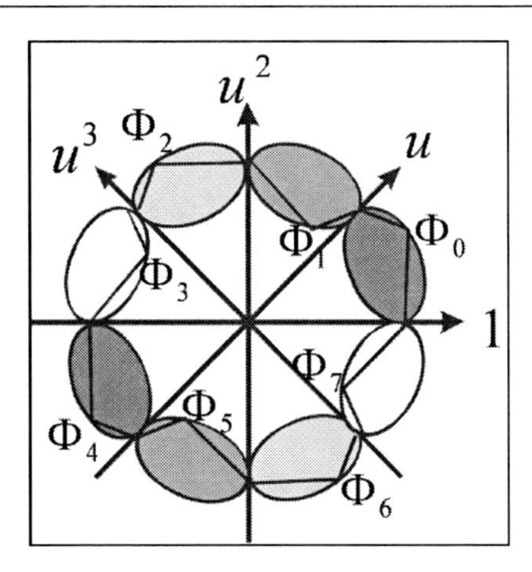


Figure 2.10. Transformation courses of four pairs of electrons along four closed circles in quaternity coordinates with each pair orbiting along both semicircles of the same shaded circle.

$$-\frac{\partial\Phi_{00}}{\partial t}=\frac{\partial\Phi_{10}}{\partial l}, \tag{2.36}$$

where $\Phi_{00}$ and $\Phi_{10}$ refer to the first terms of $\Phi_0$ and $\Phi_1$ in equation (2.18) respectively. The rate of time component reduction of one electron from a full time dimension to vanishing is equivalent to the rate of space component change of its adjacent electron from a full space dimension to vanishing. The equation may be expressed in trigonometry as

$$\cos\Psi=\sin\psi, \tag{2.37}$$

where variables $\Psi$ and $\psi$ constitute two acute angles of a right triangle so that rotatory operation is an alternative expression of Pythagorean theorem. Since both $\Psi$ and $\psi$ represent complementary time and space radian angles respectively, together they determine the course of an electron along a semicircle in dimension diagram (Figure 2.11). As $\Psi$ increases from 0 to $\pi/2$, time component $A_1\cos\Psi$ transforms into $-A_1\dot{\Psi}\sin\Psi$; and as $\psi$ reduces from $\pi/2$ to 0, space component $A_2\cos\psi$ transforms into $-A_2\dfrac{1}{\psi'}\sin\psi$. Spherical quantities of 2s2p electrons are derived from the coordination of time and space components, $A_1\cos\Psi$ and $A_2\cos\psi$. At any moment, Pythagorean theorem synchronizes the rotation of radian variables $\Psi$ and $\psi$ for the electron in harmonic oscillation. Figure 2.11 shows dimension diagram of electronic transformation from $\Phi_0$ to $\Phi_1$. Dimension diagrams for other electrons follow similar pattern with semicircles as the courses of dynamic calculus. These courses have diverse correspondences in geometric paths. We shall discuss four typical cases as follows.

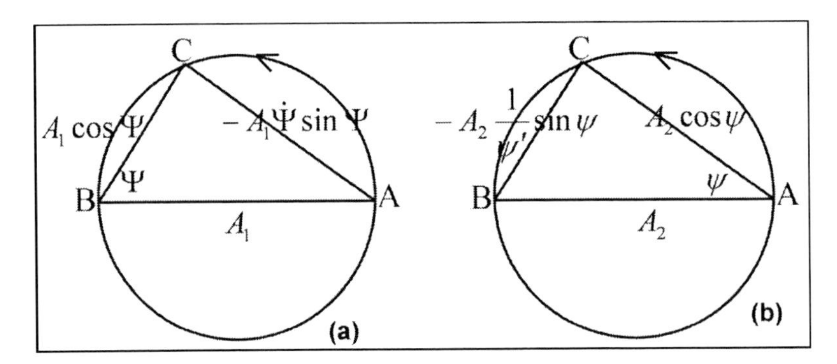


Figure 2.11. Dimension diagram of synchronized differential and integral operations of equation (2.36) in (a) time and (b) space.

## 2.3.1. Harmonic Oscillation in One Dimension

Spherical quantity for a 1s electron is $-C_1C_2v\sin\alpha\sin\beta$ ; and spherical quantity for the initial 2s electron is $A_1A_2\cos\Psi\cos\psi$ . Both are concentric spheres with the same radius geometrically at the phases of ( $\beta=\pi/2,\psi=0$ ), but the 1s sphere is a two-dimensional sphere with variable radius only while the 2s sphere is a four-dimensional one with four variable orientations. At this moment, both electrons are contiguous. The 1s electron reaches maximum space and begins to contract while the 2s electron arrives at the center and begins to skew into a polor. The dimension parameters have relationships of ( $A_1=-\omega C_1, A_2=rC_2$ ). The initial spherical quantity $A_1A_2\cos\Psi\cos\psi$ built from helium shell represents a sphere of four-dimensional time and zero-dimensional space in neon shell. Sphere O is zero-dimensional in space because its center is located at the nucleus; it is also four-dimensional in time because of its maximum size.

As shown in Figure 2.12(a), electron cloud of sphere O starts to condense and its center displace from the nucleus. The 2s electron, which initially permeates the entire sphere as electron cloud, gradually contracts its activity sphere as its spherical center moves from point O towards point B, which represents a solid particle. The polarization corresponds to the rotatory operation of spherical quantity $A_1A_2\cos\Psi\cos\psi$ . As the electron gradually appears from misty cloud to a solid particle, time reduces from four dimensions to three while space increases from zero dimension to one, i.e., from center O to position B with one-orientational displacement from the nucleus. We characterize the electron by electron cloud because it spreads over a large region of space within the atom even though it is always a particle when actually detected physically.

At any specific moment, the temporal radius of sphere O reduces from OB to AB as its center O moves to center A, which is the vertical projection of point E onto OB. Hypothetical point E is tracing along a smooth semicircle OEB. In other words, spherical center A is traveling along straight line OB but follows harmonic oscillation principle as if it were traveling uniformly along hypothetical semicircle OEB.

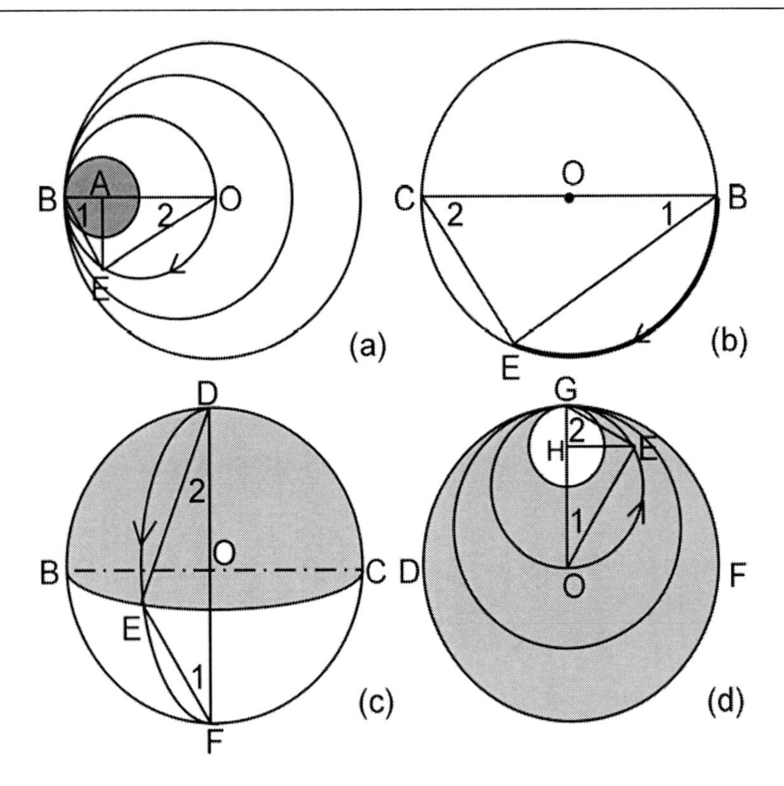


Figure 2.12. The geometric paths of electrons in various transformations (a) 2s→$2p_x$, (b) $2p_x$→$2p_y$, (c) $2p_y$→$2p_z$, and (d) $2p_z$→2s where $\angle 1$ denotes radian angle $\Psi$ and $\angle 2$ denotes radian angle $\psi$ .

In right triangle OEB, side EB measures the time component of the electron, corresponding to the product of hypotenuse BO and $\cos\Psi$ whereas side EO is space displacement from the origin O, corresponding to the product of hypotenuse OB and $\cos\psi$ . Hypotenuse OB represents a time dimension as well as a space dimension. With the rotation of $\Psi$ angle from 0 to π/2, $\psi$ rotates from π/2 to 0 so that time component BO $\cos\Psi$ reduces while space component OB $\cos\psi$ increases. From point O to point B, spherical quantity $A_1A_2\cos\Psi\cos\psi$ changes from $A_1A_2$ dimension to $-A_1A_2u$ dimension. Spherical quantity $-uA_1A_2$ is located at the down-stream of $A_1A_2$ at a full velocity dimension interval. Semicircle OEB is the imaginary course for the electron to traverse diameter OB harmonically, a full 1D space element. The course represents electron transformation from initial 2s state to full $2p_x$ state. At point B, the electron has one-dimensional space OB and three-dimensional time as a solid particle B.

## 2.3.2. Harmonic Oscillation in Two Dimensions

As shown in Figure 2.12(b), semicircle BEC represents the course as well as the geometric path of an electron transforming from $2p_x$ state to $2p_y$ state. A full $2p_x$ electron is a point particle at B whereas a full $2p_y$ electron is a semicircle of BEC in geometry. As the electron stretches along the semicircle from points B to C, radian angle $\Psi$ increases from π/2 to π (i.e., decreases from π/2 to 0 in right triangle BEC). The geometrical metamorphosis reflects the rotatory operation of spherical quantity $-uA_1A_2 \sin\Psi \sin\psi$. The initial point B denotes spherical quantity at $-uA_1A_2$ dimension. After point E reaches the end point C, the spherical quantity becomes $u^2A_1A_2$. Semicircle BEC is a full 2D space element for the electron to attain. The physical electron might manifest as magnetic flux during the process.

At any specific moment along the course, chord CE indicates time component of the electron while chord BE is space component of it. In right triangle BEC, side CE equals CB $\sin\Psi$ and side BE equals BC$\sin\psi$. Time component is decreasing while space component is increasing with the decrease of radian angle $\Psi$ from π/2 to 0. Even though semicircle BEC is the course of the electron, time and space components of it are measured by their corresponding chord lengths, CE and BE, for any point E along the course. As the electron reaches the end point C, it has traversed the second space dimension denoted by diameter BC and has drawn the shape of a full $2p_y$ orbital as a semicircle on the flat plane. A $2p_y$ electron is two-dimensional in space because its center is away from the central nucleus in two orientations, an orientation along semicircle BEC in addition to $2p_x$ electron along radial direction. Throughout this book, a semicircle always refers to a semicircular arc.

### 2.3.3. Harmonic Oscillation in Three Dimensions

Figure 2.12(c) explains the motion of an electron transforming from $2p_y$ orbital to $2p_z$ orbital. It is the rotation of semicircle BEC around axis BC in geometry. The position of whole semicircle BEC is determined by radian angle $\Psi$ in right triangle DEF. With the increment of $\Psi$ angle from π to 3π/2 (i.e., from 0 to π/2 in right triangle DEF), semicircle BEC sweeps over a hemispherical surface, which gives the shape of a full $2p_z$ orbital. In other words, the electron spreads over from a semicircle into a hemispherical surface in geometry during the life of $u^2A_1A_2 \cos\Psi \cos\psi$. The initial position of semicircle BEC at BDC denotes the spherical quantity at dimension $u^2A_1A_2$. After the point E reaches point F, the spherical quantity becomes $-u^3A_1A_2$ in spacetime.

At any specific moment along the course of DEF, the interval between semicircles BDC and BEC, as measured by side DE in right triangle DEF, represents space displacement of the electron along the third dimension. Side DE defines the progress of a curved surface shown in Figure 2.11(c) as shaded area while side FE represents time component of the electron. DE=DF $\cos\psi$ and FE=FD $\cos\Psi$. As $\Psi$ rotates from 0 to π/2 in right triangle DEF, space increases while time dwindles. After the electron traverses through diameter DF in time and space, its spherical quantity tacks another factor of a velocity dimension. The geometric shape of a full $2p_z$ is a hollow hemisphere such as BDCFE bra, which is three-dimensional in space. A $2p_z$ electron has potential energy higher than a $2p_y$ electron because it spreads over another space dimension in DF orientation in addition to the initial $2p_y$ electron.

### 2.3.4. Harmonic Oscillation in Four Dimensions

Figure 2.12(d) indicates electronic transformation from orbital types $2p_z$ to 2s. The initial $2p_z$ electron is a hemispherical surface DF, which gradually grows into a crescent. The waxing of the crescent involves the inwards filling of a spherical surface into a solid sphere as the volume of empty sphere H diminishes from maximum sphere O into point G along the radial direction. Diameter OG is the fourth space dimension for the electron to traverse harmonically via hypothetical semicircle OEG.

At any specific moment, side OE in right triangle OEG represents space component of the electron with OE=OG $\sin\psi$ ; and side GE represents time component of the electron with GE=GO $\sin\Psi$ . As radian angle $\Psi$ increases from $3\pi/2$ to $2\pi$ (i.e., from $\pi/2$ to 0 right triangle OEG), time component reduces while space component increases. The process is the rotatory operation of spherical quantity $-u^3 A_1 A_2 \sin\Psi \sin\psi$ . As the shaded crescent develops into a full sphere, the spherical quantity transforms from $-u^3 A_1 A_2$ into $u^4 A_1 A_2$. The final 2s electron occupies the whole sphere O. It is a four-dimensional space with zero-dimensional time sphere. Sphere O is four-dimensional in space because it is the maximum spatial scale ever attained by the electron; it is also zero-dimensional in time because of its relatively sparse electron cloud. Zero-dimensional time of a full 2s orbital renders the whole solid sphere instantaneous. At this state, time and space components switch their roles so that the sphere is also a spherical quantity of four-dimensional time with zero-dimensional space.

Further oscillatory processes of $\Phi_4 \mapsto \Phi_5 \mapsto \Phi_6 \mapsto \Phi_7 \mapsto \Phi_0$ are in the negative directions of quaternity axes. They constitute another quartet of $2p_x 2p_y 2p_z 2s$ orbitals. Each pair of similar orbital types transform in the opposite directions, making up both semicircles into a closed circle as was drawn by the same shade (Figure 2.10). After eight consecutive rotatory operations, an electron returns to its original state and initiates another cycle.

The foregoing description has demonstrated the principle of right triangles, i.e., Pythagorean theorem as was expressed by equation (2.37). Four distinct diameters OB, BC, DF, and OG in Figure 2.12(a), (b), (c), and (d), respectively, represent four space dimensions for an electron to traverse during $\Phi_0 \mapsto \Phi_1 \mapsto \Phi_2 \mapsto \Phi_3 \mapsto \Phi_4$ transformation processes respectively via their subtending semicircles. We label these space dimensions as $l_3$, $l_2$, $l_1$, and $l_0$ that correspond to 1D, 2D, 3D, and 4D orientations, respectively. We may generalize $l_3$, $l_2$, $l_1$, or $l_0$ dimension as $l$, which refers to the outstanding space dimension of concern or the current space dimension in transformation. And time dimension $t$ follows similar usage. This general usage has been adopted in the text above and so below. As an electron undergoes 2s → $2p_x$ → $2p_y$ → $2p_z$ → 2s rotations, the number of time dimensions decreases from 4 to 0 cosinusoidally while that of space dimensions increases from 0 to 4 sinusoidally (Figure 2.13). Four time dimensions lost are $t_0$ , $t_1$ , $t_2$ , and $t_3$ in sequence while four space dimensions gained are $l_3$, $l_2$ , $l_1$, and $l_0$ in sequence during the rotations.

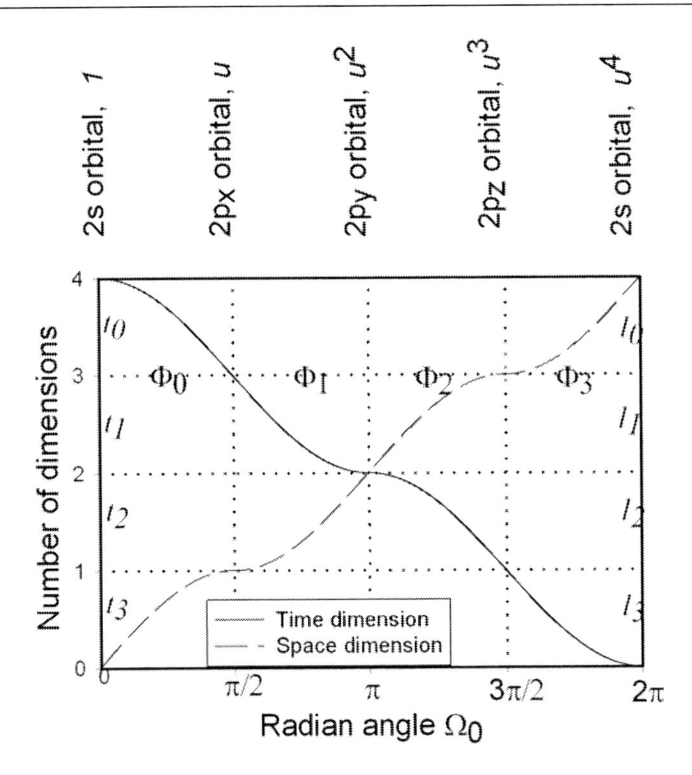


Figure 2.13. Harmonic oscillations of electrons in various space and time dimensions in correspondence with orbital types and quaternity axes.

## 2.3.5. Four Kinds of Orbital Geometries

From the above discussion on the geometries of electrons in various harmonic oscillations, it is clear that a $2p_x$-orbital is one-dimensionally oriented in space; a $2p_y$-oribital is a two-dimensional planar semicircle; a $2p_z$-electron is a three-dimensional hemispherical surface; and a 2s-orbital permeates the entire four-dimensional sphere. As shown in Figure 2.14, both $2p_x$ orbitals are represented by two symmetric points corresponding to points $P_1$ and $P_2$ in Figure 2.2; both $2p_y$ orbitals are represented by two complementary semicircles corresponding to lines $L_1$ and $L_2$ in Figure 2.2; both $2p_z$ orbitals are represented by two hemispherical surfaces corresponding to flat planes $S_1$ and $S_2$ in Figure 2.2; and both 2s electrons are represented by two concentric spheres corresponding to spheres $R_1$ and $R_2$ in Figure 2.2. The electron octet take the shapes of four geometric elements in a spherical layer.

Under quaternity coordinates, a pair of $2p_x$ electrons are located at the opposite directions of $u$-axis. They are symmetric in space and have contrary time directions. A pair of $2p_y$ orbitals are located at the opposite directions of $u^2$-axis. They constitute two complementary semicircles in space and have contrary time directions. A pair of $2p_z$ orbitals are located at the opposite directions of $u^3$-axis.

They constitute two complementary hemispherical surfaces in space and have contrary time directions. A pair of 2s orbitals are located at the opposite directions of $1$-axis. They occupy the same spatial sphere but at different time (Table 2.3). Both 2s electrons are interconnected through the tunnel of three 2p electrons. We paint one sphere to represent a 2s electron and a concentric black point to indicate another 2s sphere at the other end of the tunnel.

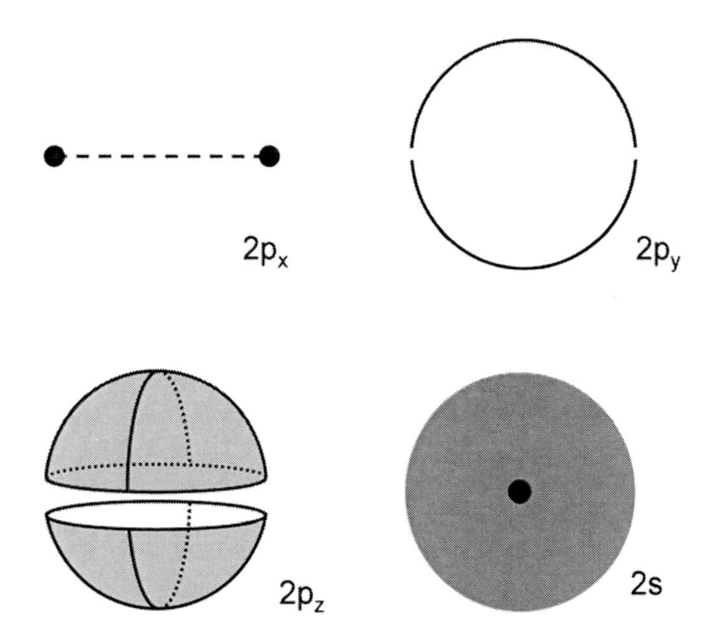


Figure 2.14. The geometric shapes of four orbital types.

There is a node between electron clouds of each orbital pair. Note that a polor includes *u*-axis, but not *1*-axis; whereas a 2s-orbital (of wave function $\Phi_0$) includes *1*-axis, but not *u*-axis. Other spherical quantities and orbital types follow the same pattern.

Under spherical polar coordinates (r, $\theta$, $\phi$) where $\theta$ is the azimuthal angle in the X-Y plane and $\phi$ is the polar angle from the Z-axis, a 2s orbital is defined as r<R where R is the maximum spatial sphere of neon shell; a $2p_z$ orbital is defined as r=R and ($0 \leqslant \phi < \pi/2$ or $\pi/2 < \phi \leqslant \pi$); a $2p_y$ orbital is defined as r=R and $\phi = \pi/2$ and ($0 < \theta < \pi$ or $\pi < \theta < 2\pi$); and a $2p_x$ orbital is defined as r=R and $\phi = \pi/2$ and ($\theta = 0$ or $\theta = \pi$). Four types of orbitals are contiguous in neon sphere. Strong evidence of these 2s2p shapes can be found in Chapter 6 as we discuss the structures of organic molecules.

**Table 2.3. Relationships between spherical quantities, space and time dimensions, and orbital types**

| Spherical quadrant | Spherical quantities | Wave Functions | Space dimension | Time dimension | Electronic orbitals |
|---|---|---|---|---|---|
| I | Polors | $\Phi_0$, $\Phi_4$ | 0→1 | 4→3 | 2s→$2p_x$ |
| II | Metors | $\Phi_1$, $\Phi_5$ | 1→2 | 3→2 | $2p_x$→$2p_y$ |
| III | Vitors | $\Phi_2$, $\Phi_6$ | 2→3 | 2→1 | $2p_y$→$2p_z$ |
| IV | Scalor | $\Phi_3$, $\Phi_7$ | 3→4 | 1→0 | $2p_z$→2s |

## 2.4. Compatibility with Classical and Quantum Mechanics

Spherical quantity in dynamic calculus is the theme of the theory of quaternity that runs through the entire book. It is a revolutionary idea, but we find that it integrates with well-established physics perfectly. First, the application of quaternity equation and rotatory operation to electronic orbitals actually indicates that electrons are oscillating among various states. Quaternity equation is a well-known one-dimensional oscillation equation in classical mechanics. But when applied to electronic orbitals in neon shell, we draw a conclusion that all electrons are undergoing oscillations around the nucleus, each of them complying with quaternity equation simultaneously and individually in its own dimension. Thus, classical physics can be applied to electronic orbitals with proper new interpretation without changing its mathematical formula. In this sense, quaternity integrates with classical mechanics and effectively extends it to the atomic world.

Second, electronic oscillations follow the principle of rotatory operation. From spherical quantities, it is easy to verify that every two adjacent electrons are orthogonal such as $\Phi_0$ and $\Phi_1$, $\Phi_1$ and $\Phi_2$, $\Phi_2$ and $\Phi_3$, $\Phi_3$ and $\Phi_4$, $\Phi_7$ and $\Phi_0$, etc. Since Faraday's law is a case of rotatory operation, electrons are interacting and transforming according to the electromagnetic law. Moreover, consider the relation of

$$\frac{\partial P(x,t)}{\partial t} = -\frac{\partial J(x,t)}{\partial x}, \tag{2.38}$$

where $P(x,t)$ represents a probability density function of an electron in a region $x$, and $J(x,t)$ denotes the flux or probability current into that region. Quantum mechanics states that a change in the density in $x$ is compensated by a net change in flux into that region. If we replace $x$ with $l$ in quaternity spacetime, then equation (2.38) is actually a special case of rotatory operation. Thus quaternity comfortably adapts to quantum mechanics in the interpretation of relationship between probability density and flux.

Third, the theory of quaternity is compatible and complementary with quantum mechanics. Harmonic oscillations of electrons within four spherical quadrants are in correspondence with four quantum numbers in quantum mechanics. The scalors correspond to the principal quantum number that accounts for energy level; the vitors correspond to the orbital quantum number that determines the magnitude of orbital angular momentum; the metors in the equatorial plane are associated with the magnetic quantum number; and the polors, having three space and one time dimensions, describe intrinsic angular momentum perfectly. In agreement with the inert property of a neon atom, quaternity theory indicates that 2s2p electrons are a set of valid wave functions fulfilling a complete cycle in their rotatory operations and hence explains satisfactorily why a full octet configuration is conservative.

However, the definitions of four various spherical quantities and four quantum numbers are not exactly the same. In neon shell, four spherical quadrants belong to the same layer of a principal quantum number 2. Quaternity defines a spherical layer between two $1$ and $u^4$ to account for this energy band. All 2p-electrons have orbital numbers and magnetic numbers. Quaternity uses dimension factors of $u$, $u^2$, and $u^3$ to account for three pairs of different orbital angular momentums, among which $u^2$ is space and time symmetry by itself corresponding to magnetic number 0 while $u$ and $u^3$ are a pair of space and time symmetric quantities corresponding to magnetic numbers of +1 and −1. The spin numbers, $\pm 1/2$, indicate the opposite electronic spins, which are distinguished in quaternity by the pairs of scalors, polors, metors, and vitors that colonize the same quaternity axes but in the opposite directions.

Finally, quaternity is advantageous over quantum mechanics in formulating wave equation for electrons because it adopts a new spacetime framework instead of Cartesian coordinates. Schrödinger's time-dependent equation, a cornerstone in quantum mechanics, takes the following form given wave function $\xi(x,y,z,t)$ and potential energy $V(x,y,z)$:

$$i\hbar\frac{\partial\xi}{\partial t} = -\frac{\hbar^2}{2m}\nabla^2\xi + V\xi \qquad (2.39)$$

By using Laplacian operator $\nabla^2$, it implies that electronic orbitals distribute equally in X, Y, and Z directions, i.e., space is homogenous and isotropic. Notwithstanding this impression, quantum mechanics handles Schrödinger's equation by transforming the isotropic Cartesian coordinates into heterotropic spherical polar coordinates pertinently and then tries to derive quantum information under the constrains that the equation must have solutions and that the solutions must be normalizable. Only via this transformation is it successful in getting much useful information on electronic orbitals in terms of discrete quantum numbers, but the information is far from complete. By this approach, even the solution for the electron in a hydrogen atom is a formidable mathematical task. This indirectly illustrates the merit of quaternity spacetime over Euclidean space. Spherical quantities in quaternity spacetime obey dynamic calculus implemented by trigonometry. Because of the close relationship between space dimensions in a spherical layer and spherical polar coordinates variables, the solutions to quaternity equation are very neat in terms of spherical polar coordinates as were described in section 2.3.5. The simplicity and beauty of quaternity reveal that it describes nature correctly.

## 2.5. Summary

We have described quaternity spacetime by geometry, trigonometry, and calculus in a coherent manner with the instantiation of electronic orbitals within neon shell. To specify, we have identified four distinct space dimensions in a spherical layer originally, discovered the mechanism for dimension curling up for the first time, examined rotatory operation as the principle of electronic transformation, and defined the geometric shapes of 2s2p orbitals in the context of harmonic oscillations. Quaternity space is spherical space; quaternity axes are dimensional axes; quaternity coordinates measure spherical quantities in dynamic calculus. A semicircle connecting two axes in quaternity coordinates system indicates the course of an electron in rotatory operation rather than the trajectory of an object in kinematic movement.

Electrons within neon shell satisfy quaternity equation. They are oscillating simultaneously and individually in harmonic ways in four various dimensions: a $2p_x$ electron is moving along a one-dimensional line segment in the radial direction; a $2p_y$ is stretching along a two-dimensional semicircle; a $2p_z$ electron is spreading around a three-dimensional hemispherical surface; and a 2s electron is permeating an entire spherical layer. A spherical layer is four-dimensional in space. This perception is remarkably different from traditional spacetime worldview. It directly challenges the traditional 2p orbital model and the theory of orbital hybridization thereon.

Quaternity description of 2s2p electrons by smooth trigonometric functions in four dimensions is an extension of classical physics, alternative to quantum mechanics for particles. Four space dimensions in neon shell, expressed as $l_3$, $l_2$, $l_1$ and $l_0$, are four diameters in dimension diagrams for electrons to traverse via their corresponding semicircles uniformly during harmonic oscillations. It is interesting that electrons trace along the semicircles in diverse geometric manifestations, but their space and time components at any moment are measured by their subtended chords consistently. The semicircular track is Newtonian time while the chord is sinusoidal time of spherical quantity. In dimension diagram, two orthogonal time or space components as two sides and the diametrical dimension as a hypotenuse always maintain a right triangle so that Pythagorean theorem indeed governs electronic motion in any cases. It is striking that Faraday's law, Pythagorean theorem, and harmonic oscillation boil down to the same principle of rotatory operation that we have prescribed for electronic motion. The unification of classical physics and particle physics is underway.

*Chapter 3*

# DYNAMIC CALCULUS IN VECTOR CALCULUS

This chapter is an exploration of adopting vector calculus to express dynamic calculus of spherical quantity. As an electron transforms from one state to another, its geometric shape evolves continuously. Vector calculus is introduced to account for the geometric evolution in Cartesian coordinates during electronic transformation. Spherical quantities are represented by vectors with variable X, Y, and Z components in Cartesian space. Spherical quantities in dynamic calculus are expressed by vector calculus with respect to line segment, arc length, spherical surface, and spherical volume. This is an advanced topic. A thorough understanding of conventional vector calculus and general Stokes's theorem is a prerequisite for studying this chapter. However, this is just a tentative work in progress. We present it with the hope that readers might get better insight into the property of spherical quantities and the structure of four-dimensional space and improve the mindset in the end. Due to its uncertainty, readers may, however, choose to skip this chapter without affecting the full comprehension of the subsequent ones.

## 3.1. DYNAMIC CALCULUS AS VECTOR CALCULUS ALONG A PATH

On Cartesian plane, exponential function $e^{-i\alpha}$ or complex function $\cos\alpha - i\sin\alpha$ defines a point tracing from coordinates A(1, 0) to B(0,1) along a quarter circle as radian $\alpha$ rotates from 0 to $\pi/2$, but in quaternity coordinates $\cos\alpha - i\sin\alpha$ represents differential transformation of a spherical quantity from $\cos\alpha$ to $-i\sin\alpha$ as $\alpha$ rotates from 0 to $\pi/2$ (Figure 3.1). The differentiation involves time dimension transformation from 1 to $-i$ (= $d\alpha/dt$) instead of position movement. Although the meanings of the complex function in Cartesian coordinates and in quaternity coordinates are different, both coordinate systems are coincided with each other, which allows us to translate dynamic calculus into vector calculus along the arc.

It has been established in section 1.5.1 that dynamic differentiation of a spherical quantity may be represented by a unit rotatory vector, u, on a two-dimensional plane. In Figure 3.1, rotatory vector u is the directed segment OC rotating around origin O. We may also use a directed curve AC starting from point A and extending along the quarter arc to represent the path of circular motion in Cartesian coordinates as well as the course of dynamic calculus in quaternity coordinates. The rotation of vector u from the abscissa orientation around the origin with vector tip along arc AC corresponds to spherical quantity $\cos\alpha$ in dynamic differentiation.

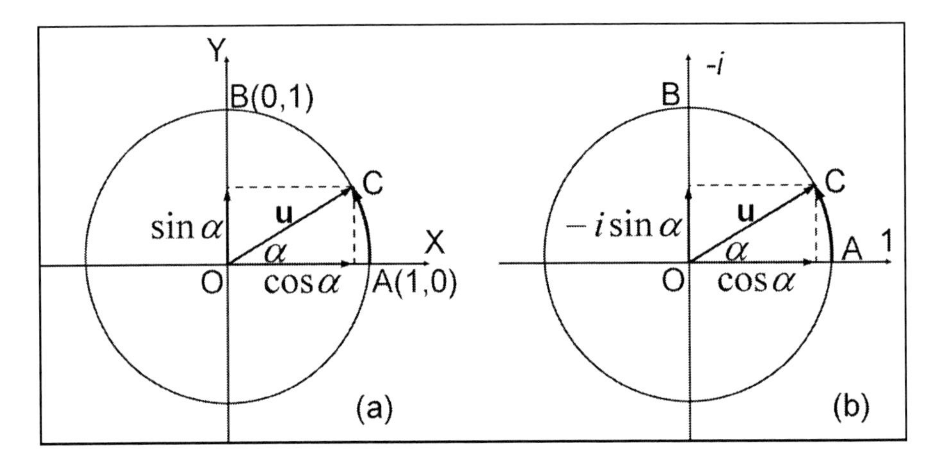


Figure 3.1. Comparison of a complex function in (a) Cartesian coordinates and (b) quaternity coordinates.

In dimension diagram, a spherical quantity in dynamic calculus is shown in Figure 3.2 where state point C traces along semicircle ACB. Diameter AB represents a time dimension as well as a space dimension during a rotatory operation. The dynamic differential process can be expressed by smooth rotation of $\alpha$ for a displacement of $\pi/2$ followed by differential chain rule:

$$-\frac{dC_1\cos\alpha}{dt} = -C_1\cos(\alpha+\frac{\pi}{2})\cdot\frac{d\alpha}{dt}. \tag{3.1}$$

As radian angle $\alpha$ increases from 0 to $\pi/2$, point C moves from A towards B so that chord $C_1\cos\alpha$ dwindles from diameter dimension $C_1$ to vanishing. Arc AC is the course of spherical quantity $C_1\cos\alpha$ in dynamic differentiation. The diameter is the dimension with respect to which dynamic differentiation is performed. The electron is tracing along the semicircle, but the spherical quantity of it is measured by chord BC at any moment. In other words, the spherical quantity in dynamic calculus defines the course of the electron along the semicircle.

This process may be expressed by a rotatory vector BC, which rotates around point B with vector tip along arc AC. Equation (3.1) may be regarded as the differentiation of vector BC with respect to time.

Because arc AC represents uniform time flow, the vector differentiation with respect to time is along the path of arc AC for proper time flow. We regard the differential path as an essential parameter parallel to an integral path in vector calculus. For example, if vector BC is a physical quantity of displacement, then the vector differentiation along the path of arc AC gives a velocity vector.

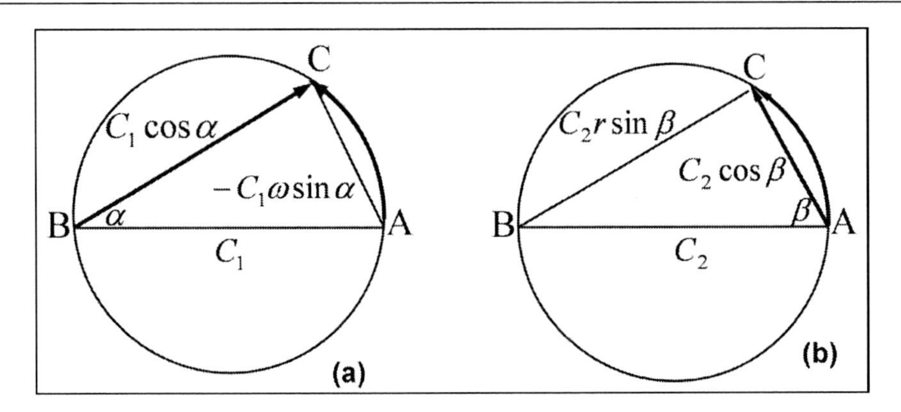


Figure 3.2. Spherical quantity in dynamic calculus as vector calculus with respect to arc length in dimension diagram.

During a rotatory operation, a differentiation is undergoing concurrently with an integral operation. As chord BC decreases, chord AC increases accordingly. The gaining of a space dimension can be represented by dynamic integral of

$$\int C_2 \cos\beta dl = C_2 \cos(\beta - \frac{\pi}{2}) \cdot r \,, \tag{3.2}$$

where

$$\frac{1}{r} = \frac{d\beta}{dl} \,. \tag{3.3}$$

As $\beta$ decreases from $\pi/2$ to 0, chord AC extends over a space dimension of diameter AB. Here radian angles $\alpha$ and $\beta$ are always complementary. Space and time components of a spherical quantity trade one for another during a rotatory operation. Changes in $\cos\alpha$ and $\cos\beta$ values as the radian angles rotate represent simultaneous differential and integral processes. Spherical quantity $C_2 \cos\beta$ is represented by chord AC and may be treated as a rotatory vector AC around point A with vector tip along arc AC. This process may be interpreted as the vector integral over a space dimension along the integral path of arc AC. For example, if vector AC is a force field, then the vector integral along the path of arc AC represents the work done by this force.

Spherical quantity in dynamic calculus may be expressed by vector calculus. In dimension diagram, we have translated spherical quantity in dynamic calculus into vector calculus. However, because spherical quantities in dynamic calculus have multifarious geometric manifestations in Cartesian coordinates, their expressions in vector calculus entail consideration case by case for electrons within helium and neon shells. In ancient Chinese Taoism, Tao refers to the general law, and De refers to individual quality, both corresponding to dynamic calculus and vector calculus, respectively. The proposition of Tao and De reflects the ancient Chinese good understanding of the natural law.

## 3.2. Vector Calculus in Two-Dimensional Space

Electrons within a helium atom transform between two dimensions. Their motion has been described in trigonometry, geometry, and probability in Chapter 1. Here we shall characterize the motion of the electrons in vector calculus. We only have to concern with space component of both 1s electrons because time component can be understood by its symmetry and complementarity to space component. As shown in Figure 3.3, spherical center O denotes the position of the nucleus, and the shaded spheres OP and ON represent the space components of both 1s electrons, $\Omega_0$ and $\Omega_2$, within a helium atom, respectively. As points P and N trace along the dashed semicircles, the radius of one sphere OP increases while that of the other ON decreases. Line segments PQ and QN are tangent to both spheres respectively so that profile NOPQ is a rectangle at any moment. Let $R_0$ represent the maximum spatial sphere of electron $\Omega_2$ so that $R_0$=OQ. As point N traces from Q towards O along the dashed semicircle, space component of $\Omega_2$ reduces sinusoidally (Figure 3.3a). This space shrinking process can be expressed in quaternity coordinates by

$$\Omega_2 = R_0 \sin\beta = R_0 \cos\alpha, \tag{3.4}$$

$$\frac{\partial \Omega_2}{\partial l} = \cos\alpha\ , \tag{3.5}$$

where $l$ denotes space dimension $R_0$ and $\alpha$ increases from 0 to $\pi/2$ smoothly. Under Cartesian coordinates, the dynamic differentiation can be expressed by vector calculus as

$$\frac{\partial \Omega_2}{\partial l} = \frac{\partial \Omega_2}{\partial x}\mathbf{i} + \frac{\partial \Omega_2}{\partial y}\mathbf{j} + \frac{\partial \Omega_2}{\partial z}\mathbf{k}\ , \tag{3.6}$$

or in concise gradient expression as

$$\frac{\partial \Omega_2}{\partial l} = \nabla \Omega_2\ , \tag{3.7}$$

where $\Omega_2$ on the left-hand side is a spherical quantity and on the right-hand side is a conventional scalar represented by a sphere. This equation establishes the relationship between a spherical quantity in dynamic calculus and a conventional vector calculus along a path. When the calculus is performed upon $l$, spherical quantity in dynamic calculus is assumed; and when the calculus is performed upon $x$, $y$, and $z$, conventional vector calculus is assumed as above, so below. To prove equation (3.6), we write vector

$$\Omega_2 = \sqrt{x^2 + y^2 + z^2}\ \cos\alpha = R_0 \cos\alpha \tag{3.8}$$

to represent the scalar sphere so that

$$\frac{\partial\Omega_2}{\partial x}=\frac{\partial\sqrt{x^2+y^2+z^2}}{\partial x}\cos\alpha=\frac{x}{R_0}\cos\alpha, \tag{3.9}$$

$$\nabla\Omega_2=\frac{x\mathbf{i}+y\mathbf{j}+z\mathbf{k}}{\sqrt{x^2+y^2+z^2}}\cos\alpha=\cos\alpha=\frac{\partial\Omega_2}{\partial l}. \tag{3.10}$$

The proof has used $R_0$ to represent a solid sphere in quaternity coordinates and as a spherical surface in Cartesian coordinate system without distinction because it is the surface boundary that defines the motion of the spatial sphere. Equation (3.6) actually establishes the connection between quaternity dimension $l$ and Cartesian dimensions $x$, $y$, and $z$. Quaternity dimension $l$ is the diameter OQ subtended by semicircular arc QNO. Spherical quantity $\Omega_2$ is the shrinking radius, which is treated as a scalar sphere.

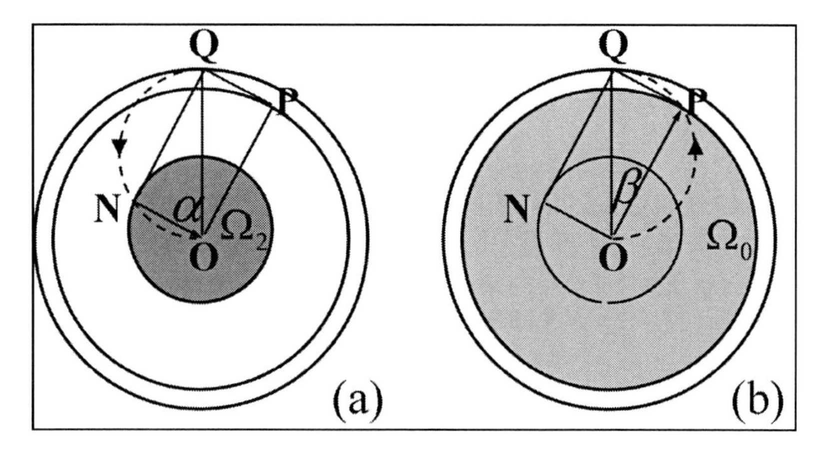


Figure 3.3. Spatial spheres of both 1s electrons within a helium atom represented by shaded circles.

While one 1s electron condenses, the other experiences spatial expansion with spherical quantity $\Omega_0$ increasing sinusoidally (Figure 3.3b). Under quaternity coordinates,

$$\Omega_0=\cos\beta, \tag{3.11}$$

where $\beta$ decreases from $\pi/2$ to 0 smoothly. This process can be expressed in dynamic integral as

$$\int\Omega_0 dl=R_0\cos\beta, \tag{3.12}$$

which is translated under Cartesian coordinates into

$$\int\Omega_0 dl=\iiint_V\left(\frac{\partial\Omega_{0x}}{\partial x}+\frac{\partial\Omega_{0y}}{\partial y}+\frac{\partial\Omega_{0z}}{\partial z}\right)dxdydz, \tag{3.13}$$

or written in divergence operator as

$$\int \Omega_0 dl = \iiint_V (\nabla \cdot \Omega_0) dxdydz \ , \tag{3.14}$$

where $\Omega_0$ on the right-hand side is regarded as a vector of

$$\Omega_0 = \Omega_{0x}\mathbf{i} + \Omega_{0y}\mathbf{j} + \Omega_{0z}\mathbf{k} \ , \tag{3.15}$$

and volume $V$ is the sphere of

$$\sqrt{x^2 + y^2 + z^2} = R_0 \cos\alpha \ . \tag{3.16}$$

To prove equation (3.14), we apply Gauss's theorem

$$\iiint_V (\nabla \cdot \Omega_0) dxdydz = \oint_S \Omega_0 dA \ , \tag{3.17}$$

where $S$ is the spherical surface enclosing the shaded sphere $V$, and the area integral on the right-hand side indicates the growing of the spherical radius by appending spherical surfaces layer upon layer as $\beta$ decreases $\pi/2$ to 0. The extending radius OP represents a vector whose tip flips through the spherical surface stack along the dashed semicircle (Figure 3.4). This is a vivid explanation of orthogonal relationship between radius and spherical surface notwithstanding the continuous variable property of the radius. The integral of radius over the area of spherical surface results in a solid sphere:

$$\oint_S \Omega_0 dA = \oint_S \cos\beta dA = R\cos\beta = \int \Omega_0 dl \ . \tag{3.18}$$

Vector $\Omega_0$ is the radius, representative of the sphere, growing from zero to $R_0$ sinusoidally. Equation (3.13) establishes the relationship between spherical quantity in dynamic integration and volume integration of the divergence of radial vector. Both spherical quantity and vector are synonymous in defining the spatial sphere in this case.

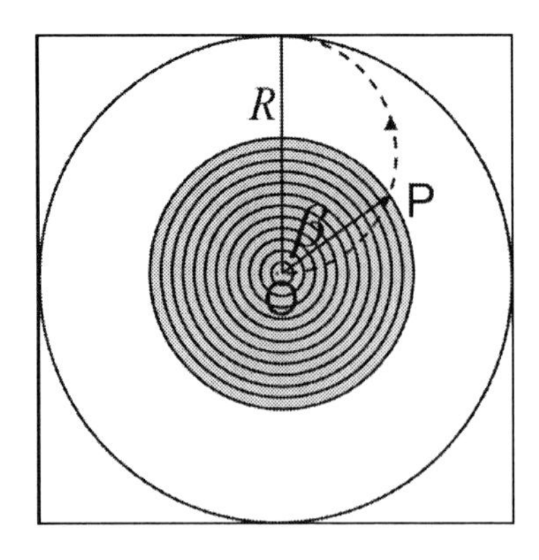


Figure 3.4. Geometric interpretation of radial vector integral over orthogonal spherical surface leading to a solid sphere.

By the way, both 1s spherical quantities are represented by dynamic radii of ON and OP spheres in Figure 3.3 respectively, one decreasing while the other increasing sinusoidally. They are orthogonal at any moment and constitute two sides of a right triangle whose hypotenuse is the radius of the maximum sphere $R_0$. The extremum of them is that one spherical quantity $\Omega_2$ equals $R_0 \cos\alpha$ $(\alpha = 0)$, the maximum sphere, while the other $\Omega_0$ equals $\cos\beta$ ( $\beta = \pi/2$ ), a dimensionless point. According to Pythagorean theorem, the total surface areas of both electrons remain constant all the time (i.e., $4\pi R_0^2 = 4\pi r_1^2 + 4\pi r_2^2$ where $r_1$ represents radius OP and $r_2$ represents radius ON). This is interesting for it is the surface area that accounts for charge activity, the inner volume being shielded mostly in electricity.

## 3.3. Vector Calculus in Quaternity Space

Four-dimensional space has been described for neon shell in dynamic calculus, trigonometry, and geometry. Here we shall characterize the motion of electrons by various vector calculus in Cartesian coordinates. As shown in Figure 3.5, spherical quantity $\Phi_0$ is a zero-dimensional sphere in space at the nuclear center. By increasing a space dimension, it displaces off the center and becomes a one-dimensional $2p_x$ orbital. By increasing another space dimension, the dislocated point $\Phi_1$ becomes a semicircular arc $\Phi_2$ , which then evolves into a three-dimensional hemispherical surface $\Phi_3$, which in turn transforms into a four-dimension solid sphere $\Phi_4$ .

At this state, the electronic space reaches the maximum and begins to dwindle by reducing the same dimensions in the reverse order. However, the roles of space and time switch after 2s orbital of $\Phi_4$ . Through four consecutive differential operations, four-dimensional time sphere $\Phi_4$ becomes zero-dimensional time point $\Phi_0$, completing a full cycle.

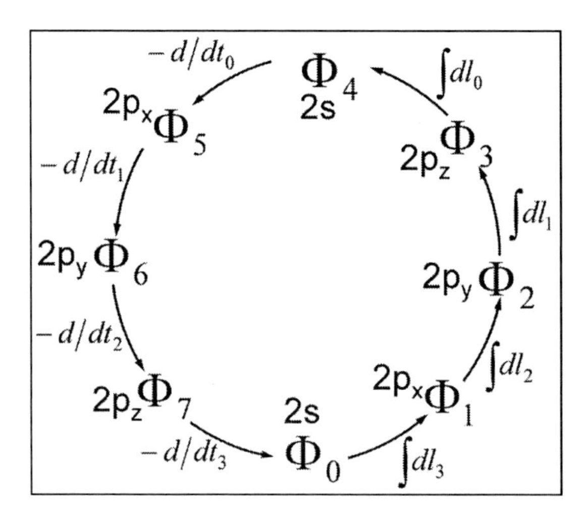


Figure 3.5. Spherical quantities and orbital types of electron octet in neon shell with dynamic calculus relationships.

At this point, the roles of space and time switch back once again to their original default states to initiate another cycle. The zero-dimensional time point $\Phi_0$ is a zero-dimensional sphere in space. We shall consider four integral operations over space on the right-hand side of Figure 3.5 and their differential reversals in vector calculus step by step as follows. Readers should refer to Figure 2.12 for the detailed description of various orbital geometries in harmonic oscillations.

### 3.3.1. Polor in the First Dimension

Electronic transformation from 2s to $2p_X$ along the first dimension is somewhat similar to harmonic oscillation of a rigid ball under the influence of springs (Figure 1.12). It also involves gradual contraction of a sphere into a point, which we disregard here for brevity. As shown in Figure 3.6, charge movement from O toward B corresponds to $\int \Phi_0 dl_3$ operation in Figure 3.5. At any specific moment, the center of the electron D is at the vertical projection of point C onto horizontal line segment OB.

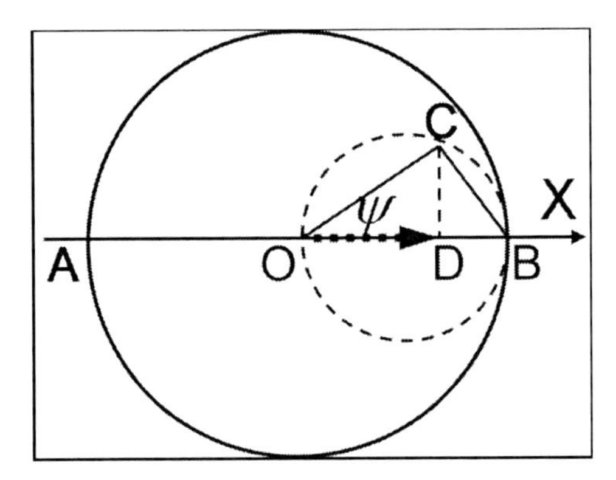


Figure 3.6. Integral of vector OD with respect to $x$ along the path of line segment OB, the first dimension of quaternity space.

As point C traces from O toward B along the dashed semicircle smoothly, point D gradually shifts away from nuclear center O. Given OB= $R$, then the distance of the electron from the nucleus is a unidirectional vector OD= $\Phi_{0x} = R\cos^2 \psi$ with $\psi$ rotating uniformly from $\pi/2$ to 0. If we set up X-axis along OB with its origin at point O, then the vector is along X-axis, measuring the displacement of the electron from the nucleus. Dynamic integral of spherical quantity $\Phi_0$ can be expressed by the integral of vector OC with respect to $x$:

$$\int \Phi_0 dl_3 = \int \Phi_0 dx = \int_R \Phi_{0x} dx \ . \tag{3.19}$$

Conversely, electronic transformation from $2p_X$ to 2s may be characterized by a vector differentiation with respect to $x$ variable:

$$\frac{\partial \Phi_1}{\partial l_3} = \frac{\partial \Phi_{1x}}{\partial x}, \tag{3.20}$$

which describes electronic movement from periphery to the nucleus. This dimension sets a reference for the other three dimensions of quaternity space in Cartesian coordinates. It is the displacement of the electron away from and centering to the nucleus along X-axis. The path of vector integral is straight line OB along X-axis.

### 3.3.2. Metor in the Second Dimension

The second dimension stretches a point into a semicircular arc. As shown in Figure 3.7, the stretching process from point A into semicircle ACB is characterized by $\int \Phi_1 dl_2$ operation that corresponds to electron transformation of $2p_x \rightarrow 2p_y$ whereas the shrinking process from the semicircle to point A is characterized by dynamic differentiation $\partial \Phi_2 / \partial l_2$ that corresponds to electron transformation of $2p_y \rightarrow 2p_x$. Spherical quantity in dynamic calculus can be expressed trigonometrically as chord AC= AB $\cos\psi$ with $\psi$ rotating from $\pi/2$ to 0 smoothly. Under Cartesian coordinates, it is a line integral of

$$\int \Phi_1 dl_2 = \int_C \Phi_{1x} dx + \Phi_{1y} dy, \tag{3.21}$$

where path $C$ is semicircle ACB defined by $x^2 + y^2 = R^2$, $R$ being the radius of semicircle ACB.

The line integral can be derived by traditional concept of vector integral calculus. The metamorphosis of the electron from a point to an arc increases kinetic energy equivalent to the work done by the electron during the process. The vector of chord AC as $\Phi_1 = \Phi_{1x}\mathbf{i} + \Phi_{1y}\mathbf{j}$ in the direction of unit tangential vector $\mathbf{T} = \sin\vartheta\,\mathbf{i} + \cos\vartheta\,\mathbf{j}$ in the direction of increasing arc length $s$ can be expressed by

$$\phi_{\mathbf{T}} = \Phi_1 \cdot \mathbf{T} = \Phi_{1x}\sin\vartheta + \Phi_{1y}\cos\vartheta, \tag{3.22}$$

so that

$$\int_C \phi_T ds = \int_C (\Phi_{1x}\sin\vartheta + \Phi_{1y}\cos\vartheta) ds = \int_C \Phi_{1x} dx + \Phi_{1y} dy. \tag{3.23}$$

This is a classical case of a line integral as an integral of a vector. If the vector is regarded as a force field, then the line integral represents the work done by this force in stretching the electron along the semicircle. Electronic orbital along the second dimension transforms from a point into a semicircle in geometry. We mark this transform by curved arrow AC, the integral path. Spherical quantity $\Phi_1$ is treated as a vector under Cartesian coordinates as

$\Phi_1 = \Phi_{1x}\mathbf{i} + \Phi_{1y}\mathbf{j} + \Phi_{1z}\mathbf{k}$, but in this special case $\Phi_{1z}$ component is zero so that the vector is on X-Y plane. As point C traces along the semicircle ACB, the direction and magnitude of the vector vary continuously. Although equation (3.21) establishes the connection, as best as we can, between a dynamic integral and an vector integral with respect to arc length, the latter describes a point tracing the semicircle while the former expresses the metamorphosis of object stretching rather than single point displacement.

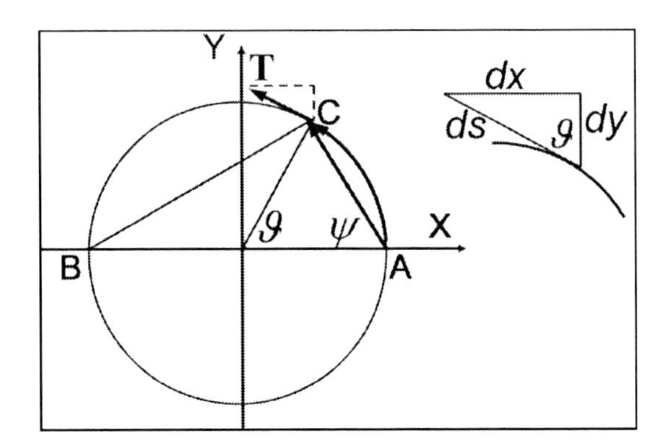


Figure 3.7. Vector integral with respect to arc length along the path of a semicircle, the second dimension of quaternity space.

Conversely, the differential process of $2p_y \rightarrow 2p_x$ transformation can be expressed by a vector in differentiation

$$\frac{\partial \Phi_2}{\partial l_2} = \frac{\partial \Phi_{2y}}{\partial x} - \frac{\partial \Phi_{2x}}{\partial y}, \tag{3.24}$$

which is a two-dimensional *curl* operation that measures angular motion of the electron along a semicircle. The electron shrinks uniformly from a semicircle to a point, which principally involves angular motion around the origin. Both terms of the partial differentiation are coordinated so as to ensure semicircular path of the differentiation, i.e., $x^2 + y^2 = R^2$.

### 3.3.3. Vitor in the Third Dimension

Orbital $2p_z$ takes the shape of a hemispherical surface as semicircle PA'AQ rotates uniformly around axis PQ with point C sweeping through semicircle ACB, point C' through semicircle A'C'B', and so on (Figure 3.8). Every point on semicircle PA'AQ stretches from a point into a semicircle during the process. Under Cartesian coordinate system, the spreading of the electron over the hemispherical surface can be expressed by area integral of

$$\int \Phi_2 dl_1 = \iint_H \Phi_{2x} dydz + \Phi_{2y} dzdx + \Phi_{2z} dxdy, \tag{3.25}$$

where $H$ represents hemispherical surface PAQBC defined by $x^2 + y^2 + z^2 = R^2$, $R$ being the radius of the hemisphere. The integral path is marked in Figure 3.8 by a representative curved arrow AC.

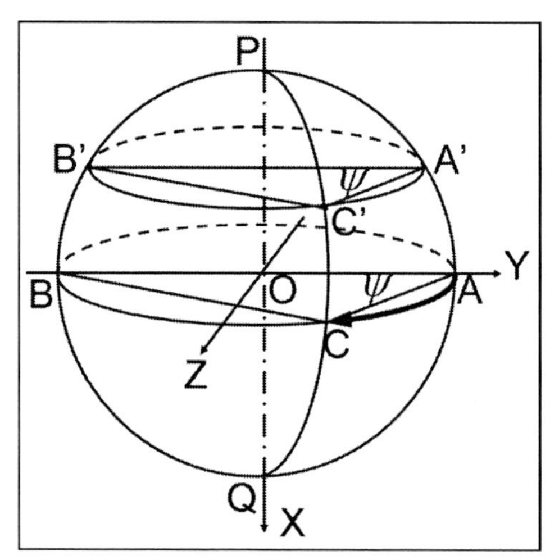


Figure 3.8. Vector integral over a hemispherical surface, the third dimension of quaternity space.

In contrary to space growth, electron transformation of $2p_z \rightarrow 2p_y$ decreases a space dimension from a hemispherical surface to a semicircle, which can be characterized by

$$\frac{\partial \Phi_3}{\partial l_1} = \left(\frac{\partial \Phi_{3z}}{\partial y} - \frac{\partial \Phi_{3y}}{\partial z}\right)\mathbf{i} + \left(\frac{\partial \Phi_{3x}}{\partial z} - \frac{\partial \Phi_{3z}}{\partial x}\right)\mathbf{j} + \left(\frac{\partial \Phi_{3y}}{\partial x} - \frac{\partial \Phi_{3x}}{\partial y}\right)\mathbf{k}, \tag{3.26}$$

where three vector components are coordinated during the vector differentiation to ensure hemispherical path of $x^2 + y^2 + z^2 = R^2$.

### 3.3.4. Scalor in the Fourth Dimension

The fourth dimension describes electron transformation of $2p_z \rightarrow 2s$ from a hemispherical surface to a full solid sphere geometrically. As shown in Figure 3.9, the original hemispherical surface is PAQ. As the hemispherical center O moves toward B, the radius of the empty sphere reduces gradually. Line segment OB is the diameter of the dashed circle. As point C moves uniformly from point O towards point B along semicircle OCB, its vertical projection onto horizontal line segment OB is point D, the center of the empty sphere. At this moment, the electron grows from the hemispherical surface into the shaded crescent volume, complementary to empty sphere D. When O passes through points D and E and finally reaches terminal point B, the empty sphere diminishes into a point at B so that the electron occupies the whole sphere. This process can be expressed under Cartesian coordinates as

$$\int \Phi_3 dl_0 = \iiint_V \left(\frac{\partial \Phi_{3x}}{\partial x} + \frac{\partial \Phi_{3y}}{\partial y} + \frac{\partial \Phi_{3z}}{\partial z}\right) dxdydz, \tag{3.27}$$

where $V$ is the crescent volume defined by equations $x^2+y^2+z^2 \le R^2$ and $(x-R\cos^2\psi)^2+y^2+z^2 \ge (R\sin^2\psi)^2$ where $\psi$ rotates from $\pi/2$ to 0 smoothly. Three partial differentiations in parenthesis indicate the progressive gradient of the waxing crescent while the triple integration indicates filling of the electron inwards over the spherical volume, both being performed simultaneously. The integral path is marked in Figure 3.9 by directed arrow AO.

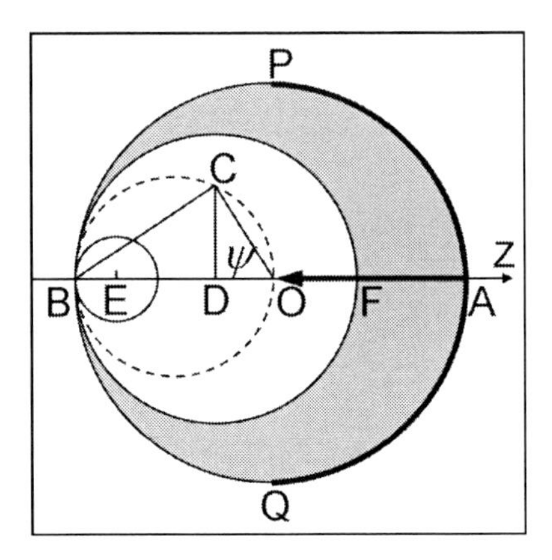


Figure 3.9. Vector integral over a spherical volume, the fourth dimension of quaternity space.

In contrary to the expansion process, electron transformation of 2s→$2p_z$ experiences flattening of a sphere into a hemispherical surface sinusoidally, which may be expressed by

$$\frac{\partial \Phi_4}{\partial l_0} = \frac{\partial \Phi_4}{\partial x}\mathbf{i} + \frac{\partial \Phi_4}{\partial y}\mathbf{j} + \frac{\partial \Phi_4}{\partial z}\mathbf{k}, \tag{3.28}$$

where i, j, k orientations are coordinated so as to ensure that the resulting hemispherical surface satisfies $x^2+y^2+z^2=R^2$ . The geometric significance of the differentiation can be explained by the waning of a shaded crescent. The process initiates by opening up an orifice at point B, which then grows into a cavity inside the spherical volume. And the cavity or empty sphere expands gradually until it eventually occupies the whole spherical volume, compressing the electron into a hemispherical surface at PAQ.

The motion of electrons in four dimensions within neon shell is summarized in Figure 3.10.

Arrows are used to represent the movements of orbital geometric centers, the black ones indicating four integral paths of vector calculus and the gray ones indicating four differential paths of vector calculus. Eight paths constitute a full cycle within the spherical layer in correspondence with Figure 3.5. Every two adjacent paths are orthogonal to each other in geometry, i.e., the downstream arrow always concatenates with the upstream one but turns to a direction perpendicular to it.

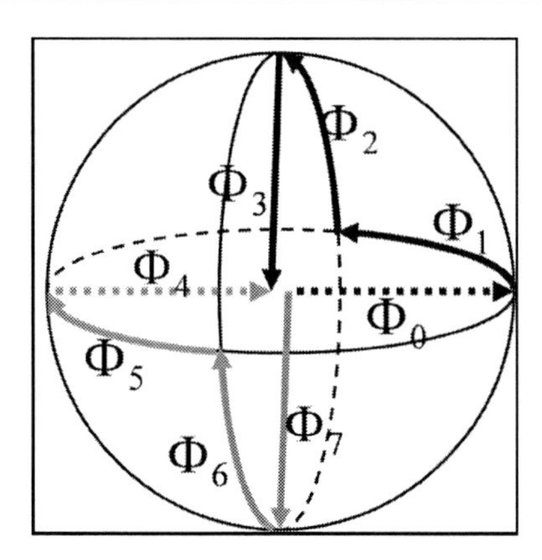


Figure 3.10. Geometric paths of vector calculus for electron octet within neon shell.

## 3.4. Orbital Synchronizations

As described above, spherical quantities for various electrons obey dynamic calculus with various geometric manifestations. In Figure 3.6, the polor in dynamic integral is a sphere converging into a point along X-axis; In Figure 3.7, the metor in dynamic integral is an arc stretching from the point into a semicircle on X-Y plane; In Figure 3.8, the vitor in dynamic integral is a curved surface extending from the semicircle towards a hemispherical surface; and in Figure 3.9, the scalor in dynamic integral is a waxing crescent progressing from the hemispherical surface into a full sphere. As radian angle $\psi$ decreases from $\pi/2$ to 0, the electrons in their various dimensions transform from one state to another. Note that the denotations of $\psi$ in Figures 3.7 and 3.9 are different from those in Figure 2.12(b)(d). At any moment along the integral paths, their spherical quantities satisfy

$$\int \Phi dl = A\cos\psi , \tag{3.29}$$

where $A$ is a dimension factor orthogonal to the trigonometric term and $l$ denotes the various dimensions to be traversed by various electrons $\Phi_0$, $\Phi_1$, $\Phi_2$, and $\Phi_3$. Radian angle, $\psi$, keeps four electrons at the same phase at any moment. It automatically synchronizes the paces of four spherical quantities. We shall explain such synchronizations in details as follows.

### 3.4.1. Green's Theorem

Electronic transformations of $2p_x \rightarrow 2p_y$ and $2p_y \rightarrow 2p_x$ draw two semicircles on X-Y plane while transformations $2s \rightarrow 2p_x$ and $2p_x \rightarrow 2s$ draw a diameter across them (Figure 3.11). Together they form two closed loops AOBCA and AOBDA. Because the centers of the electrons are never in the interior of the loop on the plane during the transformations, the area integral of zero vectors over the interior of both loops is zero. Thus Green's theorem indicates that the vector integral along the curve BCAOBDA is zero. Because of the symmetries between both $2p_x$ orbitals and between both $2p_y$ orbitals, we only have to consider the black arrows in Figure 3.11.

$$u\int \Phi_{0x} dx - \int_C \Phi_{1x} dx + \Phi_{1y} dy = 0 , \tag{3.30}$$

where the first term indicates vector integral along the radius, the second term indicates vector integral along the semicircle, and $u$ is a dimension compensator. Hence

$$u\int \Phi_{0x} dx = \int_C \Phi_{1x} dx + \Phi_{1y} dy , \tag{3.31}$$

$$u\int \Phi_0 dl_3 = \int \Phi_1 dl_2 , \tag{3.32}$$

The last equation indicates the synchronization of a $2p_x$ electron and a $2p_y$ electron in rotatory operations.

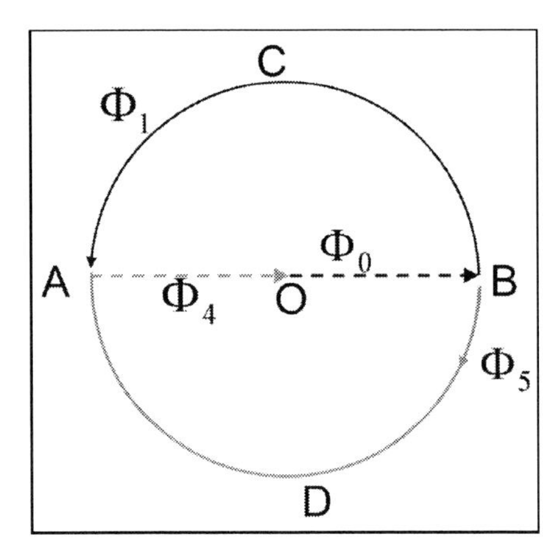


Figure 3.11. Line integral relationship between $2p_x$ and $2p_y$ orbitals within neon shell.

### 3.4.2. Stokes's theorem

Electronic transformations of $2p_y \rightarrow 2p_z$ and $2p_z \rightarrow 2p_y$ paint two hemispherical surfaces ADBCP and ADBCQ while electronic transformations $2p_x \rightarrow 2p_y$ and $2p_y \rightarrow 2p_x$ form two semicircles ADBC (Figure 3.12). Both semicircles form the boundary of both hemispherical surfaces. Because of the symmetries between both $2p_y$ orbitals and between both $2p_z$ orbitals, we only have to consider the relationship between a $2p_y$ orbital and a $2p_z$ orbital. Stokes's theorem governs the relationship between the surface integral and the line integral around the boundary:

$$u\int_C \Phi_{1x} dx + \Phi_{1y} dy = \iint_H \Phi_{2x} dydz + \Phi_{2y} dzdx + \Phi_{2z} dxdy , \tag{3.33}$$

$$u\int \Phi_1 dl_2 = \int \Phi_2 dl_1 , \quad (3.34)$$

The motion of a $2p_z$ electron along the surface is counterbalanced by the flow of a $2p_y$ electron around the periphery. Stokes' theorem synchronizes the rotatory operations of a $2p_y$ electron and a $2p_z$ electron.

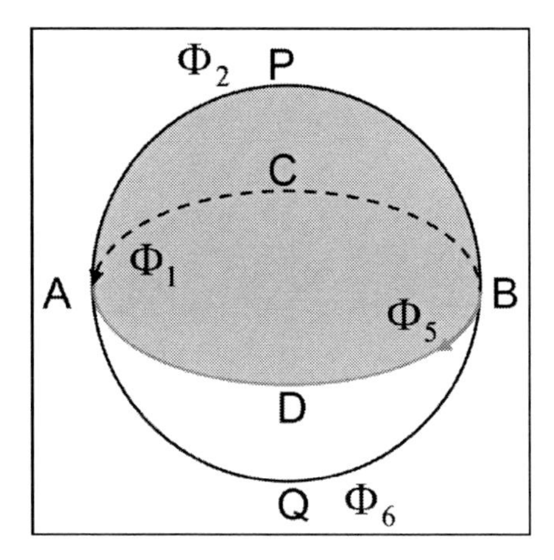


Figure 3.12. The relationship between surface integrals of both $2p_z$ electrons and line integrals of both $2p_y$ electrons around the surface boundary.

### 3.4.3. Gauss's Theorem

Both 2s orbitals are solid spheres enclosed by two hemispherical surfaces of $2p_z$ orbitals. Because of orbital symmetries, a hemispherical surface may be regarded as the boundary of a sphere. Gauss's theorem indicates an intimate relationship between the volume integral of $2p_z \rightarrow 2s$ electron and the surface integral of $2p_y \rightarrow 2p_z$ electron around the sphere:

$$u\iint_H \Phi_{2x} dydz + \Phi_{2y} dzdx + \Phi_{2z} dxdy = \iiint_V \left(\frac{\partial \Phi_{3x}}{\partial x} + \frac{\partial \Phi_{3y}}{\partial y} + \frac{\partial \Phi_{3z}}{\partial z}\right) dxdydz , \quad (3.35)$$

$$u\int \Phi_2 dl_1 = \int \Phi_3 dl_0 . \quad (3.36)$$

Gauss's theorem is commonly known as divergence theorem. Physically, it usually implies that the density change of charges within a region of space is balanced by the electric flux into or away from that region through its spherical surface boundary. Here we give it a new interpretation that the motion of $2p_z \rightarrow 2s$ within the sphere is shielded electromagnetically by the motion of $2p_y \rightarrow 2p_z$ around the globe in an equilibratory manner. Gauss's theorem synchronizes the rotatory operations of a $2p_z$ electron and a 2s electron.

## 3.5. Electronic Interaction Through Virtual Photon

We have characterized electronic orbitals in four various dimensions by vector calculus and demonstrated their tight synchronizations in motion by Green's theorem, Stokes's theorem, and Gauss's theorem. But how electrons interact to achieve at the same phase of radian angle $\psi$? What is the physical mechanism underlying the apparent phenomena? We shall borrow Feynman diagram on virtual photon and give it a fresh interpretation here.

By the principle of space and time symmetry, we have

$$u\frac{\partial\Phi_1}{\partial l_3} = -\frac{\partial\Phi_1}{\partial t_1}, \qquad (3.37)$$

where $u$ is a velocity dimension. Hence the relationship between the space and time dimensions satisfies $l_3 = -ut_1$, which implies that space dimension $l_3$ directly comes from time dimension $t_1$. Physically, we may interpret this relation as the transfer of a virtual photon between electrons $\Phi_1$ and $\Phi_0$. As shown in Figure 3.13, $\Phi_1$ emits a photon carrying a time dimension $t_1$ away while $\Phi_0$ receives that photon as a space dimension $l_3$. Because a photon travels as an electromagnetic wave, electronic transformation through photon transfer manifests sinusoidal wave behavior, consistent with the principle of dynamic calculus.

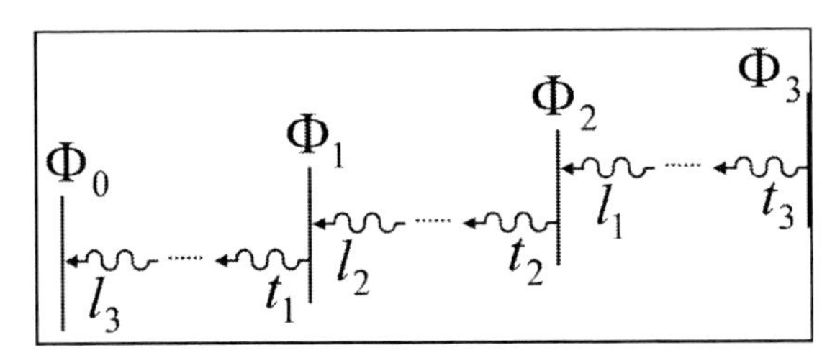


Figure 3.13. Electronic interaction through virtual photon transfer.

We draw an outgoing wave arrow as photon emission and an incoming wave arrow as photon reception. In the case of $l_3 = -ut_1$, the photon wave emitted by electron $\Phi_1$ can be expressed by $-ut_1 \cos\Psi$ carrying a time dimension away whereas the photon wave received by electron $\Phi_0$ can be expressed by $l_3 \cos\psi$ bringing a space dimension back.

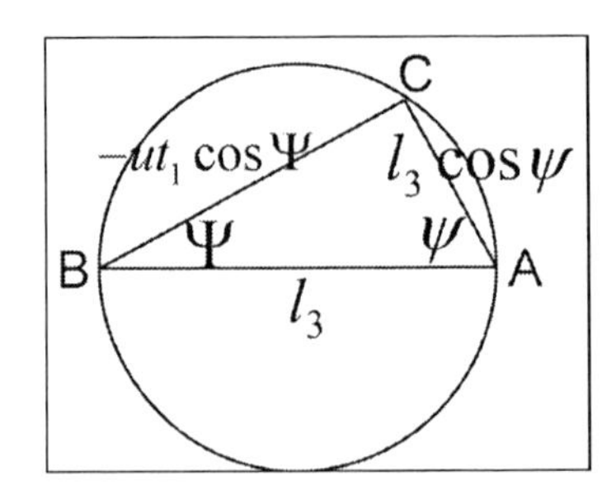


Figure 3.14. Trigonometric relationship between waves of photon emission and reception in dimension diagram.

Home axis shift signifies a rotatory operation (see Figure 2.4) During the exchange, radian angle $\psi$ decreases from $\pi/2$ to 0 while radian angle $\Psi$ increases from 0 to $\pi/2$, both being complementary. Electron $\Phi_1$ passes a photon to electron $\Phi_0$ losing a time dimension whereas $\Phi_0$ receives the photon from electron $\Phi_1$ gaining a space dimension as point C traces from A towards B along semicircle ACB (Figure 3.14). Chord BC reduces while chord AC gains sinusoidally. Similarly, electron $\Phi_2$ passes a photon to electron $\Phi_1$ while electron $\Phi_3$ delivers a photon to electron $\Phi_2$, and so forth. Hence we have implemented dynamic calculus by virtual photon transfer physically. As long as photon waves between the electrons maintain the same phase, all electrons in the transformation cycle are in good synchronization within neon shell.

## 3.6. SUMMARY

We started this chapter by expressing spherical quantity in terms of rotatory vector under dimension diagram. A spherical quantity in dynamic calculus represents the transformation of an electron along a semicircle that manifests as various geometric paths depending on the type of the electron. A spherical quantity may be treated as a conventional vector, such as $L\mathbf{i}+M\mathbf{j}+N\mathbf{k}$, in Cartesian coordinates with varying direction and magnitude in space. In this chapter, we have tried to establish the relationship between spherical quantity in dynamic calculus and vector calculus in Cartesian coordinates with the instantiation of both 1s electrons in two-dimensional helium atom and the instantiation of 2s2p electron octet in four-dimensional neon shell. The path of vector calculus could be a line segment, a semicircular arc, a hemispherical surface, or a spherical volume. We were able to specify the paths, but left the vectors undefined algebraically. We portrayed the vectors in geometry (see Figure 2.12) instead of algebra.

In quaternity space, we have defined four spherical quantities in dynamic calculus in terms of vector calculus. The polor, the metor, the vitor, and the scalor characterize the motion of electrons in point displacement, arc elongation, surface extension, and crescent volume expansion over various space dimensions. Vector integral and differential operations in four dimensions under Cartesian coordinates are summarized as follows:

$$\begin{cases} \int \Phi_0 dl_3 = \int\limits_R \Phi_{0x} dx, \\ \int \Phi_1 dl_2 = \int\limits_C \Phi_{1x} dx + \Phi_{1y} dy, \\ \int \Phi_2 dl_1 = \iint\limits_H \Phi_{2x} dydz + \Phi_{2y} dzdx + \Phi_{2z} dxdy, \\ \int \Phi_3 dl_0 = \iiint\limits_V (\frac{\partial \Phi_{3x}}{\partial x} + \frac{\partial \Phi_{3y}}{\partial y} + \frac{\partial \Phi_{3z}}{\partial z}) dxdydz. \end{cases} \tag{3.38}$$

$$\begin{cases} \dfrac{\partial \Phi_1}{\partial l_3} = \dfrac{\partial \Phi_{1x}}{\partial x}, \\ \dfrac{\partial \Phi_2}{\partial l_2} = \dfrac{\partial \Phi_{2y}}{\partial x} - \dfrac{\partial \Phi_{2x}}{\partial y}, \\ \dfrac{\partial \Phi_3}{\partial l_1} = (\dfrac{\partial \Phi_{3z}}{\partial y} - \dfrac{\partial \Phi_{3y}}{\partial z})\mathbf{i} + (\dfrac{\partial \Phi_{3x}}{\partial z} - \dfrac{\partial \Phi_{3z}}{\partial x})\mathbf{j} + (\dfrac{\partial \Phi_{3y}}{\partial x} - \dfrac{\partial \Phi_{3x}}{\partial y})\mathbf{k}, \\ \dfrac{\partial \Phi_4}{\partial l_0} = \dfrac{\partial \Phi_4}{\partial x}\mathbf{i} + \dfrac{\partial \Phi_4}{\partial y}\mathbf{j} + \dfrac{\partial \Phi_4}{\partial z}\mathbf{k}. \end{cases} \tag{3.39}$$

The equation quartets represent dynamic transformations of four types of electrons: $2p_x$, $2p_y$, $2p_z$, and 2s. The space components of these electrons are synchronized to the same radian phase of $\psi$ at any moment within a neon atom. Each electronic orbital is in dynamic equilibria with its adjacent ones during rotatory operation. We have hypothesized that the synchronization for all electrons should be physically realized through virtual photon transfer between adjacent electrons, i.e., one electron emits a photon wave as a time dimension while another electron receives the photon wave as a space dimension. The coupling of the electrons manifests as Green's theorem, Stokes's theorem, or Gauss's theorem in vector calculus. These theorems synchronize various electrons in rotatory operations. Moreover, because these theorems have broad physical implications in dynamics, fluid dynamics, electromagnetism, heat conduction, thermodynamics, etc., one may explain electronic behavior in diverse physical meanings. For example, $\int \Phi_1 dl_2$ is an integral around a semicircle describing the angular momentum of a $2p_y$ electron and can be regarded as the circulation of a velocity field in fluid dynamics. Thus, the general Stokes's theorem opens up prospects for unifying classical mechanics with particle physics.

*Chapter 4*

# QUATERNITY, RELATIVITY, AND QUANTUM MECHANICS

After formulating wave functions for electrons within helium and neon shells, we have to stop to make comments on quantum mechanics because there are claims that quantum mechanics has already given a complete description of electronic behaviors, and particles cannot be described precisely. How can quaternity theory characterize electrons in more details and better precision? Besides, what is the relationship between quaternity and relativity when both are dealing with spacetime? Does quaternity predict the same consequences such as length contraction and time dilation as special relativity or even more? In this chapter, we shall answer these questions from quaternity perspective with an effort to unify both relativity and quantum mechanics, two active fields of modern physics.

## 4.1. Scope of Views

Relativity and quantum mechanics excel at different arenas. The special and general theories of relativity mainly deal with space and time in the universe where astronomical distance is normally involved. For example, we commonly adopt light-year as a distance unit and millions of years as time units. These magnitudes are much larger than human's accustomed rulers, say, meter and hour. On the other hand, quantum mechanics is successful in describing particles on the Planck scale. For instance, an electron has a radius of 2.818 x $10^{-15}$ m and rest mass of 9.109 x $10^{-31}$ kg. The observation of the tiny size is beyond human naked eye capability. The difference in the yardsticks of relativity and quantum mechanics relative to humans gives rise to serious deviations not only in experimental approaches but also in conceptual formation in physics. We discover this issue.

Man is the intrepid observer in both fields of relativity and quantum mechanics. This is the basic requirement of experimental physics, in which human sensible evidence must be gathered to prove or disprove a proposition. Here human sensible evidence includes, of course, data that are collected by devices made by humans. In studying the behavior of electrons, man is a monster standing far away of the atomic sphere. He probes the electrons by inducing them to detectors for measuring parameters, such as angular and spin momentums. Such a perturbation is remarkable from the viewpoint of an electron, which normally results in irreversible changes of electronic states after measurements. This situation is created because human's sensible scales are so macro to the electrons. This means that electrons are violently captured out of their primitive caves for the purpose of experiments. Man intrudes abruptly. Though the proper role of physical science is to find out what the world would be like unobserved, this goal has not been achieved for particles so far. To date, our knowledge of particles is mostly obtained by bombarding atoms with high energy protons or other charges. In contrast, when examining celestial bodies, man is a tiny point inside the universe. With all instruments and apparatuses he invented, he has only penetrated a small fraction of the universe simply because his rulers are dwarfed by the vast space. He never masters the universe as a whole entity like an atomic sphere, nor does he know whether or not there are multifarious spacetime structures within it.

In both relativity and quantum mechanics, we view entities from human perspective disregarding what they are, and consequently we are actually examining them from different perspectives or angles relative to the objects. Specifically, we adopt a huge ruler relative to a particle to measure the particle in quantum mechanics; and we open our near-sighted eyes to look at the vast universe in relativity. This may lead to different conclusions. In the former case, we get a bird's-eye view over the entire particle with large uncertainty; and in the later case, we see only a fraction of the universe like a blind man touching a part of an elephant by chance. An entity as a whole and a small part of an entity are of course different. Even though both views are real, they cannot be mentioned in the same breadth due to perspective difference.

The fact that man is the center of the world in experimental physics, objects being his surroundings only, is a dilemma of science. We have to gain our knowledge based on human perception and judgments, yet this man-centered view actually biases and limits scientific progress. Since the discovery of heliocentric theory by Copernicus, everyone has known that human races are not at the center of the universe. We have spent so long getting away the general human conceit, but we haven't quite escaped this trap yet. To tackle this problem, here we bring out quaternity perspective.

Quaternity perspective is the perspective of observing an entity from outside its activity sphere and at a proper distance and time interval that match the size and oscillating rhythm of the entity. The imaginary observer uses a proper measuring rod that matches the size of the observed object (Figure 4.1). For example, an ant carries a clay-sized particle; a man rides a bicycle; and the earth interacts with the moon. There are comparable scales between the actors and the objects in these examples. From such a proper perspective, Newton observed an apple falling down from a tree that sparked a brilliant thought in his mind about gravity. When applied to our living surroundings, quaternity perspective is the perspective of classical mechanics. But the term of quaternity perspective is especially reserved for examining electrons within inert atoms on a proper scale. The theory of quaternity is essentially founded on new space and time concepts that extend our human senses in a logical way. Because the proper observer is hypothetical, the system of truth has to be derived from our rational capacities. Rigorous mathematical derivation is the key to establishing the theory.

Quaternity perspective leads to spherical view. After characterizing electronic orbitals within atoms by spherical view successfully, we may apply the same methodology to characterizing the structures of cosmic space on a proportional scale. The entire universe, when observed from a hypothetical standpoint outside the universe, is comparable to an atomic sphere under quaternity perspective. However, we do not suggest that the whole universe is similar to an atom with regard to spacetime structures.

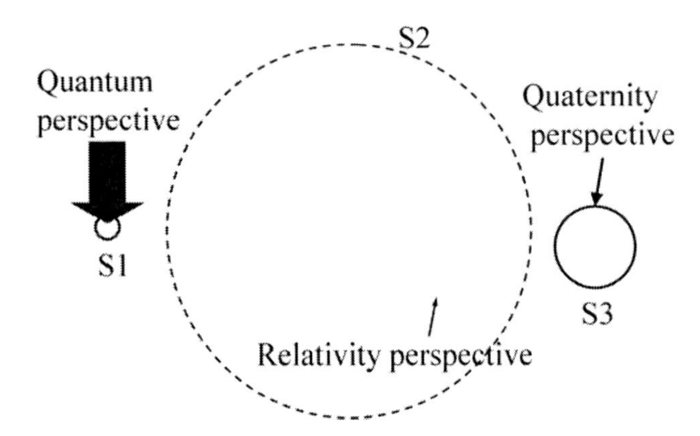


Figure 4.1. Comparison of quantum perspective, relativity perspective, and quaternity perspective where S1, S2, and S3 refer to the spheres of concerned objects and the arrows mean observing perspectives. There are certain mismatches between the arrows and spheres under quantum perspective (too large an arrow for a tiny sphere) and relativity perspective (too small an arrow for an enormous sphere).

The spacetime structures of the whole universe might be more complex than those of a single atom. The reason that we have not experienced spacetime structures other than three-dimensional space with one-dimensional time is because of our limited reach in the universe. Or we might have already detected strange media such as black holes that are radically different from other regions, but we still don't regard them as distinct spacetime structures. It is fairly to say that we are still lacking a general picture of the entire universe.

On the other hand, relativity may be applied to quantum mechanics in studying particles. We shall discuss special relativity in terms of dynamic calculus in section 4.3. Dynamic calculus is the proper way to express electrons within the inert atoms. This is why relativistic quantum mechanics is directing to the right direction.

Granting that quaternity perceives the world from a wider scope than human direct experience does not warrant it to escape the confirmation by man-centered experimental physics. On the contrary, the results that quaternity yields must be able to withstand human scrutiny. For instance, looking at a tree from different angles, one sees various shapes and shadows. The observer could be an ant that resides at the tree trunk, or it could be a bird flying over the tree. Both animals would certainly provide different pictures about the tree, but there is still a way to unify them if we take their perspectives into proper consideration. To gain more information on the tree, a good photographer should try to capture the snapshot of it from various angles, such as from inside and outside, and from far and near. The best photo to describe the whole tree is perhaps one taken at a distance that matches the tallness of the tree. However, a photo taken from a new angle, though having different looks and feels, must be able to justify its silhouette of the tree in consistent with the previous photos. Specifically, quaternity must be compatible with relativity and quantum mechanics in certain way and should not contradict with established experimental evidence.

Thus, readers cannot help asking whether quaternity agrees with quantum mechanics in the description of electrons. We shall proceed to answer this question through the discussion of EPR paradox.

## 4.2. QUATERNITY ON EPR PARADOX

Albert Einstein, Boris Podolsky, and Nathan Rosen (EPR) published a paper in 1935 to question the completeness of quantum mechanical description of particle behavior [1]. They said: "every element of the physical reality must have a counterpart in the physical theory... If, without in any was disturbing a system, we can predict with certainty the value of a physical quantity, then there exists an element of physical reality corresponding to this physical quantity." This statement formed the theme of the EPR paradox. However, when applied to the behavior of an electron, this statement makes a big assumption that the electron is an unchanging particle, i.e., it possesses certain physical quantities at all space and time. This is flaw by the judgement of spherical view.

As was discussed in Chapter 2, an electron must be described by a series of wave functions instead of a single one. For example, within neon shell, each of eight electrons is circulating and switching among eight states, obeying quaternity equation. Each state has its peculiar physical quantities associated with its space and time dimensions. Since each electron in its unperturbed atomic shell is changing between states continuously, it cannot be described by a fixed physical quantity. The concept of spherical quantity was introduced to account for electrons under cyclic rotatory operations in Chapters 1 and 2.

EPR's statement, of course, has something to do with timescale. Within a second, an electron might change states millions of times according to rotatory operation. You cannot associate a specific physical quantity with it due to its dynamic nature. But if you were an observer with a time ruler of $10^{-100}$ second, then the slow changing rate of electrons relative to the observer could be negligible. In the latter case, Einstein is correct that physical reality corresponds to the physical quantities. This problem is caused by adopting an improper scale inadvertently. For example, we cannot use a time scale of millennium to measure the growth rate of a baby. Nor should we try to characterize electronic behavior using a measuring rod that is intended for a child.

Early in the 20$^{th}$ century, physicists found that the actions of elementary particles could not be absolutely predicted, but their actions could only be forecast to a certain level of inexactness through the use of probability functions. This is not only because the measurement techniques are too heavy-handed, but also because particles are changing states intrinsically along their tracks unknown to quantum mechanics. When we use our human standard, such as position and momentum, to measure an electron, the electron is forced to stop its normal track of rotation and response to the measuring agency. At that moment, the electron transforms into positive or negative property of demand from its current state. This gives two impressions: a) the property of an electron is not real until observed; and b) the property of an electron is not permanently associated with it. These two points formed the main idea of the Copenhagen interpretation of quantum mechanics [2].

Spherical view provides deep insight into the Copenhagen interpretation. The measurement process is really an event. If not under the interaction of the measuring agency, an electron would behave as it normally does by dynamic calculus, i.e., it would not detour its track unless forced to do so. An electron within atoms must be described by spherical quantities instead of physical quantities. A spherical quantity represents dimension transformation of an electron from one physical quantity to another. An electron within an atom does not possess the property of a certain physical quantity. Any physical quantities of an electron do not exist at all until we measure them. The point here is that we are not taking proper approaches to describing electrons. We don't know the language of the electronic world at all when imposing the concept of physical quantity to electrons. And as a result, we only got statistical data in most cases in quantum mechanics.

How continuous spherical quantity collapses into discrete physical quantities has something to do with experimental system to test the electrons. To explain this, suppose that we, humans, are at the mercy of an extraterrestrial being who wants to test a property of us, and the test requires us to stand on our head at a testing machine. We shall hypothetically call the demanded property "QRate". Assuming that those people who are at sleep fail at the test while others pass, thus the number of people who pass the test has to do with whether they are at sleep or not. But the extraterrestrial being does not know the concept of "asleep", he only adopt probability to analyze the results. If on average, people sleep eight hours in a day, then the "QRate" obtained by the extraterrestrial being would be close to 8/24 in probability after a large sample is tested. But the problem is that humans sleep and wake up every day whereas an individual is sampled every thousand years in the test. Because of the large timescale of the extraterrestrial being, the test is about the stability of a human society over long history rather than the property of a person. Thus the extraterrestrial being has confused the property of a person with the property of a population in probability test. "QRate" is the property of a person but it is statistically obtained from the population. This situation is analogous to the case of our examining electrons. The timescale of human is too large to measure electrons. As a result, we are testing the stability of an experimental system rather than an electron at motion.

When a pair of electrons in the singlet state with total spin equal to zero are separated, if one electron has spin up in the X-component, the other must have spin down in the X-component. Actual measurements did show the expected approximately 100% correlation. In this case, the electronic states are predetermined by the separation process because they are no longer characterized by spherical quantities in dynamic calculus after being separated. It is totally nonsense to talk about superluminal communication between electrons.

Concerning with the second impression, whether the property of an electron is permanently associated with it or not depends on the interpretation of the words "property" and "permanently". If "permanently" refers to a timescale of a thousand years, then a baby has an uncertain growth rate because her life span is only at best a hundred year. If "permanently" refers to an interval of $10^{-100}$ second, then an electron has a fixed set of physical quantities even though it is changing states continuously. One might argue that "permanently" refers to the whole lifetime of the object. But is "asleep" a human property? Humans change states between "asleep" and "awake" everyday, and hence "asleep" should be considered to be a property inherently associated with humans. There must be some physical quantities associated with the human circadian cycle. But the extraterrestrial being can only grasp the probability of "QRate", not the "asleep" property. Whether the property is existed or not for an individual remains undecided until tested at the sampling interval of the extraterrestrial being. Although "QRate" probability happens to correspond to "asleep" property assuming population uniformity over history in this case, not every probability tested on populations in other cases may reflect a certain property of an individual! Thus the probability methodology by quantum mechanics is either incomplete in describing the electrons or simply beside the point in describing them.

The theory of quaternity has clearly demonstrated that there are precise ways to describe electronic orbitals, and hence that the description of atomic structures by quantum mechanics is incomplete. Does this violate Heisenberg's principle of uncertainty? Not at all. Being hampered by the principle of uncertainty does not preclude us from characterizing electronic behavior accurately in alternative ways. Heisenberg's principle of uncertainty ( $\Delta p \cdot \Delta \lambda \geq \hbar/2$ ) is based upon the assumptions of discrete physical quantities with linear algebra. Note that those assumptions are often unnoticed. Modern physicists even claim that there are not any assumptions underlying the principle of uncertainty and regard it as an absolute law for particles. This is a terrible mistake. The history of science tells us an appalling fact. After an assumption is repeated a million times, it becomes a law because the confidence of the speakers is assured; and after a law is repeated a million times, it becomes an obstacle to scientific progress for the confidence has developed into prejudice and fashion. The principle of uncertainty belongs to this case. Extending an assumption beyond its valid zone leads to a bad principle. Spherical view is not limited by those assumptions. It is based upon spherical quantity in dynamic calculus and hence may describe electronic behavior in an alternative manner. When we adopt a proper timescale to study human behavior, the human circadian cycle is easily understandable. Quaternity wave functions indicate that the behavior of electrons is entirely deterministic, i.e., the electron follows its own track traveling between space and time within atomic shells, just as a man sleeps at night and works at daytime having a daily living pattern, which is cognizable. There is not a so-called principle of uncertainty in spherical view.

## 4.3. Special Relativity in Quaternity

Under the principle of rotatory operation as was introduced in Chapters 1 and 2, a physical entity gaining a dimension in space is accompanied by losing a time dimension, and vice versa. For example, an electron may switch its state from a polor to a metor, i.e., changing its state from one-dimensional space with three-dimensional time to two-dimensional space with two-dimensional time. Such a switch is so rapidly within a neon atom that we have only considered its full consequence. However, in the case of a slow switch, we have to consider its partial effect during the transformation process. If a spherical quantity is undergoing a clockwise rotation from states M to N, then contraction in space dimension and dilation in time dimension proceed simultaneously. We may express such a process as

$$\int -\frac{\partial M}{\partial l} dt = N , \tag{4.1}$$

where the differential and integral operations are carried out simultaneously during the space and time dimensions tradeoff. The consequence of the process is

$$-\frac{\partial M}{\partial l} = \frac{\partial N}{\partial t} . \tag{4.2}$$

How can two operations be carried out simultaneously? Dynamic calculus can be implemented by trigonometric functions. When a spherical quantity M with space component $P$ and time component $Q$ has undergone a clockwise rotatory operation, the resultant spherical quantity N is composed of space component $-\partial P/\partial l$ and time component $\int Qdt$ (Figure 4.2). The clockwise rotatory operation can be interpreted by the rotation of radian angle $\vartheta$. As $\vartheta$ increases from 0 to π/2 continuously and smoothly, space component of M decreases a full dimension from $P$ to $-\partial P/\partial l$ while time component of M increases a full dimension from Q to $\int Qdt$.

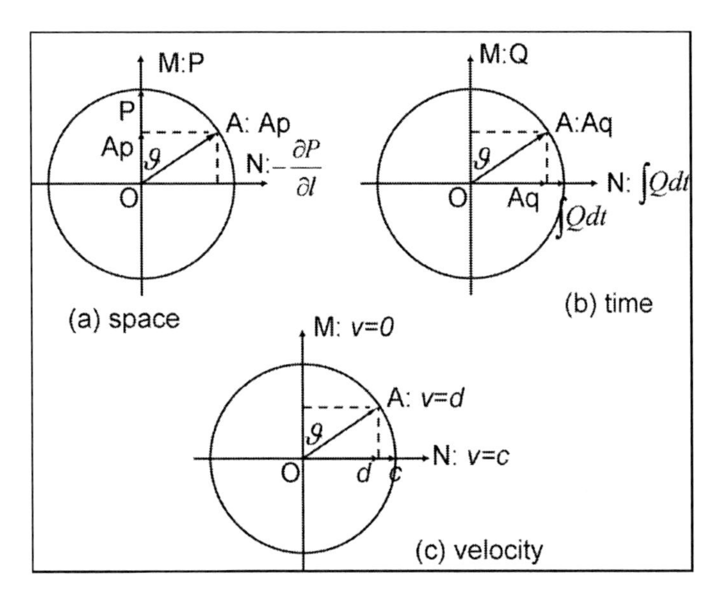


Figure 4.2. Illustration of spherical quantity changes in relation to velocity $d$ in a partial clockwise rotation.

As shown in Figure 4.2, if the velocity of a physical entity is zero ($v=0$), then $\vartheta$ equals zero and no rotation occurs; and if the physical entity travels at the speed of light ($v=c$), then $\vartheta$ equals π/2 indicating that a clockwise rotation has proceeded completely, i.e., spherical quantity M transforms into N. Light represents a clockwise rotation or space to time dimension conversion. If the velocity of the physical entity equals $d$ in the range of $0<d<c$, then the value of $\vartheta$ is between 0 and π/2. We shall call the process a partial clockwise rotatory operation, indicating that M partially transforms towards N with a velocity $d$, less than the speed of light $c$. If A is the middle state in the clockwise rotation pathway, then from Figure 4.2(a), the radius of the circle is P and space component of A is $A_p$; and from Figure 4.2(b), the radius of the circle is $\int Qdt$ and time component of A is $A_q$.

$$A_p = P\cos\vartheta, \tag{4.3}$$

$$A_q = \int Qdt \cdot \sin\vartheta. \tag{4.4}$$

These are so because spherical quantity in quaternity coordinates coincides with rotatory vector around a central point on Cartesian plane (see section 1.5.1). From Figure 4.2(c), the radius of the circle is $c$, the hypotenuse of a right triangle:

$$\sin \vartheta = \frac{d}{c}, \tag{4.5}$$

whence

$$A_p = P\sqrt{1-\left(\frac{d}{c}\right)^2}. \tag{4.6}$$

It should be understood that when $\sin \vartheta = 0$, $A_q$ declines into $Q$, reducing a dimension from $\int Qdt$. Because $A_p$ is less than $P$ and $A_q$ is larger than $Q$, state point A experiences space contraction and time dilation during the process of partial clockwise rotation. The larger the velocity $d$ is, the larger the radian angle $\vartheta$ is, and the more severe the transformation is. Thus special relativity boils down to a problem of partial clockwise rotation in quaternity. Equation (4.6) is actually a Lorentz's transformation with regard to length contraction where P is object length at rest and $A_p$ is object length at velocity $d$. The relationships between spherical quantities in full and partial clockwise rotatory operations are summarized in Table 4.1.

**Table 4.1. The transformation of a spherical quantity under partial clockwise rotation and full clockwise rotation in reference to Figure 4.2**

| Clockwise rotation | Rotating angle | Speed | Space component | Time component |
|---|---|---|---|---|
| Original state | $\vartheta=0$ | 0 | P | Q |
| Partial rotation | $0<\vartheta<\pi/2$ | $D$ | $A_p$ | $A_q$ |
| Full rotation | $\vartheta=\pi/2$ | $C$ | $-\partial P/\partial l$ | $\int Qdt$ |

## 4.4. Summary

Quantum mechanics and relativity are two cornerstones of modern physics, yet as they are applied to the atomic sphere and the universe, both have generated a lot of controversies that testify their limitations in theory and applications. These limitations are not due to the construction defect of these theories but due to the imperfection of their undergirding foundation that we have taken for granted. This foundation is physical quantity in Euclidean geometry. The limitations of quantum mechanics and relativity echo its incompetence in dealing with space and time within the atoms and the universe. When the basement of space and time deviates from the reality, the entire establishment built upon it is flaw. However, since physical quantity and Euclidean space are deeply ingrained in people's mind and imagination and have dominated the knowledge base for thousands of years, changing them seems out of the question. This makes it difficult to introduce spherical view, especially when it is not full-fledged. The scientific communities are reluctant to accept such a fresh proposal that does not fall into any existing categories of research fields.

Relativity and quantum mechanics describe the world from two extreme perspectives and in very different scopes relative to the studied objects. In relativity, man observes the universe in a way similar to a tiny ant, who resides at the trunk of a big tree, seeing a small portion of the tree from inside the tree. In quantum mechanics, man studies the atoms from outside its sphere. He gets a vague bird's-eye view of the object from a relatively far away distance. Thus the unification of relativity and quantum mechanics lies in correctly evaluating their perspectives and kens of vision. By restoring their pictures to their proper perspectives, we may reconcile these two seemingly inconsistent accounts that have confused so many physicists to date.

Besides the difference in the scope of views, one significant property of spherical quantity is that it describes an entity as a dynamic oscillating one. Within the atomic shell, electrons are changing space and time states millions of times within a second continuously and harmonically. The principle of rotatory operation is valid for describing electronic behavior. As for the entire universe, the space and time oscillation process is going on at an agonizingly slow pace due to its enormous scale, and even the partial clockwise rotation rule seems negligible from the perspective of an earthly life. However, the red shift of peculiar light spectra from distant galaxies has confirmed this dynamic nature of the universe.

Quaternity perception of the world is based on its spacetime concept and has been showed to be advantageous over quantum mechanics in explaining electronic orbitals. Although the validity of quaternity remains to be tested, experimental evidence that has supported quantum mechanics and relativity so far is all compatible with quaternity. The theory of quaternity is built upon alternative spacetime concept and is a significant progress in physics. It is expected that quaternity will become the mainstream science in the decades and centuries yet to come.

With regard to space dimensions, we are still in the Stone Age for we are able to count up to three only, namely X dimension, Y dimension, and Z dimension, as in the ancient Chinese history three was the maximum number. Where is the fourth space dimension? It is a riddle that no one really knows. Although we are taught in linear algebra about n-dimensional hyperspace, geometrical shapes of more than three dimensions have never been delineated satisfactorily. As I work on this treatise, I convince myself that we are walking out of the Stone Age as the shining light of quaternity gradually pierces the darkness on the eastern horizon---the age of quaternity is coming!

## REFERENCES

[1] Einstein, A.; Podolsky, B.; and Rosen, N. Can quantum-mechanical description of physical reality be considered complete? *Physical Review*, 1935, 47, 777-780.

[2] d'Espagnat, B. The quantum theory and reality, *Scientific American*, 1979, 241(5), 158-175.

*Chapter 5*

# THE ATOMIC STRUCTURE

We have described electronic orbitals within helium and neon shells in Chapters 1 and 2. After digression in Chapters 3 and 4, we shall resume our discussion on electronic orbitals within inert atoms. The discovery of dimension encapsulation mechanism allows us to further construct spherical quantities for electrons beyond electron octet in neon shell. Step by step, we shall build mathematical structures to account for various electrons in large inert atoms such as radon and Uuo. The elucidation of 3d- and 4f-orbitals is one of the most exciting and enthralling parts of the story. The splendid beauty of dynamic calculus convinces us that it describes the nature squarely. Of particular importance in this chapter is the recognition of orbital complements in geometry due to clockwise direction shift and due to standpoint shift, which provides vital clues to cellular model in Chapter 8.

## 5.1. ELECTRONIC ORBITALS IN ARGON SHELL

Argon shell and neon shell share the same spherical layer within an argon atom as if both 1s electrons would share helium shell. We have characterized 2s2p electron octet in reasonable details under quaternity spacetime. Here we shall present its counterpart, 3s3p electron octet, within the same framework. Space and time symmetry and orbital complementarity are the keys to understanding both 2s2p and 3s3p octets in the spherical world. Dimension factors as calculus operators and electronic ropes as conceptual abstractions demonstrate symmetry of spherical quantities about space and time. Electronic transformation due to clockwise direction shift produces pairs of complementary orbitals in geometry. The order of electrons within an argon atom is by the rule of dynamic calculus, never chaotic.

### 5.1.1. Spherical Quantities for 3s3p Electrons

Electrons within argon shell still rely upon both radian angles, $\Psi$ and $\psi$ . They are spherical quantities of 1s electrons in time and space, but both are radian angles when viewed from argon shell. Basically spherical quantities for 2s2p electrons have been constructed by eight consecutive counterclockwise rotations from $A_1A_2\cos\Psi\cos\psi$ ; and spherical quantities for 3s3p electrons will be built by clockwise rotations from it instead. A counterclockwise rotation transforms a spherical quantity in the direction of reducing a time dimension while increasing a space dimension; and a clockwise rotation transforms a spherical quantity in the direction of increasing a time dimension while reducing a space dimension. Both are spacetime symmetric in dynamic calculus. All rotatory operations that we have encountered in Chapters 1 to 3 are counterclockwise. A rotation normally refers to a counterclockwise rotation unless qualified as clockwise. Since 3s3p electrons are built by clockwise rotations, they are in symmetry to 2s2p electrons in spacetime.

Upon an initial spherical quantity $\varphi_{00} = A_1 A_2 \cos\Psi \cos\psi$, we may perform a clockwise rotation as follows:

$$\begin{aligned}\varphi_{00}^{-0} &= -\frac{dA_2 \cos\psi}{dl} \int A_1 \cos\Psi dt \\ &= (A_2 \frac{d\psi}{dl} \sin\psi) \cdot (-A_1 \frac{1}{\dot{\Psi}} \int \cos\Psi d\Psi) \\ &= -A_1 A_2 (-\frac{1}{\dot{\Psi}} \psi') \sin\psi \sin\Psi \\ &= -A_1 A_2 n \sin\psi \sin\Psi,\end{aligned} \tag{5.1}$$

where

$$\psi' = -\frac{d\psi}{dl}, \tag{5.2}$$

$$n = \frac{1}{u} = -\frac{1}{\dot{\Psi}} \psi' = \Omega_0^{-0}. \tag{5.3}$$

Variables $\Psi$ and $\psi$ are time and space radian angles, respectively. Through consecutive clockwise rotations, we obtain four spherical quantities for $3s^1 3p_z^1 3p_y^1 3p_x^1$ electrons in the negative directions of quaternity axes:

$$\begin{pmatrix} \varphi_0 \\ \varphi_1 \\ \varphi_2 \\ \varphi_3 \end{pmatrix} = \begin{pmatrix} \varphi_{00} - \dfrac{\partial \varphi_{00}}{\partial l} \\ \varphi_{00}^{-0} - \dfrac{\partial \varphi_{00}^{-0}}{\partial l} \\ \varphi_{00}^{-02} - \dfrac{\partial \varphi_{00}^{-02}}{\partial l} \\ \varphi_{00}^{-03} - \dfrac{\partial \varphi_{00}^{-03}}{\partial l} \end{pmatrix}, \tag{5.4}$$

where $\varphi_{00}^{-02}$ and $\varphi_{00}^{-03}$ denote performing clockwise rotations twice and thrice upon $\varphi_{00}$ respectively. In terms of trigonometric functions, the matrix becomes

$$\begin{pmatrix} \varphi_0 \\ \varphi_1 \\ \varphi_2 \\ \varphi_3 \end{pmatrix} = A_1 A_2 \begin{pmatrix} \cos\psi\cos\Psi - \psi'\sin\psi\cos\Psi \\ -n(\sin\psi\sin\Psi + \psi'\cos\psi\sin\Psi) \\ n^2(\cos\psi\cos\Psi - \psi'\sin\psi\cos\Psi) \\ -n^3(\sin\psi\sin\Psi + \psi'\cos\psi\sin\Psi) \end{pmatrix}. \tag{5.5}$$

After carrying out four consecutive clockwise rotations upon $\varphi_{00}$, we arrive at function

$$\varphi_{40} = A_1 A_2 n^4 \cos\psi\cos\Psi . \tag{5.6}$$

At this moment, time and space radian angles switch their roles, i.e., $\Psi$ represents space radian angle while $\psi$ denotes time radian angles for subsequent clockwise rotations. The calculus operators exchange their operands so that

$$\begin{aligned} \varphi_{40}^{-0} &= -n^4 \frac{dA_1\cos\Psi}{dl} \int A_2 \cos\psi \, dt \\ &= A_1 A_2 n^4 (\frac{d\Psi}{dl}\sin\Psi)\cdot(-\frac{1}{\dot{\psi}}\int\cos\psi \, d\psi) \\ &= -A_1 A_2 n^4 (-\frac{1}{\dot{\psi}}\Psi')\sin\Psi\sin\psi \\ &= -A_1 A_2 n^5 \sin\Psi\sin\psi, \end{aligned} \tag{5.7}$$

where

$$\Psi' = -\frac{d\Psi}{dl}, \tag{5.8}$$

$$-\frac{1}{\dot{\psi}}\Psi' = \frac{1}{u} = n . \tag{5.9}$$

By the same manner, we may derive spherical quantities for another electron quartet $3s^1 3p_z^1 3p_y^1 3p_x^1$ in the positive directions of quaternity axes from

$$\begin{pmatrix} \varphi_4 \\ \varphi_5 \\ \varphi_6 \\ \varphi_7 \end{pmatrix} = \begin{pmatrix} \varphi_{40} - \dfrac{\partial\varphi_{40}}{\partial l} \\ \varphi_{40}^{-0} - \dfrac{\partial\varphi_{40}^{-0}}{\partial l} \\ \varphi_{40}^{-02} - \dfrac{\partial\varphi_{40}^{-02}}{\partial l} \\ \varphi_{40}^{-03} - \dfrac{\partial\varphi_{40}^{-03}}{\partial l} \end{pmatrix}, \tag{5.10}$$

which expands into trigonometric terms as

$$\begin{pmatrix} \varphi_4 \\ \varphi_5 \\ \varphi_6 \\ \varphi_7 \end{pmatrix} = A_1 A_2 n^4 \begin{pmatrix} \cos\Psi\cos\psi - \Psi'\sin\Psi\cos\psi \\ -n(\sin\Psi\sin\psi + \Psi'\cos\Psi\sin\psi) \\ n^2(\cos\Psi\cos\psi - \Psi'\sin\Psi\cos\psi) \\ -n^3(\sin\Psi\sin\psi + \Psi'\cos\Psi\sin\psi) \end{pmatrix}. \tag{5.11}$$

Like electrons in neon shell, there are clockwise rotation relationships between every two adjacent spherical quantities, e.g., between $\varphi_0$ and $\varphi_1$, $\varphi_1$ and $\varphi_2$, $\varphi_2$ and $\varphi_3$, $\varphi_7$ and $\varphi_0$, and so on. Each electron is transforming through eight states continuously. After eight consecutive clockwise rotations, each electron returns to its original state. The octet cycle gives remarkable stability for argon shell. Spherical quantities $\varphi_0$ and $\varphi_4$ in the octet are $3s^2$ electrons at the quaternity dimensions of *1* and $n^4$; and the other six spherical quantities make three pairs of 3p-orbitals. Spacetime dimensions of 3s3p electrons are reciprocal to those of 2s2p electrons, one by one.

$$un = 1; n^8 = 1. \tag{5.12}$$

By the way, we have assumed that electrons within neon shell rotate counterclockwise whereas those within argon shell rotate clockwise invariably. Either octet alignment constitutes a conservative cycle that renders a neon atom or an argon atom inert. As a matter of fact, neon and argon atoms are noble gases. However, within a fluorine or a chlorine atom, the order of electrons is not so neat. Depending on the position of the unfilled void in the atomic sphere, an electron may rotate either counterclockwise or clockwise in order to fill the vacancy. Thus each electron has the potential of rotating back and forth. For example, if Φ represents a metor at a dimension of two-dimensional space with two-dimensional time, then it may transform into either a polor or a vitor depending on the position of the vacancy. One question arises as to whether or not two electrons collide at a particular dimension when both tend towards the same vacancy at the same moment. The existence of a vacancy is the source of instability of an unsaturated atom such as fluorine and chlorine.

### 5.1.2. Complementary Orbitals Due to Direction Shift

The geometric relationship between 2s2p and 3s3p electrons deserves further scrutiny. Here we shall first present a typical case of orbital complement and then extend it to more cases. As shown in Figure 5.1, we compare the process of $2s \rightarrow 2p_x$ transformation with that of $3s \rightarrow 3p_z$ transformation. Transformation of $2s \rightarrow 2p_x$ involves the sinusoidal contraction of a solid sphere and the displacement of the spherical center rightwards. The whole sphere eventually converges into a point at the periphery. In contrast, transformation of $3s \rightarrow 3p_z$ involves the waxing of a hemispherical surface at the left hemisphere into a crescent (shaded volume). The entire empty sphere is eventually compressed rightwards until it disappears at the periphery. Both orbitals under transformations are complementary in geometric shapes, i.e., the solid sphere in the former exactly coincides the empty sphere in the latter. Both orbitals share the same spatial sphere but their electron cloud distributions at any moment are complementary like a pair of plug and socket.

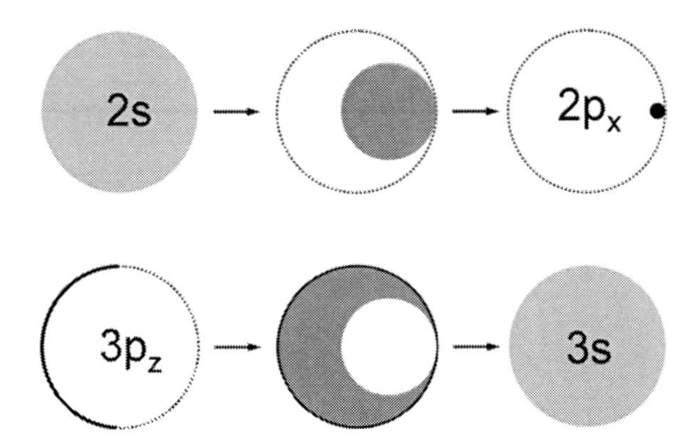


Figure 5.1. Profile diagrams of two complementary orbitals in transformations.

The complementarity is achieved because $3p_Z$ orbitals are oriented along the opposite direction of $2p_X$ orbitals. Within neon shell, eight electrons are built by consecutive counterclockwise rotations from dimensionless quaternity axis *1*; and within argon shell, eight electrons are constructed by consecutive clockwise rotations from the same notch. Clockwise direction shift results in anti-parallel matches between both octets. As shown in Figure 5.2, the space dimensions of every orbital pair in the same row are complementary in considering that the time flow direction of 3s3p orbitals is contrary to that of 2s2p.

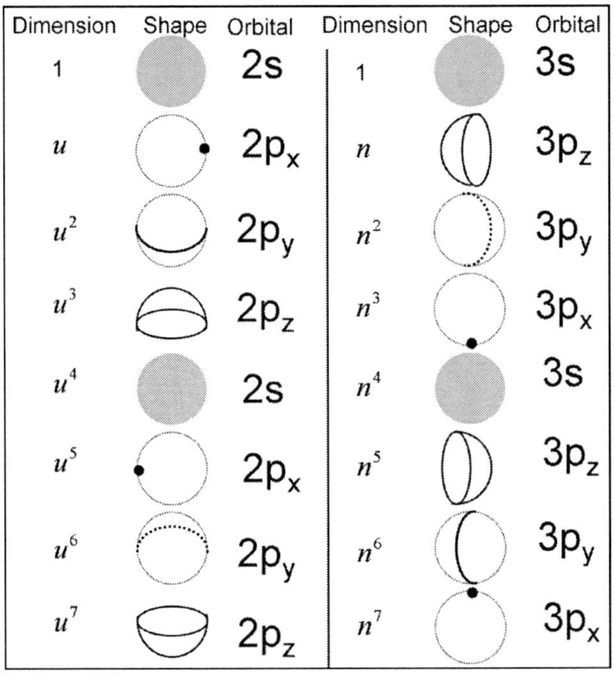

| Dimension | Shape | Orbital | Dimension | Shape | Orbital |
|---|---|---|---|---|---|
| 1 | | 2s | 1 | | 3s |
| $u$ | | $2p_x$ | $n$ | | $3p_z$ |
| $u^2$ | | $2p_y$ | $n^2$ | | $3p_y$ |
| $u^3$ | | $2p_z$ | $n^3$ | | $3p_x$ |
| $u^4$ | | 2s | $n^4$ | | 3s |
| $u^5$ | | $2p_x$ | $n^5$ | | $3p_z$ |
| $u^6$ | | $2p_y$ | $n^6$ | | $3p_y$ |
| $u^7$ | | $2p_z$ | $n^7$ | | $3p_x$ |

Figure 5.2. Complementary dimensions and orbital shapes of 2s2p and 3s3p electrons due to clockwise direction shift.

We have illustrated complementary processes of $1 \rightarrow u$ and $n \rightarrow 1$ orbital transformations in Figure 5.1. From Figure 5.2, we may find other complementary pairs such as $u^3 \rightarrow u^4$ and $n^4 \rightarrow n^3$, $u^4 \rightarrow u^5$ and $n^5 \rightarrow n^4$, and $u^7 \rightarrow 1$ and $1 \rightarrow n^7$. Each pair are complementary in a unique spacetime match. Note that Figure 5.1 is a profile diagram where $3p_z$ orbital is drawn as a semicircle; but Figure 5.2 is a stereo sketch where $p_z$-orbitals are drawn as hemispheres. Although $u^7$ dimension is equivalent to $n$ dimension in considering equation (5.12), the $2p_z$ orbital is different from the $3p_z$ orbital in orientation. The $2p_z$ orbital is a hemispherical surface with a mouth opening upwards whereas the $3p_z$ orbital is a hemispherical surface with a mouth facing right. Dimension series of $u^i$ define the orientations of 2s2p electrons while dimension series of $n^i$ indicate the orientations of 3s3p electrons. By the way, the relationship between $2p_y$ and $3p_y$ orbitals is also complementary in geometry (Figure 5.3). In these ways, both 2s2p and 3s3p shells occupy the same spherical layer to complement each other in maximum.

It is also noteworthy that clockwise shift in rotations is different from time and space roles switch between radian angles. As radian angles $\Psi$ and $\psi$ switch their time and space roles, the differential and integral operations of a rotation switch their operands without changing the operator. A clockwise direction shift changes the operator rather than its operands, from $\int -(\partial/\partial t)dl$ to $\int -(\partial/\partial l)dt$ for counterclockwise to clockwise shift. Roles switch does not involve changes between $u^i$ and $n^i$ dimension series for any spherical quantities. But clockwise rotations produce spherical quantities in $n^i$ dimension series instead of $u^i$ dimension series, which implicitly alter the geometrical orientations of p-orbitals.

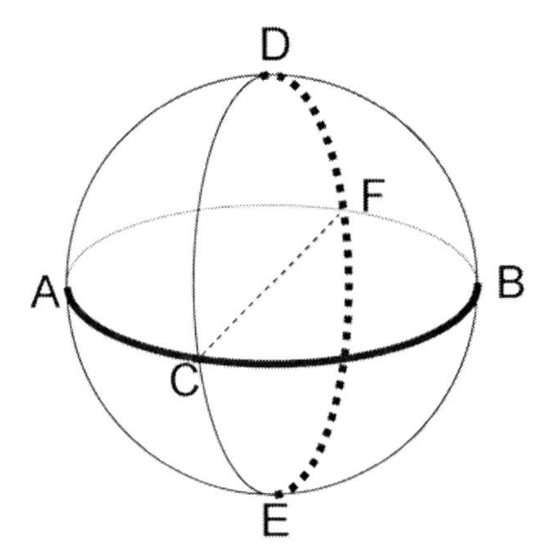


Figure 5.3. Spatial relationship between $2p_y$ and $3p_y$ orbitals represented by two interlaced arcs ACB and DFE on a sphere.

### 5.1.3. Dimension Factors as Operators

From equations (1.1) and (1.4), we may write

$$\omega\Omega = -\frac{\partial\Omega}{\partial t}, \tag{5.13}$$

$$\omega^2\Omega = \frac{\partial}{\partial t}(-\frac{\partial\Omega}{\partial t}), \tag{5.14}$$

where $\omega$ is an angular velocity dimension in function expressions $\omega\Omega$ and $\omega^2\Omega$. This dimension factor may be regarded as a differential operator with respect to a time dimension. The first derivative is negative while the second differential operator is positive $\partial/\partial t$ because there is only one dimension of time within helium shell, the second time dimension being the first one wrapped back in contrary direction. By similar logic, orbital radius $r$ may be treated as an integral operator when placed before a proper spherical quantity directly.

$$r\Omega = \int \Omega dl , \tag{5.15}$$

$$r^2\Omega = -\iint \Omega dl^2 . \tag{5.16}$$

Combining operators $\omega$ and $r$ yields

$$\omega r\Omega = v\Omega = \int -\frac{\partial\Omega}{\partial t} dl , \tag{5.17}$$

where a velocity dimension factor serves as a rotatory operator. Thus duality equation may be expressed concisely by

$$v^2\Omega = \Omega , \tag{5.18}$$

which indicates that rotating a spherical quantity twice returns it to its original state since $v^2 = 1$ in two-dimensional helium shell.

The interpretation of dimension factors as operators may be extended to four-dimensional case. Given $\dot{\Psi}$ as an angular velocity in neon shell, we may treat it as a negative differential operator:

$$\dot{\Psi}\Phi = -\frac{\partial\Phi}{\partial t}, \tag{5.19}$$

$$\dot{\Psi}^2\Phi = -\frac{\partial}{\partial t}(-\frac{\partial\Phi}{\partial t}) = \frac{\partial^2\Phi}{\partial t^2}, \tag{5.20}$$

$$\dot{\Psi}^3\Phi = -\frac{\partial^3\Phi}{\partial t^3}, \tag{5.21}$$

$$\dot{\Psi}^4\Phi = -\frac{\partial}{\partial t}(\dot{\Psi}^3\Phi) = \frac{\partial^4\Phi}{\partial t^4} . \tag{5.22}$$

Given $\psi'$ as a dimension factor in equation (5.2), we may also write

$$-\frac{1}{\psi'}\Phi = \int \Phi dl \, , \quad (5.23)$$

$$\frac{1}{\psi'^2}\Phi = \iint \Phi dl^2 , \quad (5.24)$$

$$-\frac{1}{\psi'^3}\Phi = \iiint \Phi dl^3 \, , \quad (5.25)$$

$$\frac{1}{\psi'^4}\Phi = \int\iiint \Phi dl^4 \, . \quad (5.26)$$

Combining the differential and integral operators yields

$$-\frac{1}{\psi'}\dot{\Psi}\Phi = u\Phi = \int -\frac{\partial \Phi}{\partial t} dl \, , \quad (5.27)$$

where $u$ is a velocity dimension in four-dimensional spacetime serving as a rotatory operator upon function $\Phi$. From equations (2.16) and (2.23), we have $u = v\Omega$. Since both helium and neon shells are coupled together, the rotation of a 1s electron is a rotatory operator upon a 2s2p electron. In other words, the cycling of 1s electrons drives the cycling of 2s2p electron octet. After radian angles $\Psi$ and $\psi$ switch their time and space roles, we have another pair of differential and integral operators instead:

$$-\frac{1}{\Psi'}\dot{\psi} = u \, . \quad (5.28)$$

Thus quaternity equation may be expressed in a concise way of

$$u^4\Phi = \Phi \, , \quad (5.29)$$

where $u$ is a counterclockwise rotatory operator upon a spherical quantity. In this way, 2s2p octet may be compactly expressed as $\sum_{i=0}^{7} u^i \Phi_0$ where index $i$ indicates the number of rotatory operations to perform upon the spherical quantity and plus signs indicate continuity of orbitals in spacetime. Likewise, because $n$ is a clockwise rotatory operator, 3s3p electron octets may be expressed as $\sum_{i=0}^{7} n^i \varphi_0$ as well where index $i$ indicates the number of clockwise rotations to perform upon the spherical quantity and $\Sigma$ indicates spacetime continuity.

Treating dimension indicators as operators is neat. However, we must still treat dimension factors that associate with trigonometric functions as dimension factors and complex number identifiers to avoid any confusion. These are the cases as were expressed in equations (2.18), (2.25), (5.5), and (5.11). Only when a dimension indicator is placed before by a spherical quantity directly should we regard it as a calculus operator.

### 5.1.4. Electronic Ropes

A counterclockwise rotation transforms a spherical quantity in the direction of reducing a time dimension while increasing a space dimension. In two-dimensional helium, we have

$$\Omega_0^0 = \Omega_2 ; \ \Omega_2^0 = \Omega_0 . \qquad (5.30)$$

The rotation of one electron results in another electron, and rotating an electron twice results in itself after a transformation cycle. Mechanically, when you twist a strand of fiber or hemp in a fixed direction and then let it fold back, a rope is formed. If we regard rotatory operator $\int(-\partial/\partial t)dl$ upon an electron as a kind of twisting tension, then electron transformations of equation (5.30) constitute the art of making a rope with $\Omega_0$ and $\Omega_2$ as two participating strands (Figure 5.4). If one strand represents time component, the other strand is space component, both being intertwined together. If one strand represents momentum $p$, then the other strand denotes energy $E$ so that the rotatory relationship indicates the balance of forces binding both strands.

$$-\frac{\partial p}{\partial t} = \frac{\partial E}{\partial l} . \qquad (5.31)$$

In formulating spherical quantities for 2s2p electrons, we performed eight consecutive counterclockwise rotations upon function $A_1 A_2 \cos\Psi \cos\psi$ . The head and tail of 2s2p electrons are interconnected so that the octet is a two-strand rope.

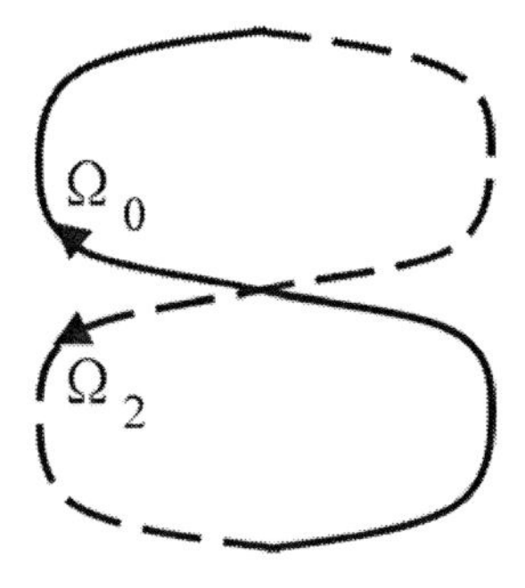


Figure 5.4. Conceptual diagram of both 1s electrons forming a rope structure.

One strand is composed of four electrons in the positive directions of quaternity axes, the other strand is composed of another four electrons in the negative directions of quaternity axes. Four pairs of electrons ($\Phi_0$ and $\Phi_4$, $\Phi_1$ and $\Phi_5$, $\Phi_2$ and $\Phi_6$, and $\Phi_3$ and $\Phi_7$) are located in the opposite directions of the same quaternity axes with anti-parallel alignments so as to form a rope properly (see Figure 2.10).

Likewise, spherical quantities for 3s3p were obtained through clockwise rotations from function $A_1A_2\cos\Psi\cos\psi$ so that 3s3p octet constitutes a rope as well. The counterclockwise rotations of 2s2p electrons and clockwise rotations of 3s3p electrons within a argon atom do not counteract each other, but rather twist the orbitals in a fixed direction because they are performed upon both ends of 1s sphere. They are in a similar chiralty. To illustrate, if we stand at a certain point and look right and forward at a helical cord (Figure 5.5), the helix is wound counterclockwise; and if we face left but look backward from the same standpoint, the helix is wound clockwise. Dimension factor $n$ in 3s3p electrons indicates a clockwise rotation upon $\Omega_0$ while dimension factor $u$ in 2s2p electrons indicates a counterclockwise rotation upon $\Omega_0$. Because $\Omega_0^0$ and $\Omega_0^{-0}$ form a rotatory cycle, 3s3p and 2s2p octets twist 1s sphere in both ends (forward and backward) but in a fixed direction (Figure 5.5).

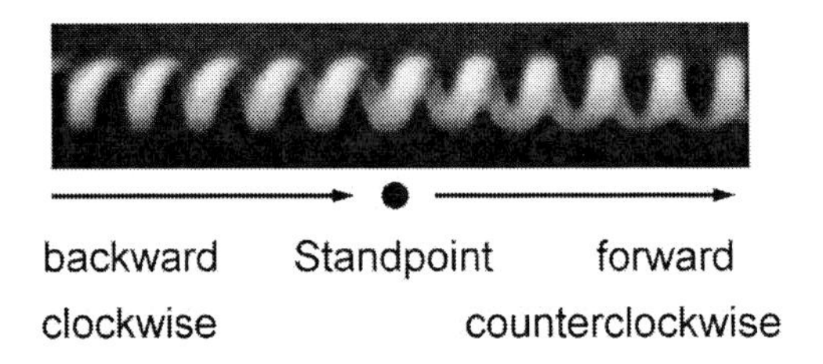


Figure 5.5. Perspective shift accounting for clockwise shift of a helical structure. Arrows represent direction of time flow and hence observation direction.

There are two junctions between 2s2p and 3s3p octets at dimensions $\Phi_{00}$ and $\Phi_{40}$, known alternatively as $\varphi_{00}$ and $\varphi_{40}$. Since a rotatory operation is a differential operation with a simultaneous integral operation, we may purposely confuse the expressions between $\Phi_{00}$ and $\Phi_0$, and between $\Phi_{40}$ and $\Phi_4$, and so on. At the junctions, we have

$$\Phi_0^0 = \Phi_1, \Phi_7^0 = \Phi_0; \tag{5.32}$$

$$\varphi_0^{-0} = \varphi_1, \varphi_7^{-0} = \varphi_0; \tag{5.33}$$

$$\Phi_0 = \varphi_0 . \tag{5.34}$$

$$\Phi_4^0 = \Phi_5, \Phi_3^0 = \Phi_4; \tag{5.35}$$

$$\varphi_4^{-0} = \varphi_5, \varphi_3^{-0} = \varphi_4; \tag{5.36}$$

$$\Phi_4 = \varphi_4 . \tag{5.37}$$

Thus there are four electrons converge at each junction. As shown in Figure 5.6, the junctions under quaternity coordinates are dimension positions of A and B. Both 2s2p and 3s3p octets constitute an electronic rope comprising four strands, each strand having four electrons in unique alignments. They are intertwined and complementary within an argon atom.

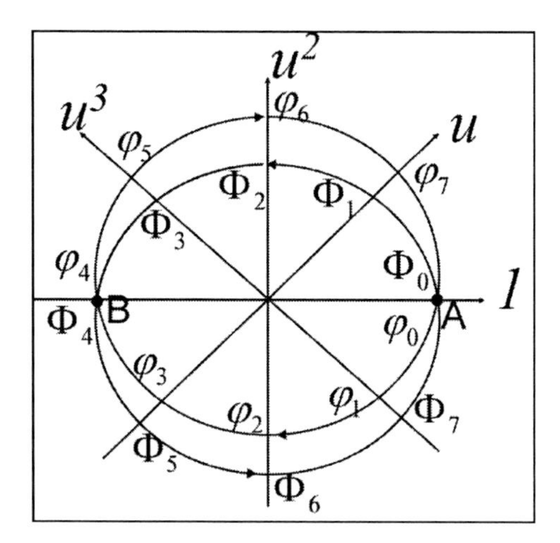


Figure 5.6. Dimensional relationships between 2s2p and 3s3p electron octets having two junctions. The arrows indicate rotation directions of the electrons under quaternity coordinates.

## 5.2. Electronic Orbitals in Krypton Shell

Because electronic orbitals within krypton shell occupy a spherical layer outside argon shell, spherical quantities for 4s4p may be constructed by rotatory operations upon an initial function encapsulating 2s2p and 3s3p shells. The dimension encapsulation involves casting two circular complex functions into complementary radian angles by cosine operators. Both circular complex functions are obtained by dividing spherical quantities for 2s2p and 3s3p electrons into time and space components. Before doing that, we have got to define circular complex functions and learn the properties of radian angles as circular complex functions and as exponential functions.

### 5.2.1. Coordination of Circular Complex Functions

Because sine and cosine functions are one-dimensional projections of uniform circular motion, they are called circular functions in mathematics. Here we introduce the concept of circular complex function. A circular complex function is a complex function composed of circular functions. A circular complex function represents a complete set of time or space dimensions of spherical quantities within an inert atomic shell. For example, both $\Psi$ and $\psi$ functions are circular complex functions within helium shell. Function $\Psi$ contains both time dimensions $C_1 \cos\alpha$ and $-C_1\omega \sin\alpha$ ; and function $\psi$ contains both space dimensions $C_2 \cos\beta$ and $C_2 r \sin\beta$ in the two-dimensional system. Each function expresses a full cycle of a spherical quantity in time or space. Because of this, performing a differential or integral operation upon a circular complex function shifts the sequence of its trigonometric terms, but does not change the function itself. Dimension wrapping around in a circular complex function determines its invariant calculus property.

$$\Psi = C_1(\cos\alpha - \omega\sin\alpha)\,, \tag{5.38}$$

$$\omega\Psi = -\frac{d\Psi}{dt} = C_1(-\omega\sin\alpha + \cos\alpha) = \Psi\,. \tag{5.39}$$

$$\psi = C_2(\cos\beta + r\sin\beta)\,, \tag{5.40}$$

$$r\psi = \int\psi\, dl = C_2(r\sin\beta + \cos\beta) = \psi\,. \tag{5.41}$$

From equations (1.8) and (1.9), we obtain a complete set of dimensions in the two-dimensional spacetime by combining time and space components:

$$\Omega = C_1(\cos\alpha - \omega\sin\alpha)\cdot C_2(\cos\beta + r\sin\beta)\,, \tag{5.42}$$

where the dot operation should be regarded as a coordination between time and space components instead of conventional multiplication. The rule of coordination operation is that only adjacent terms between both functions are associated (Figure 5.7). The coordination of time and space components in equation (5.42) results in four terms as were expressed in equation (1.9). These four terms correspond to four pairs of associations indicated by zigzag lines. Although the result in this case happens to coincide with the consequence of conventional multiplication, coordination is not multiplication.

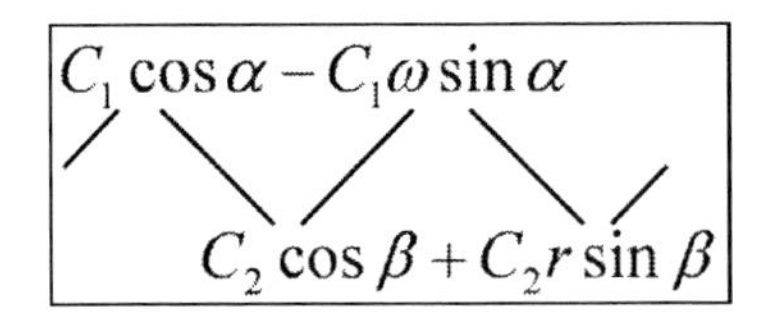


Figure 5.7. Association pairs related by lines between adjacent terms of circular complex functions in coordination. The last half line connects with the first half to form an association pair.

In quaternity spacetime, we may partition spherical quantities of 2s2p electrons into time and space components as follows:

$$\begin{aligned}\Theta &= A_1(\cos\Psi - \dot{\Psi}\sin\Psi - \dot{\Psi}^2\cos\Psi + \dot{\Psi}^3\sin\Psi)\\ &+ A_1\dot{\Psi}^4(\cos\Psi - \frac{1}{\Psi'}\sin\Psi - \frac{1}{\Psi'^2}\cos\Psi + \frac{1}{\Psi'^3}\sin\Psi),\end{aligned} \tag{5.43}$$

$$\theta = A_2(\cos\psi - \frac{1}{\psi'}\sin\psi - \frac{1}{\psi'^2}\cos\psi + \frac{1}{\psi'^3}\sin\psi)$$
$$+ A_2\frac{1}{\psi'^4}(\cos\psi - \dot{\psi}\sin\psi - \dot{\psi}^2\cos\psi + \dot{\psi}^3\sin\psi), \tag{5.44}$$

Eight terms in function $\Theta$ are obtained by performing four consecutive differential operations upon $A_1\cos\Psi$ with respect to time dimensions followed by four consecutive integral operations over space dimensions; and eight terms in function $\theta$ are similarly obtained by performing four consecutive integral operations upon $A_2\cos\psi$ over space dimensions followed by four consecutive differential operations with respect to time dimensions. Because four space dimensions in function $\Theta$ are actually four time dimensions in sequence upon switching space and time roles, there are

$$\dot{\Psi}^4\frac{1}{\Psi'^4} = 1, \tag{5.45}$$

and similarly,

$$\frac{1}{\psi'^4}\dot{\psi}^4 = 1. \tag{5.46}$$

Functions $\Theta$ and $\theta$ are four-dimensional versions of circular complex functions because each of them comprises eight dimensions in the spherical layer. The coordination of both yields eight electrons of 2s2p.

$$\Phi = \sum_{i=0}^{7}\Phi_i = \Theta\cdot\theta. \tag{5.47}$$

The association pairs of the coordination are related by 16 lines in Figure 5.8, and the calculation result can be found in equations (2.18) and (2.25). It becomes evident that coordination is different from multiplication in the four-dimensional spacetime.

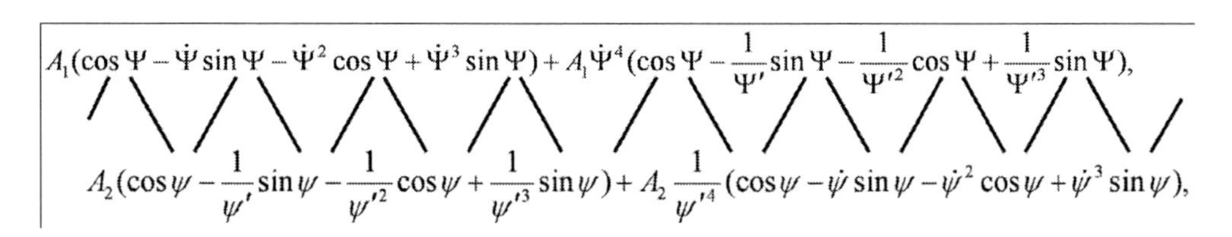


Figure 5.8. The coordination of both circular complex functions resulting in 16 pairs of spherical quantities in zigzag lines.

Both circular complex functions $\Psi$ and $\Theta$ represent time components of helium and neon shells respectively. They are geared together as was illustrated in Figure 2.9 by cogwheels in mesh. Here we further implement the cogwheels by trigonometry in Figure 5.9, where the arrows indicate electron rotation directions in opposite to cogwheel rotation direction. For example, after a rotatory operation, the coupling relation of helium and neon shells shifts from Figure 5.9(a) to 5.9(b). Given the differential chain of $-\partial\Theta/\partial t = -\partial\Theta/\partial\Psi \cdot \partial\Psi/\partial t$, function $-\partial\Theta/\partial t = -A_1 \cos\Psi$ represents a wave function of 2s electron that encapsulates the dimension of both 1s electrons; and function $-\partial\Psi/\partial t = -C_1 \cos\alpha$ is the unfolding dimension of a 1s electron. Derivation from $-\partial\Theta/\partial t$ to $-\partial\Psi/\partial t$ observes differential chain rule and derivation from $C_1 \cos\alpha$ to $A_1 \cos\Psi$ follows dimension encapsulation mechanism. While dimension encapsulation of $C_1 \cos\alpha$ leads to the formation of initial dimension $A_1 \cos\Psi$ in neon shell, which then transforms through eight states consecutively, the differential chain of $-\partial\Psi/\partial t$ alternates between transformations of $C_1 \cos\alpha \mapsto -C_1\omega \sin\alpha$ and $-C_1\omega \sin\alpha \mapsto C_1 \cos\alpha$ in gearing with the rotations of both 1s electrons. Circular complex functions of space components $\theta$ and $\psi$ are coupled together in a similar manner.

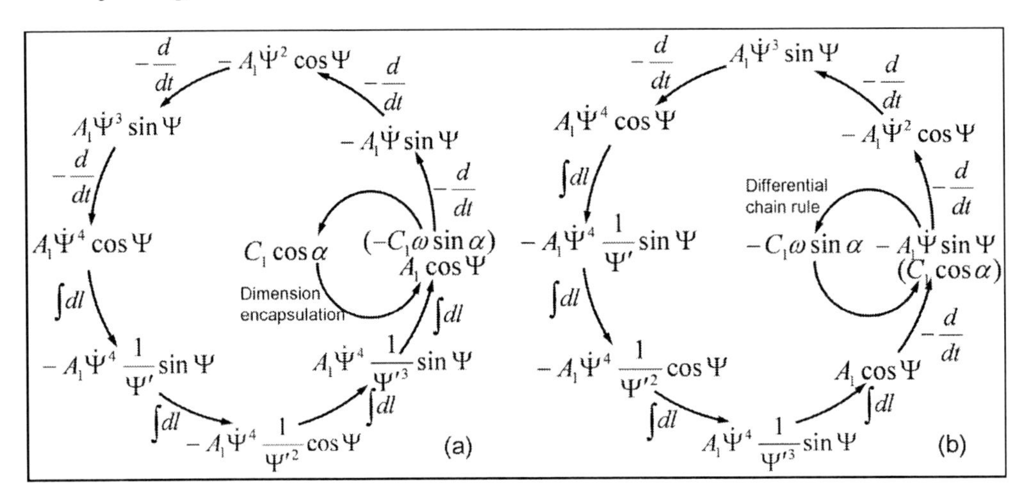


Figure 5.9. Coupling relationship between circular complex functions of helium and neon shells in trigonometry.

## 5.2.2. Circular Complex Functions as Exponential Functions

In line with their invariant calculus property, functions $\Psi$ and $\psi$ can be expressed in exponential form:

$$\Psi = C_1 e^{-\omega\alpha}, \tag{5.48}$$

$$\psi = C_2 e^{r\beta}, \tag{5.49}$$

where $\alpha$ and $\beta$ are radian angles, and $\omega$ and $r$ are complex number identifiers. Performing a differential or integral operation upon each function does not change its expression. This is an important property of an exponential function and a circular complex function.

$$
\begin{aligned}
&-\frac{d\Psi}{dt} = -C_1\frac{de^{-\omega\alpha}}{dt} \\
&= -C_1 e^{-\omega\alpha}\cdot(-\omega)\cdot\frac{d\alpha}{dt} \\
&= -\Psi\cdot\omega^2 \\
&= \Psi.
\end{aligned}
\tag{5.50}
$$

and similarly,

$$
\begin{aligned}
&\int \psi dl = \int C_2 e^{r\beta} dl \\
&= \int C_2 e^{r\beta} r d\beta \\
&= C_2 e^{r\beta} \\
&= \psi.
\end{aligned}
\tag{5.51}
$$

We applied equations (1.3) and (1.6) to the above calculations. Equations (5.50) and (5.51) agree with equations (5.39) and (5.41) in dynamic calculus, respectively.

Upon four-dimensional circular complex functions, we may perform calculus operation as follows:

$$
\begin{aligned}
&\dot{\Psi}\Theta = -A_1 d(\cos\Psi - \dot{\Psi}\sin\Psi - \dot{\Psi}^2\cos\Psi + \dot{\Psi}^3\sin\Psi)/dt \\
&+ A_1\dot{\Psi}^4\int(\cos\Psi - \frac{1}{\Psi'}\sin\Psi - \frac{1}{\Psi'^2}\cos\Psi + \frac{1}{\Psi'^3}\sin\Psi)dt \\
&= A_1(-\dot{\Psi}\sin\Psi - \dot{\Psi}^2\cos\Psi + \dot{\Psi}^3\sin\Psi + \dot{\Psi}^4\cos\Psi) \\
&+ A_1\dot{\Psi}^4(-\frac{1}{\Psi'}\sin\Psi - \frac{1}{\Psi'^2}\cos\Psi + \frac{1}{\Psi'^3}\sin\Psi + \frac{1}{\Psi'^4}\cos\Psi) \\
&= \Theta.
\end{aligned}
\tag{5.52}
$$

The above calculations disregard the sequence change of the trigonometric terms. Strictly speaking, sequence change in a circular complex function indicates home axis shift in quaternity coordinates. Function $\Theta$ is circular because its head and tail are continuous by dynamic calculus, i.e., the last trigonometric term of the function is continuous with the first term. A circular complex function represents a continuum in space or time. By the same manner, we also have

$$
-\frac{1}{\psi'}\theta = \theta. \tag{5.53}
$$

Because of their invariant property in calculus, these circular complex functions may be expressed in exponential form as well:

$$\Theta = A_1 e^{-\dot{\Psi}\Psi}, \theta = A_2 e^{-\frac{1}{\psi'}\psi}, \tag{5.54}$$

where $\Psi$ and $\psi$ are radian angles, and $\dot{\Psi}$ and $-1/\psi'$ are complex number identifiers. Performing a differentiation upon $\Theta$ with respect to a time dimension leads to the function itself.

$$\begin{aligned}
&-\frac{d\Theta}{dt} = -A_1 de^{-\dot{\Psi}\Psi}/dt \\
&= -A_1 e^{-\dot{\Psi}\Psi} \cdot (-\dot{\Psi}) \cdot \frac{d\Psi}{dt} \\
&= -\Theta \cdot (\dot{\Psi})^2 \\
&= \Theta,
\end{aligned} \tag{5.55}$$

where stand-alone factor $(\dot{\Psi})^2$ is construed as $i^2$ that traverses the same dimension to and fro. It is different from $\dot{\Psi}^2$ in equation (5.20) or (5.43) that refers to two consecutive time dimensions in four-dimensional spacetime. By the same manner, we may also write

$$\int \theta dl = \int A_2 e^{-\frac{1}{\psi'}\psi} dl = \int A_2 e^{-\frac{1}{\psi'}\psi} \cdot -\frac{1}{\psi'} d\psi = \theta. \tag{5.56}$$

Hence we have set the rule for dynamic calculus upon exponential functions with imaginary exponents and the rule for dynamic calculus upon circular complex functions. They are not only self consistent, but also consistent with infinitesimal calculus. For example,

$$e^x = 1 + x + \frac{x^2}{2!} + \frac{x^3}{3!} + \dots, \tag{5.57}$$

where the right-hand side can be regarded as a special circular complex function whose second term is the integral of the first term, $\int 1dx$, and the third term is the integral of the second term, $\int xdx$, and so forth. This is in good agreement with the property of a circular complex function that every two adjacent terms have calculus relationship. Equation (5.57) is special because it is a unidirectional ray.

### 5.2.3. Circular Complex Functions as Radian Angles

Of great interest is that the whole circular complex function can be treated as a radian angle, which permits the expression of iterated exponential function such as:

$$\Theta = A_1 e^{-\dot{\Psi} \cdot C_1 e^{-\omega\alpha}}, \tag{5.58}$$

where $\Theta$ itself is a radian angle, too. Within the scope of helium shell, both $\Psi$ and $\psi$ are circular complex functions; but within the scope of neon shell, they are radian angles whose values have been assigned in Figure 2.6. Within the scope of neon shell, $\Theta$ and $\theta$ are circular complex functions; but within the scope of krypton shell, they are radian angles. Circular complex functions are chords transforming sinusoidally by dynamic calculus while radian angles are physical quantities changing uniformly. As shown in dimension diagram (Figure 5.10), the magnitudes of radian angles $\Theta$ and $\theta$ are measured by arcs AC and BC, respectively. They are equivalent to radian angles $\Psi$ and $\psi$ respectively and are complementary at any moment as point C traces semicircle ACB.

$$\cos\Theta = \sin\theta \, . \tag{5.59}$$

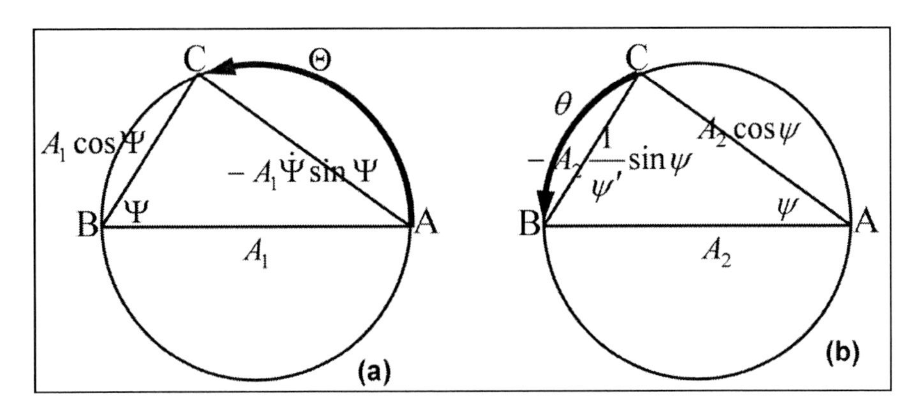


Figure 5.10. Dimension diagram showing the complement of radian angles $\Theta$ and $\theta$ in semicircle ACB.

In Figure 5.10, state point C is codetermined by chords $A_1 \cos\Psi$ and $-A_1\dot{\Psi}\sin\Psi$ in time as well as by chords $A_2 \cos\psi$ and $-A_2 \frac{1}{\psi'} \sin\psi$ in space. This indicates the transformation of spherical quantity from $A_1 A_2 \cos\Psi \cos\psi$ to $-A_1 A_2 u \sin\Psi \sin\psi$ . For subsequent rotations, state point C is codetermined by four similar chords (Figure 5.11). We shall call state point C centromere because it connects with four chords and moves from one end of the semicircle to the other during rotation transformation. The division of spherical quantity $\Phi$ into circular complex functions $\Theta$ and $\theta$ is somewhat analogous to the meiosis of cells because variables $\Psi$ and $\psi$ are separated during the process. This reduces the number of variables by half.

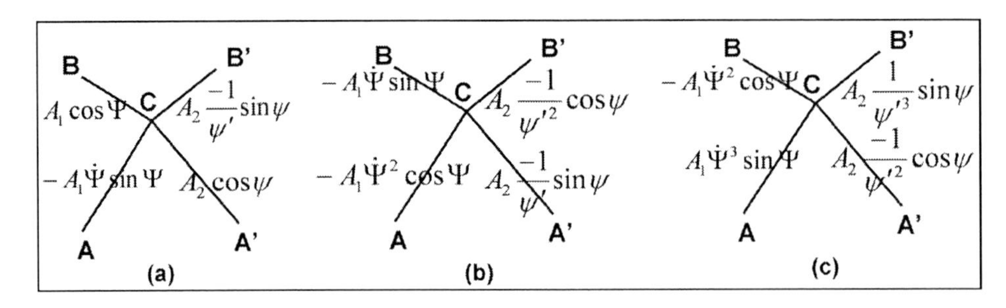


Figure 5.11. Conceptual diagram of four chords connected by a centromere in each of three consecutive rotatory operations.

By the same manner, spherical quantity $\varphi$ within argon shell may be divided into two circular complex functions $\hat{\Theta}$ and $\hat{\theta}$ as well.

$$\varphi = \sum_{i=0}^{7} \varphi_i = \hat{\theta} \cdot \hat{\Theta}\,, \tag{5.60}$$

$$\begin{aligned} \hat{\theta} = A_2 e^{-\psi'\psi} &= A_2(\cos\psi - \psi'\sin\psi - \psi'^2\cos\psi + \psi'^3\sin\psi) \\ &+ A_2\psi'^4(\cos\psi - \frac{1}{\dot{\psi}}\sin\psi - \frac{1}{\dot{\psi}^2}\cos\psi + \frac{1}{\dot{\psi}^3}\sin\psi), \end{aligned} \tag{5.61}$$

$$\begin{aligned} \hat{\Theta} = A_1 e^{-\frac{1}{\dot{\Psi}}\Psi} &= A_1(\cos\Psi - \frac{1}{\dot{\Psi}}\sin\Psi - \frac{1}{\dot{\Psi}^2}\cos\Psi + \frac{1}{\dot{\Psi}^3}\sin\Psi) \\ &+ A_1\frac{1}{\dot{\Psi}^4}(\cos\Psi - \Psi'\sin\Psi - \Psi'^2\cos\Psi + \Psi'^3\sin\Psi). \end{aligned} \tag{5.62}$$

Although $\hat{\Theta}$ and $\hat{\theta}$ as circular complex functions are different from functions $\theta$ and $\Theta$, they take the same values as radian angles.

$$\Theta = \hat{\theta};\ \theta = \hat{\Theta}. \tag{5.63}$$

The equivalence in terms of radian angles can be explained by the synchronization of counterclockwise rotation $\Phi_{00}^{0}$ and clockwise rotation $\varphi_{00}^{-0}$ within a krypton atom. The former has been illustrated in Figure 5.10 and the latter can be found in Figure 5.12. Radian angles $\hat{\Theta}$ and $\hat{\theta}$ are measured by arcs BC and AC respectively. By comparing both figures, it is easy for us to derive equation (5.63) when the counterclockwise and clockwise rotations proceed at the same pace.

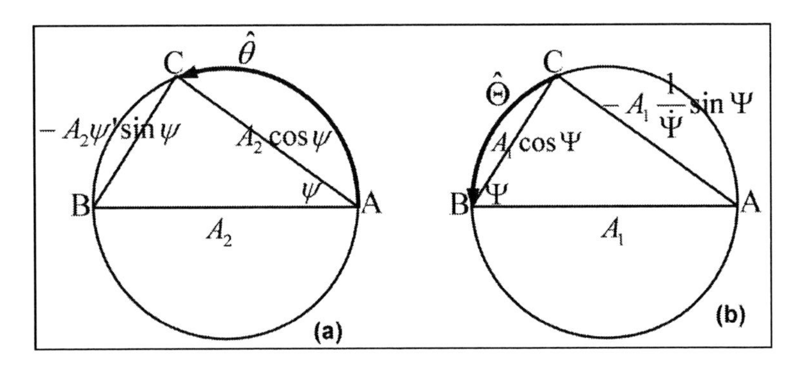


Figure 5.12. Radian angles $\hat{\Theta}$ and $\hat{\theta}$ in dimension diagram of clockwise rotation $\varphi_{00}^{-0}$ in comparison with radian angles $\theta$ and $\Theta$ in Figure 5.10.

How to explain the dimension diagrams? For counterclockwise rotation, in differential operation of Figure 5.10(a), arc BC decreases while arc AC increases. Spherical quantity $A_1 \cos\Psi$ gradually transforms into spherical quantity $-A_1\dot{\Psi}\sin\Psi$. And in integral operation of Figure 5.10(b), arc BC decreases while arc AC increases, which indicates a contrary effect that chord $A_2 \cos\psi$ gives way to chord $-A_2 \frac{1}{\psi'} \sin\psi$. For clockwise rotation, dimension diagram shows reverse trends. In integral operation of Figure 5.12(b), arc BC decreases while arc AC increases. Spherical quantity $A_1 \cos\Psi$ gradually transforms into spherical quantity $-A_1 \frac{1}{\dot{\Psi}} \sin\Psi$. And in differential operation of Figure 5.12(a), arc BC decreases while arc AC increases, which indicates a contrary effect that chord $A_2 \cos\psi$ gives way to chord $-A_2\psi' \sin\psi$. In all cases, state point C determines the values of the radian angles so that equation (5.63) is confirmed when both dimension diagrams are synchronized within a krypton atom.

## 5.2.4. Spherical Quantities for 3d Electrons

After dividing spherical quantities $\Phi$ and $\varphi$ into two radian angles $\Theta$ and $\theta$, the spacetime dimensions of 2s2p/3s3p spherical layer may be encapsulated by function:

$$\Gamma_{00} = A_3 A_4 \cos\Theta \cos\theta, \tag{5.64}$$

which serves as the starting point for building krypton shell. To assume crossing over between neon and argon shells, radian angle $\Theta$ in wave function $\Gamma_{00}$ inherits from spherical quantity $\Phi$ while radian angle $\theta$ from spherical quantity $\varphi$ ( $\theta = \hat{\Theta}$. ). We may perform rotatory operation upon $\Gamma_{00}$ as follows:

$$\Gamma_{00}^{0} = -\frac{d(A_3 \cos\Theta)}{dt}\int A_4 \cos\theta dl$$
$$= A_3 \sin\Theta \frac{d\Theta}{dt}\cdot(-A_4 \frac{1}{\theta'}\int \cos\theta d\theta) \quad (5.65)$$
$$= A_3 A_4 (-\dot{\Theta}\cdot\frac{1}{\theta'})\sin\Theta \sin\theta$$
$$= A_3 A_4 w \sin\Theta \sin\theta,$$

where $w$ is a velocity dimension in krypton shell that initiates the branch of 3d electrons via calculus chain rule.

$$\dot{\Theta} = -\frac{d\Theta}{dt}, \quad (5.66)$$

$$\theta' = -\frac{d\theta}{dl}, \quad (5.67)$$

$$w = -\frac{1}{\theta'}\cdot\dot{\Theta} = \frac{dl}{d\theta}\cdot(-\frac{d\Theta}{dt})$$
$$= \int A_2 \cos\psi \; dl\cdot(-\frac{dA_1 \cos\Psi}{dt}) \quad (5.68)$$
$$= \overline{\Phi}_0^0 = u\overline{\Phi}_0,$$

where $\overline{\Phi}_0$ is a 3d orbital in krypton shell. It is similar to $\Phi_0$ in mathematical form but has a complementary shape in geometry. Instead of building upon 1s electrons, 3d orbitals attach to 4s electrons. The dimension encapsulation of neon/argon spherical layer into $\Gamma_{00}$ and the differential chain rule upon rotation $\Gamma_{00}^{0}$ follow two different routes. In krypton shell, dimension encapsulation of 2s2p electrons leads to 4s electrons; and the differential chain rule of 4s electrons results in 3d electrons. On ten consecutive rotations of $\Gamma_{00}$, the power of $w$ is consumed to the construction of 3d electrons. Not until 3d electrons are filled does the power of $w$ raise up to two and more to produce 4p electrons. To express spherical quantities for 3d electrons clearly, we divide a rotatory operation into differential and integral steps awkwardly. Under the platform of

$$\Gamma_0 = A_3 A_4 (\cos\Theta\cos\theta - \dot{\Theta}\sin\Theta\cos\theta), \quad (5.69)$$

eight 3d electrons take the values of $\overline{\Phi}_i$ as gateway function $\dot{\Theta}$ initiates rotatory operations upon $\overline{\Phi}_0$ as follows:

$$\dot{\Theta}\cdot\frac{-1}{\theta'} = w = u\overline{\Phi}_i = \begin{pmatrix} \overline{\Phi}_0 & \overline{\Phi}_4 \\ \overline{\Phi}_1 & \overline{\Phi}_5 \\ \overline{\Phi}_2 & \overline{\Phi}_6 \\ \overline{\Phi}_3 & \overline{\Phi}_7 \end{pmatrix}, \tag{5.70}$$

which means rotations iterating through eight electronic states. Spherical quantities $\overline{\Phi}_i$ are 3d electrons similar to $\Phi_i$ in mathematical forms, so they are labeled as $3d^{2s2p}$ electrons. Another two 3d electrons are closely coupled with 1s electrons and are labeled as $3d^{1s}$. Both $3d^{1s}$ electrons attach to $\overline{\Phi}_0$ orbital as gateway function $\dot{\Psi}$ propagates into $\overline{\Omega}_0$ and $\overline{\Omega}_2$ functions respectively via differential chain rule (see equations 1.15 to 1.18). The rotation of $\Gamma_{00}$ drives the rotation of $\overline{\Phi}_i$, which in turn drives the rotation of $\overline{\Omega}_0$ and $\overline{\Omega}_2$.

$$\overline{\Phi}_0 = A_1 A_2(\cos\Psi\cos\psi - \dot{\Psi}\sin\Psi\cos\psi), \tag{5.71}$$

$$\dot{\Psi}\cdot\frac{-1}{\psi} = u = v\overline{\Omega}_i = \left(\overline{\Omega}_0 \quad \overline{\Omega}_2\right). \tag{5.72}$$

from the construction order, it is easy to see that 3d electrons are somewhat similar to $2s^2 2p^2$ and $1s^2$ electrons combined. They are similar in wave function expressions, the difference being that 1s2s2p electrons wrap around the nucleus whereas $3d^{2s2p}$ electron octet attach to the outmost 4s sphere and $3d^{1s}$ electrons in turn attach to the octet (Figure 5.13). Both 3d and 1s2s2p electrons are supposed to be complementary in spacetime. Eight $3d^{2s2p}$ electrons pass a gateway while two $3d^{1s}$ electrons have to pass two gateways in connection with the 4s electrons. They are long-range passengers that fill the niches of krypton sphere. The flexible mobility of 3d orbitals contributes to the good electrical conductivity of their corresponding transition metals especially copper and zinc with $3d^{1s}$ electrons.

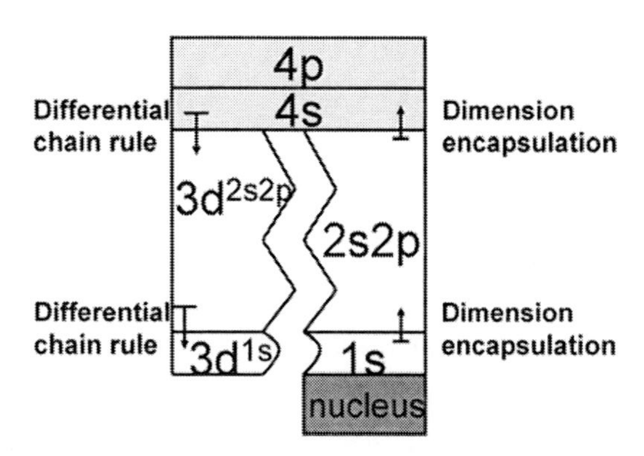


Figure 5.13. Dimension encapsulation mechanism of 2s2p building upon 1s electrons and of 4s building upon 2s2p electrons in contrast with differential chain rule of $3d^{2s2p}$ hanging down from 4s electrons and of $3d^{1s}$ hanging down from $3d^{2s2p}$ electrons.

After ten 3d electrons have been instantiated, 4s4p electrons are generated by the formulae of

$$\begin{pmatrix} \Gamma_0 \\ \Gamma_1 \\ \Gamma_2 \\ \Gamma_3 \\ \Gamma_4 \\ \Gamma_5 \\ \Gamma_6 \\ \Gamma_7 \end{pmatrix} = \begin{pmatrix} \Gamma_{00} - \dfrac{\partial \Gamma_{00}}{\partial t} \\ \Gamma_{00}^{0} - \dfrac{\partial \Gamma_{00}^{0}}{\partial t} \\ \Gamma_{00}^{02} - \dfrac{\partial \Gamma_{00}^{02}}{\partial t} \\ \Gamma_{00}^{03} - \dfrac{\partial \Gamma_{00}^{03}}{\partial t} \\ \Gamma_{00}^{04} - \dfrac{\partial \Gamma_{00}^{04}}{\partial t} \\ \Gamma_{00}^{05} - \dfrac{\partial \Gamma_{00}^{05}}{\partial t} \\ \Gamma_{00}^{06} - \dfrac{\partial \Gamma_{00}^{06}}{\partial t} \\ \Gamma_{00}^{07} - \dfrac{\partial \Gamma_{00}^{07}}{\partial t} \end{pmatrix}. \tag{5.73}$$

In trigonometric functions, the matrix takes the form of

$$\begin{pmatrix} \Gamma_0 \\ \Gamma_1 \\ \Gamma_2 \\ \Gamma_3 \\ \Gamma_4 \\ \Gamma_5 \\ \Gamma_6 \\ \Gamma_7 \end{pmatrix} = A_3 A_4 \begin{pmatrix} \cos\Theta\cos\theta - \dot{\Theta}\sin\Theta\cos\theta \\ -w(\sin\Theta\sin\theta + \dot{\Theta}\cos\Theta\sin\theta) \\ w^2(\cos\Theta\cos\theta - \dot{\Theta}\sin\Theta\cos\theta) \\ -w^3(\sin\Theta\sin\theta + \dot{\Theta}\cos\Theta\sin\theta) \\ w^4(\cos\theta\cos\Theta - \dot{\theta}\sin\theta\cos\Theta) \\ -w^5(\sin\theta\sin\Theta + \dot{\theta}\cos\theta\sin\Theta) \\ w^6(\cos\theta\cos\Theta - \dot{\theta}\sin\theta\cos\Theta) \\ -w^7(\sin\theta\sin\Theta + \dot{\theta}\cos\theta\sin\Theta) \end{pmatrix}, \tag{5.74}$$

$$\dot{\theta} = -\frac{d\theta}{dt}. \tag{5.75}$$

The order of wave function generations reflects the order of energy levels of electronic orbitals. Electrons within krypton shell corresponding to chemical elements in the fourth period have energy levels in the sequence of 4s3d4p. From 2s2p/3s3p duality sphere, the dimension encapsulation leads to the formation of 4s scalor in a larger spherical layer. Then there are two branches of rotatory operations from the established 4s platform. One branch is the upward rotatory operations according to equation (5.73), which give 4s4p electrons; the other branch is the downward rotatory operations stemming from gateway function $\dot{\Theta}$ via calculus chain rule, which result in ten 3d electrons. The downward branch has a lower energy and hence shows up before the upward branch in the periodic table of elements. Because 3d electrons attach to the platform of equation (5.69), 4s electrons must be established preceding any 3d electrons. This is the priority order of wave functions in krypton shell.

In general, after filling all available dimensions in a certain spherical layer, we roll up to a larger spherical layer by dimension encapsulation. We built 2s2p electrons from 1s sphere and built 4s electron from 2s2p/3s3p sphere in this way. On the other hand, calculus chain rule hanging down from 4s to $3d^{2s2p}$ and to $3d^{1s}$ electrons constitute a symmetric loop to dimension encapsulation. Electrons $3d^{2s2p1s}$ and 2s2p1s constitute two anti-parallel strands that form a rope as they pair together (Figure 5.14). During the rolling up, we treated spherical quantities as radian angles whereas in the hanging down we converted the radian angles back into spherical quantities. Such bi-directional conversions reflect the scope shift of the observer outside and inside a certain spherical layer.

While 4s4p octet represent the primary scale of spacetime in krypton shell, the propagation of gateway function $\dot{\Theta}$ into various $\overline{\Phi}_i$ is secondary in spacetime and has a minor distinction or effect on the overall atomic properties. This accounts for the more physical similarities among transition metal elements from Sc to Zn than among elements from Ga to Kr. However, because the mathematical forms of $3d^{2s2p1s}$ electrons coincide with those of 2s2p1s electrons, there are somewhat similar properties between 3d electrons and 2s2p1s electrons. For instance, Cu and Zn ($3d^{1s}$) correspond to $1s^2$ electrons and tend to take valences of +1 and +2, respectively, during chemical reactions. Their chemical behaviors are somewhat similar to those of alkaline and alkaline earth elements, respectively.

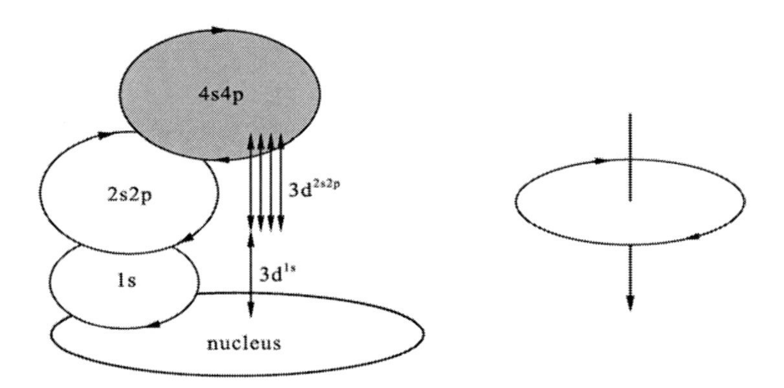


Figure 5.14. An electronic rope with two strands, one being spherical shape of 2s2p1s electrons, and the other being linear shape of 3d electrons. They are intertwined together inductively albeit separate in the imaginary diagram, which is reminiscent of the induction of circular magnetic wave by electric current flowing along a straight wire.

So far we have refrained from pointing out an important rule in wave function expression because we do not want to confuse the readers from the beginning. When we said that $\Gamma_1$ electron had a wave function designated by equation (5.74) where dimension factor $w$ took the value of equation (5.68), it was different from

$$\begin{aligned}\Gamma_1 &= -A_3 A_4 w(\sin\Theta \sin\theta + \dot{\Theta}\cos\Theta \sin\theta) \\ &= -A_1 A_2 u(\cos\Psi \cos\psi - \dot{\Psi}\sin\Psi \cos\psi) A_3 A_4 (\sin\Theta \sin\theta + \dot{\Theta}\cos\Theta \sin\theta).\end{aligned} \tag{5.76}$$

Even thought the mathematical substitution is correct, equation (5.76) suggests more waveforms for an electron than $\Gamma_1$ in equation (5.74). A 4s4p electron treats $w$ as a dimension indicator and as a whole unit that stays at that level only. The oscillation of 4s4p octet does not involve the details of $w$ expression, i.e., the physical correspondence for equation (5.68) is not a part of 4s4p electrons. Waveforms contained in $w$ represent 3d electrons. In other words, equation (5.76) indicates two electrons ( $\Gamma_1$ and $\overline{\Phi}_0$ ) rather than one. As we formulate wave functions, we must consider such nuances in physical implication. That a mathematical waveform corresponds to a physical electron is the criterion for writing a good wave function.

### 5.2.5. Complementary Orbitals Due to Standpoint Shift

The geometries of $3d^{2s2p}$ orbitals challenge your wildest space imagination. Although they are supposed to be complementary to 2s2p octet, we will not make any intellectual speculations on their geometric shapes thus far. However, we would like to touch on the geometry of $3d^{1s}$ orbitals that are complementary to 1s orbitals in spacetime. As shown in Figure 5.15, both 1s electrons are solid spheres that center at nucleus O. As radian angle $\alpha$ increases from 0 to $\pi/2$, radian angle $\beta$ reduces from $\pi/2$ to 0. One 1s electron reduces its spherical radius from QO to NO (shaded area in Figure 5.15a) and to vanish while the other 1s electron increases its spherical radius from naught to OP (shaded area in Figure 5.15b) and to OQ. Both 1s electrons oscillate cyclically, but their geometric centers are always at point O.

In contrast, both $3d^{1s}$ electrons are hollow spheres that attach to spherical surface Q. Spherical surface Q is the boundary of 1s sphere at maximum. As radian angle $\alpha$ increases from 0 to $\pi/2$, one $3d^{1s}$ electron ingrows from a spherical surface to a thick hollow (shaded ring in Figure 5.15c) sphere and eventually to a solid sphere while the other $3d^{1s}$ electron changes from a solid sphere into a hollow one (the shaded ring in Figure 5.15d) and eventually to a spherical surface. Spherical surface Q is the region that both $3d^{1s}$ electrons always connect with during harmonic oscillations. Electrons in Figure 5.15(a) and 5.15(c) are always complementary sharing the whole sphere O together whereas electrons in Figure 5.15(b) and 5.15(d) are complementary in geometry similarly. The spacetime complements arise from the fact that 1s electrons spin around the nucleus (point O) whereas $3d^{1s}$ electrons always stick to $3d^{2s2p}$ octet (spherical surface Q). The standpoint shift determines their complements in geometry. It is interesting that 1s electrons are introvert whereas $3d^{1s}$ are extrovert in behavior. Their complementarity ensures the harmony and integrity of an entire krypton atom.

We shall explain the orbital complementarity by set theory here. A set is a collection of objects or a container of objects. But here a set refers to a part of spacetime continuum of certain dimensions or a spherical quantity. Because of the standpoint shift, both $3d^{1s}$ electrons $\overline{\Omega}_0$ and $\overline{\Omega}_2$ are complementary to both 1s electrons $\Omega_0$ and $\Omega_2$ in spacetime, respectively.

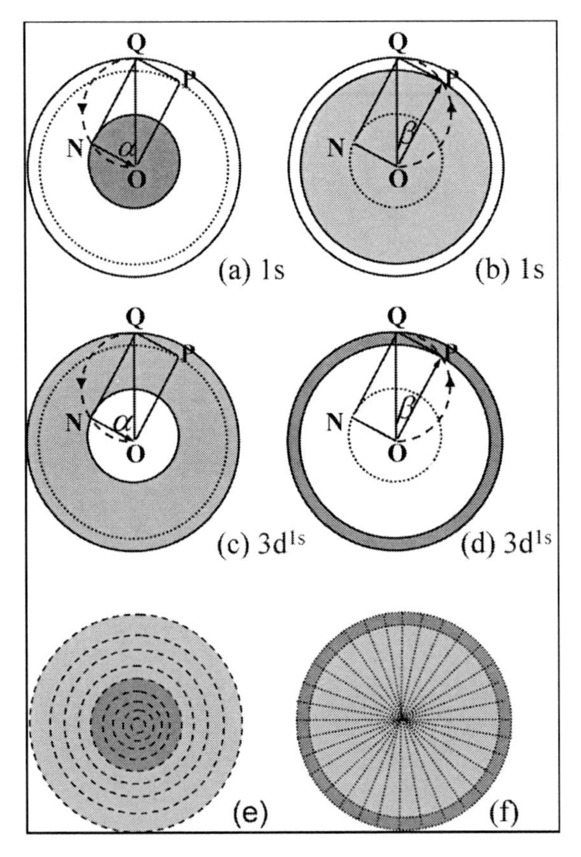


Figure 5.15. Geometric complements and electromagnetic inductions between $3d^{1s}$ orbitals and 1s orbitals.

The union of each pair represents both dimensions in a two-dimensional sphere; and the intersection of each pair is a null set.

$$\Omega_0 \cup \overline{\Omega}_0 = \Omega, \ \Omega_0 \cap \overline{\Omega}_0 = \varnothing, \tag{5.77}$$

$$\Omega_2 \cup \overline{\Omega}_2 = \Omega, \ \Omega_2 \cap \overline{\Omega}_2 = \varnothing, \tag{5.78}$$

where Ø is an empty set and $\Omega$ denotes the universal set of both dimensions in the two-dimensional sphere as was expressed in equation (1.9). Equation (5.77) indicates that $\overline{\Omega}_0$ is a complement of $\Omega_0$ in spacetime; and equation (5.78) indicates that $\overline{\Omega}_2$ is a complement of $\Omega_2$ in spacetime. The complementarity is an alternative expression of Pythagorean theorem for right triangles of Figure 1.7. Because of the standpoint shift, as $\Omega_0$ is reducing a time dimension, $\overline{\Omega}_0$ is gaining a time dimension. The counterclockwise rotation of $\overline{\Omega}_0$ has a reverse effect when viewed from the standpoint of central nucleus so that the following two transformations occur at the same pace. Time and space dimensions of $3d^{1s}$ electrons transform in the contrary directions to those of 1s electrons.

$$\int -\frac{\partial \Omega_0}{\partial t} dl = \Omega_2 \, , \tag{5.79}$$

$$\int -\frac{\partial \overline{\Omega}_0}{\partial l} dt = \overline{\Omega}_2 \, . \tag{5.80}$$

It is worthwhile to note that Figures 5.13, 5.14, and 5.15 actually express the same atomic structure from various perspectives. Figure 5.13 draws the architecture of wave function construction in krypton shell; Figure 5.14 exaggerates the intertwining relation of 3d electrons with 1s2s2p electrons; and Figure 5.15 focuses on the complementary geometries of $3d^{1s}$ and 1s electrons, especially Figure 5.15(e) (f) gives a visual aid of electromagnetic field continuum of $\Omega_0$ and $\overline{\Omega}_0$ electrons and of $\Omega_2$ and $\overline{\Omega}_2$ electrons. We ignored 3s3p electrons in these graphics for brevity.

By the way, the order of electrons within a krypton or a zinc atom provides a functional model for a eukaryotic cell. According to endosymbiotic hypothesis, a eukaryotic cell began as a large anaerobic organism. A mitochondrion arose when a small aerobic bacterium was ingested by the larger cell. Here $3d^{1s}$ electrons hang down from the outer layer of $3d^{2s2p}$ and 4s electrons like a mitochondrion coming from outside the cell. And the aerobic and anaerobic complementary roles of the mitochondrion and the larger cell are analogous to spacetime complements of $3d^{1s}$ and 1s electrons. We shall revert to such an interesting biological subject in Chapter 8.

## 5.3. Electronic Orbitals in Radon Shell and Beyond

We shall proceed to formulate spherical quantities for electrons in radon and Uuo shells. Uuo is the largest element with 118 protons. We shall arrange all elements into a quaternity periodic table of elements based on the property of their characteristic electrons. To make such an arrangement, 3d-orbitals are divided into $3d^{2s2p}$ and $3d^{1s}$ because their waveforms are similar to 2p2s1s electrons, and 4f-orbitals are divided into $4f^{4p}4f^{4s}4f^{2p}$ in correspondence with 4p4s2p electrons as well. Under the framework of quaternity spacetime, this alternative periodic table provides sharp insight into the structure of various atoms. Moreover, we consider atomic shells of hydrogen, helium, neon, argon, krypton, xenon, radon, and Uuo to be an octet configuration. The shell octet within a Uuo atom is a stable system in a certain way similar to electron octet within neon shell. This perception lets us jump to the conclusion that there are exactly 118 elements existing in the world.

### 5.3.1. Spherical Quantities for 4f Electrons

Spherical quantities for 5s4d5p can be constructed through consecutive clockwise rotations on the initial function of

$$\gamma_{00} = A_3 A_4 \cos\Theta \cos\theta \, , \tag{5.81}$$

where $\Theta$ and $\theta$ are another pair of radian angles derived from spherical quantities $\Phi$ and $\varphi$ ( $\Theta = \hat{\theta}$ ). Performing clockwise rotatory operations similar to equations (5.4) and (5.10) with substitutions of $\varphi_{00}$ with $\gamma_{00}$ yields 5s5p electron octet:

$$\begin{pmatrix} \gamma_0 \\ \gamma_1 \\ \gamma_2 \\ \gamma_3 \\ \gamma_4 \\ \gamma_5 \\ \gamma_6 \\ \gamma_7 \end{pmatrix} = A_3 A_4 \begin{pmatrix} \cos\theta\cos\Theta - \theta'\sin\theta\cos\Theta \\ -m(\sin\theta\sin\Theta + \theta'\cos\theta\sin\Theta) \\ m^2(\cos\theta\cos\Theta - \theta'\sin\theta\cos\Theta) \\ -m^3(\sin\theta\sin\Theta + \theta'\cos\theta\sin\Theta) \\ m^4(\cos\Theta\cos\theta - \Theta'\sin\Theta\cos\theta) \\ -m^5(\sin\Theta\sin\theta + \Theta'\cos\Theta\sin\theta) \\ m^6(\cos\Theta\cos\theta - \Theta'\sin\Theta\cos\theta) \\ -m^7(\sin\Theta\sin\theta + \Theta'\cos\Theta\sin\theta) \end{pmatrix}, \tag{5.82}$$

where

$$m = -\frac{1}{\dot{\Theta}}\theta' = \frac{1}{w}, \tag{5.83}$$

or after role exchange between time and space radian angles $\Theta$ and $\theta$ , we write

$$\Theta' = -\frac{d\Theta}{dl}, \tag{5.84}$$

$$m = -\frac{1}{\dot{\theta}}\Theta' = \frac{1}{w}. \tag{5.85}$$

Factor $m$ is a reciprocal velocity dimension in xenon shell that initiates the branch of 4d electrons. 4d-oribtals may be divided into $4d^{3s3p}$ and $4d^{1s}$ according to their waveforms. The first octet, $4d^{3s3p}$, bear close relationships to $\varphi_i$ and the last two $4d^{1s}$ electrons are somewhat complementary to both 1s electrons. Electrons octet $4d^{3s3p}$ attach to 5s electrons in the outer spherical layer and $4d^{1s}$ electrons in turn attach to the octet. These transition electrons are inductive with 3s3p1s electrons. Within xenon shell, there are two electronic strands, one composed of 4s3d4p twisted counterclockwise, and the other composed of 5s4d5p strained clockwise, both forming a rope.

The scalability of quaternity spacetime makes it possible to build spherical quantities for various electrons up to four spherical layers. Electrons are packed in each layer by similar orders even though the structural complexity of an outer spherical layer tends to increase as a result of more dimensions involved. Spherical quantities for 6s4f5d6p electrons in the sixth period can be derived by encapsulating dimensions within Kr/Xe spherical layer. The encapsulation regards time and space components of spherical quantities $\Gamma$ and $\gamma$ as radian angles. To save letters, we still use variables $\Gamma$ and $\gamma$ to represent those radian angles derived from the meiosis of spherical quantities $\Gamma$ and $\gamma$ . This is possible because it is reasonable to use a male to denote a sperm and a female to denote an ovum analogously. Thus dimension encapsulation of Kr/Xe spherical layer leads to a membrane function

$$\mathrm{H}_{00} = A_5 A_6 \cos\Gamma \cos\gamma \,, \tag{5.86}$$

where $A_5$ and $A_6$ are dimension constants, and $\Gamma$ and $\gamma$ serve as two complementary radian angles in radon shell. The function is called a membrane function because it encapsulates a spherical layer as if plasma membrane would enclose the cytoplasm in a cell. We shall construct electrons 6s4f5d6p starting from this membrane function. Let

$$\begin{pmatrix} 6s & 6s \\ 6p_x & 6p_x \\ 6p_y & 6p_y \\ 6p_z & 6p_z \end{pmatrix} = \begin{pmatrix} \mathrm{H}_0 & \mathrm{H}_4 \\ \mathrm{H}_1 & \mathrm{H}_5 \\ \mathrm{H}_2 & \mathrm{H}_6 \\ \mathrm{H}_3 & \mathrm{H}_7 \end{pmatrix}, \tag{5.87}$$

where spherical quantities in the right matrix correspond to orbital types in the left matrix. We may assign the values of $\mathrm{H}_i$ through consecutive counterclockwise rotations by replacing $\Gamma_{00}$ with $\mathrm{H}_{00}$ in equation (5.73) so that

$$\begin{pmatrix} \mathrm{H}_0 \\ \mathrm{H}_1 \\ \mathrm{H}_2 \\ \mathrm{H}_3 \\ \mathrm{H}_4 \\ \mathrm{H}_5 \\ \mathrm{H}_6 \\ \mathrm{H}_7 \end{pmatrix} = A_5 A_6 \begin{pmatrix} \cos\Gamma\cos\gamma - \dot{\Gamma}\sin\Gamma\cos\gamma \\ -p(\sin\Gamma\sin\gamma + \dot{\Gamma}\cos\Gamma\sin\gamma) \\ p^2(\cos\Gamma\cos\gamma - \dot{\Gamma}\sin\Gamma\cos\gamma) \\ -p^3(\sin\Gamma\sin\gamma + \dot{\Gamma}\cos\Gamma\sin\gamma) \\ p^4(\cos\gamma\cos\Gamma - \dot{\gamma}\sin\gamma\cos\Gamma) \\ -p^5(\sin\gamma\sin\Gamma + \dot{\gamma}\cos\gamma\sin\Gamma) \\ p^6(\cos\gamma\cos\Gamma - \dot{\gamma}\sin\gamma\cos\Gamma) \\ -p^7(\sin\gamma\sin\Gamma + \dot{\gamma}\cos\gamma\sin\Gamma) \end{pmatrix}, \tag{5.88}$$

where

$$\dot{\Gamma} = -\frac{d\Gamma}{dt}, \tag{5.89}$$

$$\gamma' = -\frac{d\gamma}{dl}, \tag{5.90}$$

$$p = -\dot{\Gamma} \cdot \frac{1}{\gamma'} = w\Gamma \,. \tag{5.91}$$

Factor $p$ is a velocity dimension in radon shell. Or after role exchange between time and space radian angles $\Gamma$ and $\gamma$, there are

$$\dot{\gamma} = -\frac{d\gamma}{dt}, \tag{5.92}$$

$$\Gamma' = -\frac{d\Gamma}{dl}, \tag{5.93}$$

$$p = -\dot{\gamma}\frac{1}{\Gamma'}. \tag{5.94}$$

Under the platform of spherical quantity $H_0$, gateway function $\dot{\Gamma}$ initiates the branch of 5d electrons:

$$\begin{pmatrix} 5d^{4s} & 5d^{4s} \\ 5d^{4px} & 5d^{4px} \\ 5d^{4py} & 5d^{4py} \\ 5d^{4pz} & 5d^{4pz} \end{pmatrix} = A_5 A_6 (\cos\Gamma\cos\gamma - \begin{pmatrix} \overline{\Gamma}_0 & \overline{\Gamma}_4 \\ \overline{\Gamma}_1 & \overline{\Gamma}_5 \\ \overline{\Gamma}_2 & \overline{\Gamma}_6 \\ \overline{\Gamma}_3 & \overline{\Gamma}_7 \end{pmatrix} \sin\Gamma\cos\gamma), \tag{5.95}$$

where spherical quantities in the right matrix correspond to orbital types on the left matrix. Functions $\Gamma_i$ were spherical quantities of 4s4p electrons, and functions $\overline{\Gamma}_i$ refer to $5d^{4s4p}$ electrons attaching to spherical quantity $H_0$. They have similar mathematical forms but complementary geometries. Hanging down from $5d^{4s4p}$ octet are another two spherical quantities for $5d^{2s}$.

$$\begin{pmatrix} 5d^{2s} \\ 5d^{2s} \end{pmatrix} = A_3 A_4 (\cos\theta\cos\Theta - \begin{pmatrix} \phi_0 \\ \phi_4 \end{pmatrix} \sin\theta\cos\Theta). \tag{5.96}$$

The expression means that both functions $\phi_0$ and $\phi_4$ are similar to 2s electrons in waveform, but they attach to $5d^{4s4p}$ octet instead of building upon 1s electrons. And $5d^{4s4p}$ octet attach to 6s electron $H_0$. We have labeled 5d electrons with superscripts in matrices (5.95) and (5.96) to indicate their orbital properties. These 5d electrons are all buried under 6s electrons. If we use elements to represent their characteristic electrons, then there is a one to one correspondence between characteristic waveforms and chemical elements.

$$\begin{pmatrix} 5d^{4s} & 5d^{4s} \\ 5d^{4px} & 5d^{4px} \\ 5d^{4py} & 5d^{4py} \\ 5d^{4pz} & 5d^{4pz} \\ 5d^{2s} & 5d^{2s} \end{pmatrix} = \begin{pmatrix} La & Hf \\ Ta & W \\ \mathrm{Re} & Os \\ Ir & Pt \\ Au & Hg \end{pmatrix} = \begin{pmatrix} \overline{\Gamma}_0 & \overline{\Gamma}_4 \\ \overline{\Gamma}_1 & \overline{\Gamma}_5 \\ \overline{\Gamma}_2 & \overline{\Gamma}_6 \\ \overline{\Gamma}_3 & \overline{\Gamma}_7 \\ \phi_0 & \phi_4 \end{pmatrix}. \tag{5.97}$$

Inner-transition 4f electrons are characterized by spherical quantities hanging down to even a lower level. They reflect the possible combinations of gateway functions $\dot{\Gamma}$ and $\dot{\Theta}$. Fourteen 4f electrons may be categorized into three major types: $4f^{4p}$, $4f^{4s}$, and $4f^{2p}$, which correspond to chemical elements as follows.

$$\begin{pmatrix} 4f^{4px} & 4f^{4px} \\ 4f^{4py} & 4f^{4py} \\ 4f^{4pz} & 4f^{4pz} \\ 4f^{4s} & 4f^{4s} \\ 4f^{2px} & 4f^{2px} \\ 4f^{2py} & 4f^{2py} \\ 4f^{2pz} & 4f^{2pz} \end{pmatrix} = \begin{pmatrix} Ce & \mathrm{Pr} \\ Nd & Pm \\ Sm & Eu \\ Gd & Tb \\ Dy & Ho \\ Er & Tm \\ Yb & Lu \end{pmatrix}. \tag{5.98}$$

Under the framework of function $H_0$, gateway function $\dot{\Gamma}$ may take eight values as $\overline{\Gamma}_i$; and within each value $\overline{\Gamma}_i$, gateway function $\dot{\Theta}$ (or $\dot{\theta}$) might take eight values as $\phi_j$, too.

$$\mathrm{H}_0 = A_5 A_6 (\cos\Gamma\cos\gamma - \begin{pmatrix} \overline{\Gamma}_0 & \overline{\Gamma}_4 \\ \overline{\Gamma}_1 & \overline{\Gamma}_5 \\ \overline{\Gamma}_2 & \overline{\Gamma}_6 \\ \overline{\Gamma}_3 & \overline{\Gamma}_7 \end{pmatrix} \sin\Gamma\cos\gamma), \tag{5.99}$$

$$\overline{\Gamma}_0 = A_3 A_4 (\cos\Theta\cos\theta - \begin{pmatrix} \phi_0 & \phi_4 \\ \phi_1 & \phi_5 \\ \phi_2 & \phi_6 \\ \phi_3 & \phi_7 \end{pmatrix} \sin\Theta\cos\theta), \tag{5.100}$$

where $\overline{\Gamma}_i$ means $5d^{4s4p}$ electrons in complement to 4s4p electrons because they attach to $H_0$ in the outer spherical layer. Function $\overline{\Gamma}_0$ in equation (5.100) is only an example of $\overline{\Gamma}_i$. Spherical quantities $\phi_j$ attach to $\overline{\Gamma}_i$. However, the combination of both $\overline{\Gamma}_i$ and $\phi_j$ values in matrices (5.99) and (5.100) does not produce $8\times 8$ variants of wave functions. To form valid wave functions, an s-orbital in one variable may combine with an s-orbital in another variable to form an s-nature orbital, but three pairs of p-orbitals in one variable must combine with two s-orbitals in another to form six p-nature orbitals. The combinations of p-orbitals with p-orbitals are not valid within an atom. Mathematically, we shall define a new subscript symbol "$_{\otimes}$" as a gateways combination operator. According to the above combination rule, we write $\overline{\Gamma}_{i\otimes}\phi_j$ to indicate valid combinations of both gateway variables ( $\dot{\Gamma}$ and $\dot{\Theta}$ ) and to indicate that $\phi_j$ attaches to function $\overline{\Gamma}_i$ so that

$$\overline{\Gamma}_{i\otimes}\phi_j = \begin{pmatrix} \overline{\Gamma}_0 & \overline{\Gamma}_4 \\ \overline{\Gamma}_1 & \overline{\Gamma}_5 \\ \overline{\Gamma}_2 & \overline{\Gamma}_6 \\ \overline{\Gamma}_3 & \overline{\Gamma}_7 \end{pmatrix}_{\otimes} \begin{pmatrix} \phi_0 & \phi_4 \\ \phi_1 & \phi_5 \\ \phi_2 & \phi_6 \\ \phi_3 & \phi_7 \end{pmatrix} = \overline{\Gamma}_{0\otimes}(\phi_0 \quad \phi_4) + \begin{pmatrix} \overline{\Gamma}_1 & \overline{\Gamma}_5 \\ \overline{\Gamma}_2 & \overline{\Gamma}_6 \\ \overline{\Gamma}_3 & \overline{\Gamma}_7 \end{pmatrix}_{\otimes} (\phi_0 \quad \phi_4) \tag{5.101}$$

$$+\overline{\Gamma}_{4\otimes}(\phi_0 \quad \phi_4) + (\overline{\Gamma}_0 \quad \overline{\Gamma}_4)_{\otimes} \begin{pmatrix} \phi_1 & \phi_5 \\ \phi_2 & \phi_6 \\ \phi_3 & \phi_7 \end{pmatrix},$$

where the first term on the right-hand side of the equation denotes two $5d^{2s}$ electrons as were defined in equation (5.96) previously; the second term contains six $4f^{4p}$ electrons; the third term is two $4f^{4s}$ electrons; and the last term comprises six $4f^{2p}$ electrons. In the second term, rotations iterating from $\overline{\Gamma}_1$ through $\overline{\Gamma}_7$ drive the rotations alternating between both $5d^{2s}$ electrons; and in the fourth term, the rotations alternating between both $5d^{4s}$ electrons drive the rotations iterating from $\phi_1$ through $\phi_7$. Fourteen 4f electrons are confined between two $5d^{4s}$ electrons. The outer electron of La represents the ground state of $5d^{4s}$ whereas that of Hf is the upper bound of $5d^{4s}$ energy level. Because the energy difference between these two states is trivial, fourteen 4f electrons have very close energy levels and therefore exhibit similar physical properties. And for this reason, a 4f electron may easily switch between various neighboring states. For example, electron configuration of a cerium atom may take the form of $[Xe]6s^25d^14f^1$, $[Xe]6s^25d^2$ or $[Xe]6s^24f^2$. The valence of cerium normally takes +3 or +4 in chemical compounds. Since 5d and 4f electrons are intertwined with 4s4p and 2s2p electrons, their existence brings 6s6p electrons closer to the inner spherical layers. In other words, 5d and 4f orbitals serve as expressways for 6s6p electrons to communicate with the inner 4s4p and 2s2p shells. They behave like many anchors cast deeply into the inner shells to fasten the floating ship of 6s6p electrons.

## 5.3.2. Quaternity Periodic Table of Elements

Spherical quantities for 7s5f6d7p electrons can be constructed through clockwise rotations based on radian angles $\gamma$ and $\Gamma$. Given the initial function of

$$\eta_0 = A_5 A_6 (\cos\gamma \cos\Gamma - \gamma' \sin\gamma \cos\Gamma), \tag{5.102}$$

electron octet 7s7p may be expressed by

$$\eta_i = q^i \eta_0, i = 1,2,...,7 \tag{5.103}$$

where $q$ is reciprocal to $p$ serving as a clockwise rotation operator in Uuo shell. Spherical quantities for 6d and 5f orbitals can be obtained in a similar way to those for 5d and 4f orbitals. We shall not elaborate here again. Like 4f electrons, 5f electrons pass two gateways and are inductively intertwined with 5p5s3p waveforms. They serve as mortar joints among 3s3p, 5s5p, and 6d orbitals. Their energy levels are confined to those of Ac and Rf elements.

Although d- and f-orbitals span large space and time ranges, they are still sufficiently explained by four-dimensional spacetime framework. All electrons are divided into four spherical layers. Spherical quantities for electrons in an outer spherical layer are constructed upon those in the inner layer through dimension encapsulation. When considering spherical quantities of an electron in the outer layer, the curled dimensions of the inner layer can be ignored to a certain extent. Each spherical layer is relatively independent, but there are transition messengers passing gateways between spherical layers to deliver messages for better communication.

For example, 3d electrons attach to 4s4p electrons and complementary to 1s2p2s electrons in spacetime. While s- and p-electrons stay within the same shell, d- and f-orbitals are long-range ones inductive with the inner shells of the atom. Because 3d electrons are complementary to 1s2p2s electrons and the other d-electrons have similar properties, we may treat them as composite orbitals of $s^2p^6s^2$ to a certain extent. By the same way, $f^{14}$-electrons can be regarded as $p^6s^2p^6$-like orbitals, e.g., 4f electrons form a weak rope with 4p4s2p electrons combined. In this manner, we organize all elements into a periodic table according to quaternity spacetime.

**Table 5.1. Periodic table of elements based on quaternity spacetime, which shows the enclosure of p-orbitals between s-orbitals in general except f-orbitals**

| s | | $p_x$ | | $p_y$ | | $p_z$ | | s | |
|---|---|---|---|---|---|---|---|---|---|
| H | | | | | | | | | He |
| Li | Be | B | C | N | O | F | Ne | Na | Mg |
| | | Al | Si | P | S | Cl | Ar | K | Ca |
| Sr | Ti | V | Cr | Mn | Fe | Co | Ni | | |
| Cu | Zn | Ga | Ge | As | Sn | Br | Kr | Rb | Sr |
| Y | Zr | Nb | Mo | Tc | Ru | Rh | Pd | | |
| Ag | Cd | In | Sn | Sb | Te | I | Xe | Cs | Ba |
| La | | *Ce* | *Pr* | *Nd* | *Pm* | *Sm* | *Eu* | *Gd* | |
| | *Tb* | *Dy* | *Ho* | *Er* | *Tm* | *Yb* | *Lu* | | Hf |
| | | Ta | W | Re | Os | Ir | Pt | | |
| Au | Hg | Tl | Pb | Bi | Po | At | Rn | Fr | Ra |
| Ac | | *Th* | *Pa* | *U* | *Np* | *Pu* | *Am* | *Cm* | |
| | *Bk* | *Cf* | *Es* | *Fm* | *Md* | *No* | *Lr* | | Rf |
| | | Db | Sg | Bh | Hs | Mt | Uun | | |
| Uuu | Uub | Uut | Uuq | Uup | Uuh | Uus | Uuo | | |

The periodic table shows that p-orbitals are enclosed by s-orbitals. Such an enclosure of p-orbitals ensures the stability of atoms except lanthanides and actinides that feature anti-enclosure. Of these two inner-transition metal series, only Gd, Tb, Cm, and Bk have s-orbital properties, the other twenty-four being p-orbitals in nature. If we regard s-orbitals as round spheres, then p-orbitals are oval, d-orbitals are narrow oval, and f-orbitals are even narrow oval. Because of their severe deviations from spheres, f-orbitals are so narrow and long-ranged that they are susceptible to decay like the instability of a comet surrounding the sun along a narrow ellipse.

Table 5.1 is not intended to replace the traditional periodic table of elements, but it is an alternative view in quaternity spacetime. Some elements have two possible arrangements in the table. For example, Cu and Zn can also be placed in the slots under K and Ca. This agrees with their chemical properties in those columns. Cu and Zn often take +1 and +2 valances respectively in chemical compounds. Their chemical properties are somewhat similar to those of K and Ca, but not exactly the same due to their d-orbital nature.

Despite the primary difference between d-orbitals and s- and p-orbitals, elements within the same column still show certain similarity in their behaviors. For example, manganese is a useful alloying element to achieve high strength stainless steels. It has been found that the addition of manganese increases the solubility of nitrogen in austenitic steels [1], somewhat reflecting the attraction of nitrogen by manganese. This attraction lies in the inductive effect of their complementary waveforms because Mn and N are in the same column in Table 5.1. This partly uncovers the secret of good solubility with similar molecules as the solvent and the solute. Electronic attraction between complementary orbitals is a key factor to good solubility.

### 5.3.3. Forces and Electronic Waves

Even though we have formulated electronic orbitals for all kinds of electrons, the forces that bind them together in atoms are not clear yet. They deserve further explanations. To clarify, a force is not a mysterious quantity or the power of an invisible hand. In its primitive form, a force requires certain way of message exchanges between acting particles. When two people exchange information or goods, they are likely to make friends if the exchange benefits both and is meaningful to their livelihoods. The interdependence of entities is a manifestation of forces. Message exchange is an essential aspect of force expression or mechanism. Without messengers or force carriers to deliver messages, there would be no forces. We previously introduced virtual photon as the vehicle of electron interaction in section 3.5. A virtual photon delivers an electronic wave. Here we shall characterize message exchanges within atoms by waves, the themes that organize various electrons in atoms in harmonious order. Forces are implemented by the wave behavior of electrons.

Within the atomic sphere, the most significant force is perhaps the interaction between the nucleus and electrons, or more specifically the force between the nucleus and both 1s electrons because these two electrons form the inner core that delivers wave signals towards outer spherical layers. The interaction between 1s electrons and the nucleus might involve the continual annihilations and re-creation of electrons at the nucleus (see Figure 1.15). The electron disappears and is immediately created again with a fresh electronic state that carries messages from the nucleus. Imagine throwing a pebble into a still pond and watching the circular ripples moving outward. This initial signal must be strong enough to generate waves that eventually propagate over the entire atomic sphere.

In a neon atom, the waves generated by both 1s electrons are soon delivered to 2s2p electrons through dimension encapsulation mechanism. Wave functions of 2s2p are built upon 1s wave functions. As was shown in equation (2.16), dimension factor $u$ of 2s2p wave functions denotes the rotation of 1s wave functions ($u = v\Omega_0$). In other words, the rotation of 1s electrons serves as a dimension factor and a rotation operator for 2s2p electrons. Electronic waves in helium shell drive electronic waves in neon shell, and 2s2p wave functions harness wave energy from 1s electrons. The wave signals then reflect back from 2s2p towards 1s through differential chain rule, reinforcing the original 1s waves. Thus 1s electronic waves propagate outwards into 2s2p waves and echo back as 1s waves again cyclically forming a resonant medium within a neon atom.

In a larger atom such as radon, 2s2p and 3s3p electronic waves are driven by 1s waves, and 4s3d4p and 5s4d5p are in turn generated by 2s2p and 3s3p waves, and finally the outskirt 6s4f5d6p electrons are built upon 4s4p/5s5p spherical layer. Since electronic orbitals are waves in nature, the entire atom is a dynamic multi-dimensional wave that propagates out from central nucleus to its outer electrons and reflects backward through d- and f-orbitals. Dimension encapsulation represents waves delivered outwards; and calculus chain rule represents waves reflected inwards as d- and f-orbitals. For example, 3d-orbitals are wave signals that are reflected from 4s electrons towards inner neon and helium shells. $3d^{2s2p}$ waves synchronize with 2s2p waves; and $3d^{1s}$ waves are coupled with 1s waves.

A significant role of 3d-orbitals is the reinforcement of contact between electrons in the outer and inner spherical layers.

Wave functions of 4s4p electrons are based on dimension encapsulation upon 2s2p/3s3p electrons. The force induced by message exchanges between 4s4p and the inner spherical layer through dimension encapsulation is not strong enough to bind all 4s4p electrons to the atom. That is why 3d-orbitals are involved. On one end, 3d electrons directly connect with 4s4p sphere; and on the other end, they interact with electrons in neon and helium shells. A 3d-orbital is like a tendril dropping down from 4s4p spherical branch and tightly grasping the inner cores of 2s2p1s electrons. This helps stabilize the system.

Compared with d-orbitals that deliver strong messages, f-orbitals implement forces that are relatively weak. The weakness is revealed by the instability of inner-transition metals that are susceptible to decaying.

Atoms in the last period of the periodic table have maximum space volume and hence minimum time stability. They are all radioactive due to their need for delicate balances that are often broken.

### 5.3.4. Atomic Shell Octet

We shall extend the significance of a spherical quadrant to an atomic shell as a whole. Within the atom of Uuo with 118 protons, the atomic shells in four spherical layers can be organized into an octet as shown in Table 5.2. This octet has the characteristics of polors, metors, vitors, and scalors in certain manifestations. Eight shells split into two groups with H, Ne, Kr, and Rn in one group having counterclockwise rotations property and He, Ar, Xe, and

Uuo in another group featuring clockwise rotations, both groups forming a rope structure within an Uuo atom (Figure 5.16). Note that hydrogen is not an inert atomic shell, but it is a representative atomic shell of a 1s electron that is in symmetry to another 1s electron in helium shell.

Helium shell has s-orbitals only. Dimension encapsulation upon helium shell leads to wave functions of neon shell, which contains s- and p-orbitals. Dimension encapsulation upon neon shell leads to wave functions of krypton shell, which contains s-, p-, and d-orbitals. And dimension encapsulation upon krypton shell brings f-electrons in addition to s-, p- and d-electrons in radon shell. Thus it is clear that each dimension encapsulation leads to one increment of orbital types in the outer spherical layer. Helium shell is the most stable atomic shell. The stability of the atomic shell is relaxed stepwise with adding orbital types upon dimension encapsulations.

**Table 5.2. Atomic shell octet within the last atom of $_{118}$Uuo**

| Spherical quadrant | Spherical layer | Electronic Shells | Electronic orbitals | Rotation directions |
|---|---|---|---|---|
| Polor | 1 | H, He | $1s^1$, $1s^2$ | ↑,↓ |
| Metor | 2 | Ne, Ar | $2s^22p^6$, $3s^23p^6$ | ↑,↓ |
| Vitor | 3 | Kr, Xe | $4s^23d^{10}4p^6$, $5s^24d^{10}5p^6$ | ↑,↓ |
| Scalor | 4 | Rn, Uuo | $6s^25d^14f^{14}5d^96p^6$, $7s^26d^15f^{14}6d^97p^6$ | ↑,↓ |

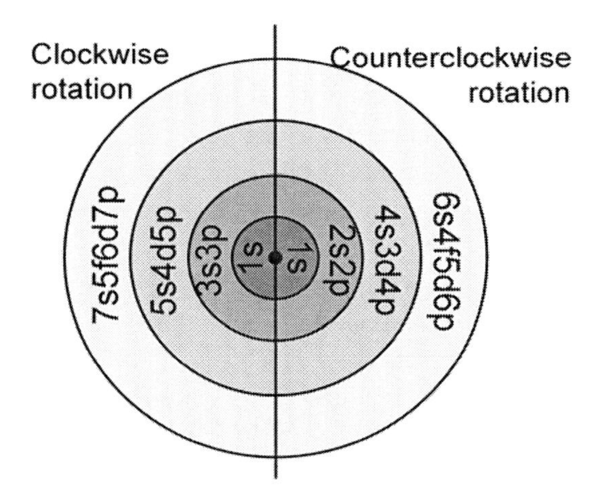


Figure 5.16. Atomic shell octet within an Uuo atom.

If we regard dimension encapsulation as a rotatory operation, then there are rotation relationships between four shells on the right-hand side of Figure 5.16. For example, $A_1A_2 = -vC_1C_2$, where velocity dimension $v$ is a rotation operator. Of course, after four consecutive rotations, time and space switch roles: four-dimensional space (spherical layer 4) becomes four-dimensional time; and zero-dimensional time (minimum stability with s-, p-, d-, and f-electrons) becomes zero-dimensional space. Further rotations are performed in the negative directions of quaternity axes. The rotation cycle of the atomic shells within an Uuo atom is like that of equation (2.29).

$$H \mapsto Ne \mapsto Kr \mapsto Rn \mapsto He \mapsto Ar \mapsto Xe \mapsto Uuo \mapsto H, \quad (5.104)$$

where each arrow $\mapsto$ indicates a rotatory operation from one atomic shell to another. Shell octet is a conservative system similar to the configuration of electron octet within neon shell. Due to the octet arrangement, various shells can cooperate with each other harmonically in a Uuo atom. This implies that there are only 118 possible chemical elements because the octet arrangement is a full and stable configuration.

## 5.4. Concluding Remarks

We have formulated wave functions for all electrons in their primitive inert atoms in quaternity spacetime in a relentless effort. Each electron is unique in terms of the waveform and role in delivering force. Not any two electrons are exactly the same. This is in good agreement with the Pauli exclusion principle that no more than one electron can occupy a single quantum state. The organization of electrons shows some patterns: p-electrons are enclosed by s-electrons in a spherical layer; and d- and f-electrons pass gateways to deliver messages between electrons in the outer and inner spherical layers. All electrons are interwoven, bringing about harmonic oscillating waves over the whole atomic sphere. They interact with each other in such a way that keeps electronic waves harmonic and total energy low.

The theory of quaternity provides a unique observation on the atomic structure. It has given us much more detailed explanations on electronic arrangement and motion than quantum mechanics. Among other things, it is revealing to see that chemical elements corresponding to their outmost characteristic electrons can be tabulated under the fields of s-orbitals, $p_x$-orbitals, $p_y$-orbitals, $p_z$-orbitals, and 2-orbitals into a quaternity periodic table of elements. This alternative approach is fully compatible with established chemistry and physics and provides further insight into elemental behaviors.

The elucidation of electronic pattern within inert atoms is perhaps one of the greatest achievements of mathematical derivation combining calculus, trigonometry, and geometry together. While dimension encapsulation reflects layered structure of electrons, the principle of rotatory operation on spherical quantities brings order and harmony to electrons within each spherical layer. Orbital wave functions follow a pattern of self-similarity within every spherical layer. No matter how many spherical layers an atom may have, all layers are spacetime continuous via calculus chain rule. The theory of quaternity is so elegant and wonderful in mathematical expression and conceptual interpretation of electrons. The striking beauty that it reveals is the best proof of its correctness. Only truth has so unparalleled beauty and exactness.

## REFERENCES

[1] Lula, R. A. High manganese austenitic steels: past, present and future, *Conference Proceedings of high manganese high nitrogen austenitic steels*, edited by R. A. Lula. ASM International, 1987, pp. 1-80.

# PART II. THE NATURAL PATTERN

We shall set out from the theoretical basis of Part I to further explore the unknown spacetime realm in Part II. Part I has defined the motion of electrons in various atoms with spherical quantities in dynamic calculus; and Part II will continue to pursue spherical view for the characterization of molecules, cells, organisms, and the earth. For example, both 1s electrons within helium shell constitute a two-dimensional system, which is a common characteristic of life phenomena. A couple of husband and wife are two dimensions of a family; plants and animals are two kingdoms of the advanced lives. The interplay and transformation between both dimensions are the eternal theme of culture as well as nature. Spherical view describing the atomic structure may be applied to studying the biological pattern. Spherical quantity in dynamic calculus is the law of nature.

Starting from quaternity spacetime in neon shell, Chapter 6 explains the configurations and reaction mechanisms of several organic molecules by the anisotropy of 2s2p orbitals within carbon atoms forming the molecules. The explanation is more natural than orbital hybridization. The anisotropy of 2s2p orbitals is the origin of a chiral carbon bonded to four different substituents. It is deduced that a chiral carbon may transfer electrons along a selective pathway from a bonded substituent to another. Based on the chiral property, Chapter 7 characterizes a DNA molecule as stepwise *LC* oscillatory circuits. Charge transfer from one base pair to another through the phosphate bridges observes harmonic oscillation. If the electrical characteristic of a DNA molecule obeys dynamic calculus, then a cell determined by the DNA molecule must also obey dynamic calculus. Chapter 8 examines the relationship between biochemical cycles and the relationship between cellular organelles. It is suggested that these relationships within a cell are similar to those between electronic orbitals within atoms. They are governed by the same natural law. Chapter 9 further explores the possibility of unifying planetary orbits with electronic orbitals under the framework of uniform circular motion. It also establishes the connection between central force and wave function.

After defining spherical quantity in dynamic calculus as the law of nature, it is a logical step to arrange physical quantities into a periodic table according to their spacetime dimensions. From the intricate relationships between various physical quantities, Chapter 10 tries to clarify the spacetime dimensions of each physical quantity so as to make a periodic table of physical quantities. This rudimentary work provides sharp insight into the laws of physics, the rules and properties of one or more physical quantities, or the relationships that govern two or more physical quantities. Last but not least, Chapter 11 defines the core concepts of traditional Chinese medical theory in strict geometry and dynamic calculus. The mathematical definition with certainty and precision is revolutionary as compared with traditional Chinese medical theory in philosophy that invariably produces doubts and disputes. Spherical view coincides with traditional Chinese medical theory and is expected to revitalize the tradition effectively in generations to come.

The discovery of the atomic structures provides a model for describing the natural pattern in general. Compared with Part I, this extensional research is more radical and speculative, hiking towards the unification of the world order. Hypothetical as it may be occasionally, it is nevertheless valuable because it furnishes fresh methodology and sharp vision, broadening our conventional mindset. After all, science is the unreserved communication between scientists. Part I focused on mathematics and physics; and Part II is mainly about chemistry and biology. So far, readers are advised to peruse Part I before continue.

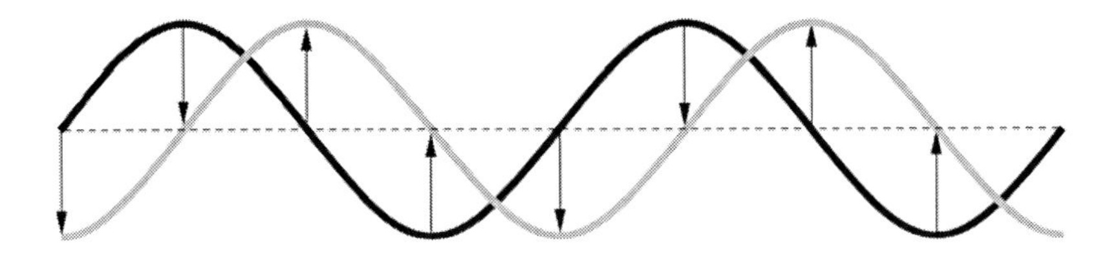

Motif 2. Oscillations, waves, rotations, circulation, rope, and trigonometry.

*Chapter 6*

# ANISOTROPIC 2P ORBITALS AND CHIRAL CARBON

The outer electrons of an atom determine the chemical property of the atom in forming inorganic and organic molecules. This chapter extends the theory of quaternity from the atomic structure to the molecular level. The discovery of $2s2p_x2p_y2p_z$ geometries corresponding to four dimensions in quaternity space has enormous significance in organic chemistry. By analyzing the structure of carbon atoms in organic molecules as basic as methane, ethane, ethyne, and benzene, we provide strong evidence of anisotropic electron configuration $2s^1 2p_x^1 2p_y^1 2p_z^1$ of the carbon atoms that form the molecules, alternative to orbital hybridization. The anisotropic electron configuration of carbon atoms explains well the reaction mechanisms of nucleophilic substitutions of alkyl halides and of electrophilic substitutions of mono-substituted benzenes. On further analyses, we find that anisotropic electron configuration $2s^1 2p_x^1 2p_y^1 2p_z^1$ gives the characteristic of an asymmetrical carbon for chirality. A chiral molecule is a molecule that is nonsuperposable on its mirror image. A pair of chiral enantiomers rotate the plane of polarization of plane-polarized light an equal amount, but in opposite directions. We believe that the spacetime differences of 2s, $2p_x$, $2p_y$, and $2p_z$ orbitals are responsible for the optical rotations. Light bends due to density gradient of electrons within the atomic spacetime. By the same principle, chiral carbon centers also play an important role in directing electron flow through itself in biomolecules. We illustrate the last point by the examples of C1', C3', and C4' stereocenters in a deoxyribonucleoside.

## 6.1. EVIDENCE OF ANISOTROPIC 2P ORBITALS

Quantum mechanics depicts 2p orbitals as being equally distributed in the X, Y, and Z directions in Euclidean geometry. Such an intuitive assumption violates true orthogonal principle and has escaped scientific scrutiny for nearly a century. For quantities to be orthogonal, they must belong to different spacetime dimensions as were distinguished by the polor, the metor, the vitor, and the scalor (Table 6.1). Three quantities being equally distributed in the X, Y, and Z orientations do not constitute orthogonality. Assuming that electron cloud distributes in three spatial orientations equally and statically cannot explain the orientation of 2p orbitals in the carbon atoms that form methane, ethene, ethyne, and benzene. Hypothetical orbital hybridization has to be supplemented for those accounts. Starting from the definition of the geometries of anisotropic 2p orbitals, this section tries to explain quaternity space in carbon atoms that form various molecules. For examples, how to explain the chemical structure of basic organic compounds, such as methane, ethane, ethyne, and benzene, without appealing to orbital hybridization? How does quaternity space in carbon atom facilitate $S_N2$ reaction mechanism in nucleophilic substitution of alkyl halides? Does quaternity theory agree with established organic chemistry in predicting the consequence of electrophilic substitutions of mono-substituted benzenes? By applying quaternity space to carbon atoms, we shall answer these questions in a coherent manner. Successful applications are the best evidence of anisotropic 2s2p orbitals.

**Table 6.1. Wave functions and geometrical shapes of eight electrons in neon shell from Chapter 2**

| Spherical quadrants | Wave functions | Space dimension | Final orbitals | Final shapes radiating from central nucleus |
|---|---|---|---|---|
| Polors | $\Phi_0$, $\Phi_4$ | 0→1 | $2p_x^2$ | Two poles |
| Metors | $\Phi_1$, $\Phi_5$ | 1→2 | $2p_y^2$ | Two flat sectors |
| Vitors | $\Phi_2$, $\Phi_6$ | 2→3 | $2p_z^2$ | Two hemispheres |
| Scalors | $\Phi_3$, $\Phi_7$ | 3→4 | $2s^2$ | Two concentric spheres |

### 6.1.1. Geometries of 2p Orbitals

In traditional organic chemistry, 2p orbitals are known to be dumbbell in shape aligned along the X, Y, and Z axes. So they are assigned $2p_x$, $2p_y$, and $2p_z$ orbitals (Figure 6.1). It is believed that the 2p orbitals have a higher energy than 2s orbital, and $2p_x$, $2p_y$, and $2p_z$ orbitals have equal energy. According to aufbau principle, electrons will fill up lower energy atomic orbitals before entering higher energy orbitals; according to Pauli exclusion principle, each orbital is allowed a maximum of two electrons of opposite spins; and according to Hund's rule, when orbitals of equal energy are available, electrons will occupy separate orbitals before paring up. These principles determine that electron configuration of a carbon atom is $1s^2 2s^2 2p_x^1 2p_y^1$.

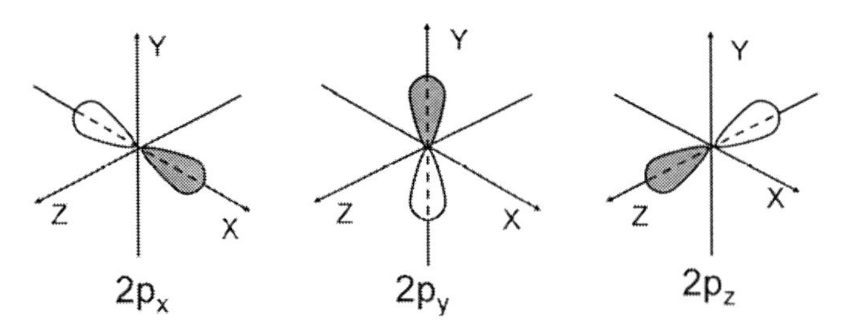


Figure 6.1. Schematic diagram of 2p orbital geometries in traditional opinion under Cartesian coordinates.

In contrast to the traditional view, quaternity theory describes 2p orbitals by two polors, two metors, and two vitors, which are of different dimensions in spacetime. As shown in Figure 6.2, $2p_x$, $2p_y$, and $2p_z$ orbitals possess different space dimensions in Euclidean space. A $2p_x$ is a polor of a line segment pointing from the nucleus to either pole; a $2p_y$ is a metor of a planar sector spreading out from the nucleus to a semicircular arc; and a $2p_z$ is a vitor of hemispherical shape. The shapes plotted here are those geometrical shapes in Figure 2.14 connecting to the central nucleus (Table 6.1). Electronic orbitals along the X, Y, and Z directions are different in geometry, so they are termed anisotropic 2p orbitals. The $2p_x$ orbital has a smaller volume with denser electron cloud than the $2p_y$ orbital whereas the $2p_z$ orbital has a larger volume with sparser electron cloud than the $2p_y$ orbital. A 2s orbital is spherical in shape, which has a maximum volume with minimum density. According to quaternity theory, electrons within neon shell are in dynamic motion. Each electron is evolving through various states cyclically in the arrow direction shown in Figure 6.3. Although 2s orbital is the first orbital of the second shell (neon shell), electrons will not occupy the orbital statically all the time. Four outer electrons of a carbon atom typically occupy 2s, $2p_x$, $2p_y$, and $2p_z$ orbitals, respectively. Thus, it is rational to believe that electron configuration of a carbon atom is $1s^2 2s^1 2p_x^1 2p_y^1 2p_z^1$.

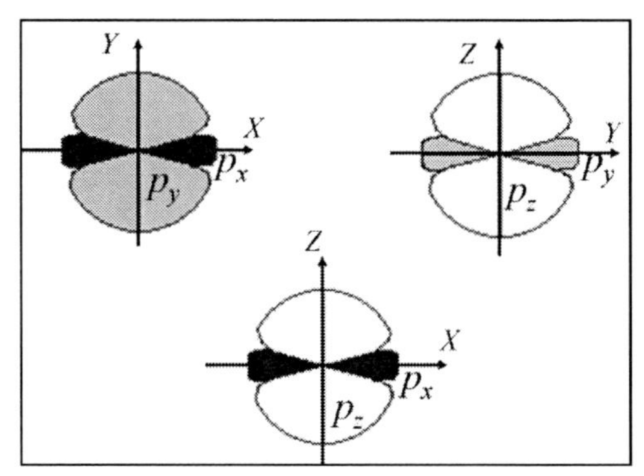


Figure 6.2. Profile diagrams of the shapes of 2p orbitals in three-dimensional Euclidean space where $2p_x$ are two one-dimensional sticks (black); $2p_y$ are two flat sectors (gray); and $2p_z$ are two hemispheres (white).

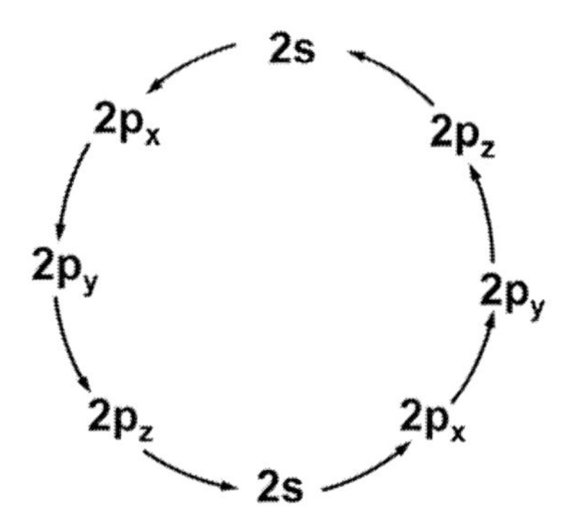


Figure 6.3. Electronic transformation cycle within neon shell.

The different perceptions of 2p orbitals in a carbon atom lead to different accounts on the nature of covalent bonds in molecules such as methane, ethene, ethyne, and benzene. Electron configuration of carbon $1s^2 2s^2 2p_x^1 2p_y^1$ implies that carbon should form two bonds. However, it is known that carbon forms four bonds. The traditional opinion of carbon electron configuration cannot explain the geometrical structures of those basic molecules. Orbital hybridization is traditionally hypothesized to describe the organic molecules case by case.

Specifically, $sp^3$ hybridization is fabricated to explain the tetrahedral shape of methane; $sp^2$ hybridization is fabricated to explain the planar shape of alkene; and sp hybridization is fabricated to explain the linear shape of ethyne. Although orbital hybridization has been written in the textbook of elementary organic chemistry, we don't think it is correct. In the following sections, we shall apply electron configuration of carbon atom, $1s^2 2s^1 2p_x^{\ 1} 2p_y^{\ 1} 2p_z^{\ 1}$, in quaternity space to the explanation of various molecular structures and reaction mechanisms. The purpose is to provide an alternative, simple, and unified account of carbon atoms in organic chemistry.

## 6.1.2. Covalent Bonds in Methane, Ethene, and Ethyne

As an alternative to orbital hybridization, here we describe molecular structures by quaternity space. As was mentioned previously, electronic orbitals are dynamic ones, whose geometries change continuously as radian angles $\Psi$ and $\psi$ vary. Four electrons in the outer shell of a carbon atom tend to occupy orbitals of $2s^1 2p_x^{\ 1} 2p_y^{\ 1} 2p_z^{\ 1}$. Because four orbitals are different in space dimension, there are various spatial flexibilities associated with them. To specify, a 2s orbital is in a spherical shape and hence may move around the entire sphere; a $2p_x$ orbital is rectilinear radiating from the nucleus to either pole and is restricted to the unidimensional line in space; a $2p_y$ orbital is a flat sector and therefore may move along the semicircle; and a $2p_z$ orbital is a hemisphere and has the highest degrees of freedom in space among three types of 2p electrons. In addition, a free electron may evolve dynamically from 2s to $2p_x$, to $2p_y$, and to $2p_z$ in sequence (Figure 6.3). These restrictions and flexibilities determine their possible orbital orientations in forming covalent bonds between atoms.

The anisotropic shapes of 2p orbitals were derived from dimensional analyses of a spherical layer. They have broad supports from organic chemistry. The shapes explain the σ-bond and π-bonds in methane, ethene, and ethyne reasonably well. In a methane molecule, a carbon atom uses a 2s electron, a $2p_x$ electron, a $2p_y$ electron, and a $2p_z$ electron to form covalent bonds with four hydrogen atoms (Figure 6.4). Because of the flexibility of $2p_y$, $2p_z$, and 2s electronic orbitals in space, each bond is restricted to a corner of a tetrahedron under electrical repulsion. But the four covalent bonds are all different due to different nature of the four electrons in carbon atoms. However, because the electrons may evolve dynamically from 2s to $2p_x$, to $2p_y$, and to $2p_z$ in sequence, each covalent bond switches among four types of bonds cyclically. This explains why four covalent bonds have equal lengths upon rough measurements.

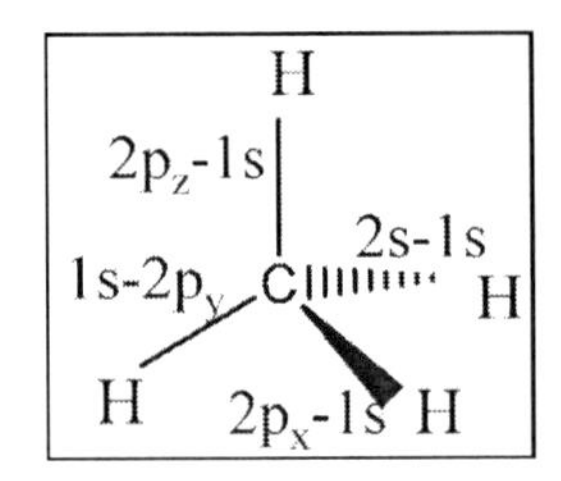


Figure 6.4. A methane molecule with four types of C-H covalent bonds at any moment.

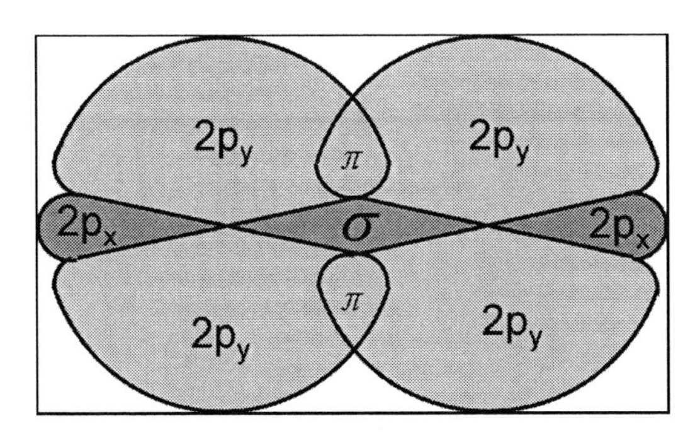


Figure 6.5. In an ethene molecule, a σ-bond is in the middle and a π-bond involves two electrons colonizing a pair of overlapping regions above and below the σ-bond.

In an ethene molecule, two carbon atoms form a stable $2p_x$-$2p_x$ σ-bond and a $2p_y$-$2p_y$ π-bond. Geometrically, the π-bond is the partial overlap of two $2p_y$ flat sectors in one atom with two $2p_y$ flat sectors in another atom (Figure 6.5). This is slightly different from traditional concept of π-bond. The π-bond has less overlap than the σ-bond and is an unsaturated covalent bond because only one electron occupies each $2p_y$-$2p_y$ overlapping region. The geometrical orientation of $2p_y$ electrons to accommodate for π-bond and the geometrical flexibility of 2s and $2p_z$ under electrical repulsion of the σ-bond and π-bond explain its planar molecular conformation. If the σ-bond and π-bond are on a vertical plane, then 2s-1s bond and $2p_z$-1s bond are on a horizontal plane.

In an ethyne molecule, two carbon atoms form a $2p_x$-$2p_x$ σ-bond, a $2p_y$-$2p_y$ π-bond, and a $2p_z$-$2p_z$ π-bond. These two π-bonds are directionally specific and are of different energy, one being two flat sectors partially overlapping two flat sectors, the other being two hemispheres partially overlapping two hemispheres on the edges. Both π-bonds are perpendicular in orientations. The C-H bonds are oriented towards the opposite sides of the three bonds due to electrical repulsion, giving its linear molecular structure. Overall, any organic molecular structures can be adequately illustrated by the combined effects of the original 2p orbital geometries, their degrees of freedom, and electrical repulsion. Orbital hybridization is not necessary in quaternity theory.

Because quantum mechanics cannot explain geometrical arrangements of methane, ethene, and ethyne molecules, orbital hybridization is artificially made up for those accounts, respectively. Although the hybridization explains the configuration of organic molecules reasonably well, it is purely based on observations rather than theoretical deduction. For example, one can describe the result of $sp^2$ hybridization, but cannot explain how and why it behaves in that way. The difficulty and limitation of quantum mechanics underline the success of quaternity. Four spherical quadrants match exactly with 2s, $2p_x$, $2p_y$, and $2p_z$ orbitals. The suitability of quaternity space for explaining the geometries of organic molecules is the basic evidence of quaternity theory.

### 6.1.3. Nucleophilic Substitutions of Alkyl Halides

Nucleophilic substitution of an alkyl halide involves the substitution of the halogen atom with a different nucleophile. The reaction may follow two types of mechanisms, namely $S_N1$ and $S_N2$. We shall not present the subject systematically here, but only present the context of

$S_N2$ mechanism for illustrating how electron configuration $2s^1 2p_x^1 2p_y^1 2p_z^1$ of carbon center facilitates the reaction. $S_N2$ mechanism is a concerted reaction that proceeds directly from reactant to product without forming any detectable intermediates.

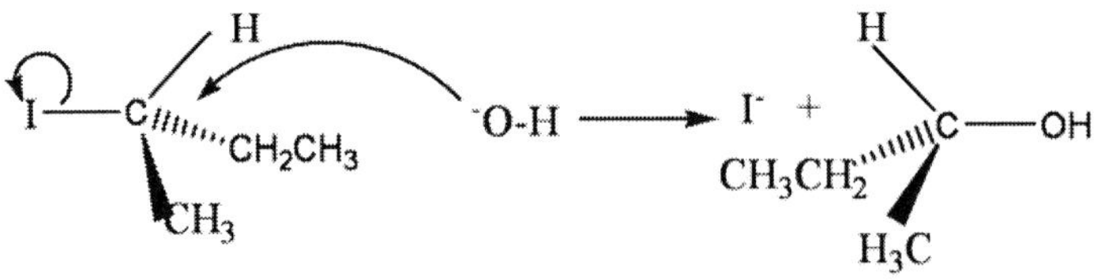

Figure 6.6. Walden inversion of an asymmetrical carbon center in $S_N2$ reaction.

The incoming nucleophile forms a bond to the carbon center at the same time as the C-X bond is broken. The incoming nucleophile attacks the carbon center from one side of the molecule and the outgoing halide departs from the other side. This is called back-side displacement. As a result, the reaction center is inverted during the process. For example, hydroxide ion displaces iodide ion in the following reaction (Figure 6.6).

If electron configuration of the carbon center in (S)-2-iodobutane is $sp^3$ orbital hybridization, then there is not any empty atomic orbital on the other side of iodine for hydroxide ion to approach. So we must expect that hydroxide ion attack the carbon center from any directions. Thus, $sp^3$ orbital hybridization in carbon atom cannot explain the reaction mechanism satisfactorily.

We now turn to quaternity viewpoint of $2s^1 2p_x^1 2p_y^1 2p_z^1$ electron configuration of the carbon center. Because iodine group is more electronegative than –H, $-CH_3$, and $-CH_2CH_3$ groups in the molecule, it bonds to the densest orbital ($2p_x$ orbital) in carbon. The other three groups bond to 2s, $2p_z$, and $2p_y$ orbitals in the carbon center, respectively. Because of electrical repulsion, four substituent groups are arranged towards tetrahedral apices. Why hydroxide ion attacks the carbon center from the other side of iodine group? Because in the molecule, there are another three empty atomic orbitals $2p_x 2p_y 2p_z$ in the carbon center on the other side of the occupied $2p_x^1 2p_y^1 2p_z^1$ orbitals respectively and an empty spherical 2s orbital around the carbon. The hydroxide ion preferably attacks empty $2p_x$ atomic orbital in the carbon because it is denser (less space dimension with more time dimensions) than any other exposed empty atomic orbitals. The reaction results in Walden inversion of the chiral carbon center from S to R configuration. Thus quaternity viewpoint of 2p orbitals explains the back-side displacement of $S_N2$ mechanism reasonably well. And the explanation is simple and natural.

### 6.1.4. Covalent Bonds in Aromatic Rings

It is traditionally believed that all carbon atoms in benzene undergo $sp^2$ hybridization. Each carbon forms three σ-bonds (C-H, C-C, and C-C) and a π-bond (C-C) with the other atoms. There is also a $2p_y$ orbital left over in each carbon. The unhybridized $2p_y$ oribtals of six carbons overlap together, each contributing one electron for a total of six π electrons. These six π electrons are said to be delocalized around the aromatic ring, increasing the stability of it.

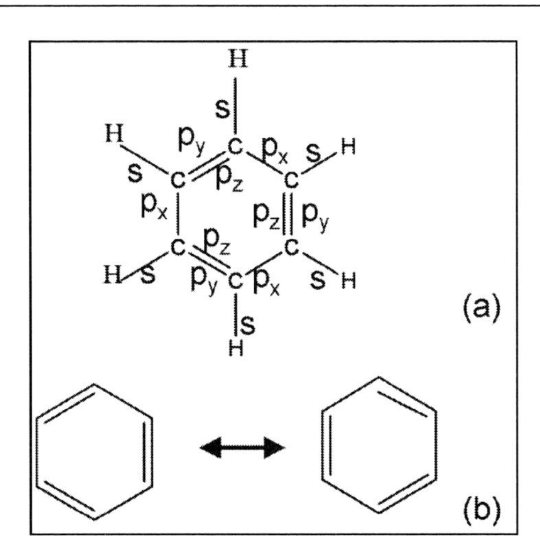


Figure 6.7. The nature of covalent bonds in benzene and its resonance structures.

Quaternity provides an alternative explanation on the structure of benzene. With electron configuration of carbon $2s^1 2p_x^1 2p_y^1 2p_z^1$, each carbon in benzene forms a 2s-1s covalent bond with hydrogen atom, a $2p_x$-$2p_x$ covalent bond with carbon on the one side, and two ($2p_y$-$2p_y$ and $2p_z$-$2p_z$) π-bonds with carbon on the other side (Figure 6.7a). The $2p_x$-$2p_x$ covalent bonds alternate with the $2p_y$-$2p_y$ and $2p_z$-$2p_z$ double bonds around the ring, giving its aromatic stability. However, because each carbon is bonded by similar atoms, the double bonds and $2p_x$-$2p_x$ covalent bond may switch their positions rapidly, giving two resonance structures as electrons transform dynamically and cyclically (Figure 6.7b). The resonance structures render equivalent carbon-carbon bond lengths of six sides around the ring.

### 6.1.5. Electrophilic Substitutions of Mono-Substituted Benzenes

Substituents on an aromatic ring can activate or deactivate the ring towards further electrophilic substitution. They can also direct substitution either to the *meta* position or to the *ortho* and *para* positions. In this section, we shall explain these phenomena by quaternity space in carbon atoms. Aniline and nitrobenzene serve as two typical examples of mono-substituted benzenes. In an aniline molecule, because nitrogen is more electronegative than carbon, N-C bond is a $2p_x$-$2p_x$ covalent bond. This disturbs the original structure of the aromatic ring, around which $2p_x$-$2p_x$ bond alternates with double π bonds of $2p_y$-$2p_y$ and $2p_z$-$2p_z$. As shown in Figure 6.8(a), after $2p_x$ orbital of C1 has connected to N, there is only a 2s orbital available for C6.

The covalent bond between C1 and C6 (2s-$2p_x$) is mismatched. Only orbitals of similar type form a stable covalent bond. Thus the amino group destabilizes the ring especially at *ortho* position. Moreover, the nitrogen atom of aniline is also able to donate lone-pair electron density to the intermediate cations arising from electrophilic aromatic substitution at *ortho* and *para* positions. Thus the amino group activates the ring towards further electrophilic substitution at *ortho* and *para* positions.

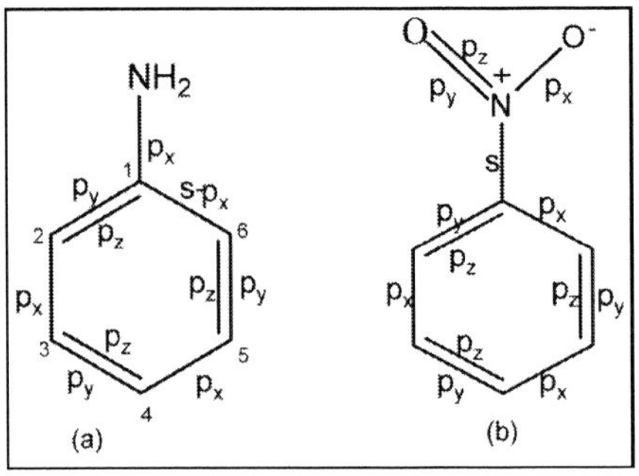

Figure 6.8. The nature of covalent bonds in aniline and nitrobenzene.

A nitro group attaches to the ring to form nitrobenzene through a nitrogen atom with a formal positive charge. As shown in Figure 6.8(b), because $2p_x$, $2p_y$, and $2p_z$ orbitals of nitrogen are bonded to oxygen, which is more electronegative than carbon, the bond of N-C is a 2s-2s covalent bond. Comparing Figure 6.7(a) with Figure 6.8(b), we find that the nitro group does not disturb the original bonding structure of the aromatic ring. Instead, the positive charge absorbs electron density around the ring and hence deactivates further electrophilic substitution.

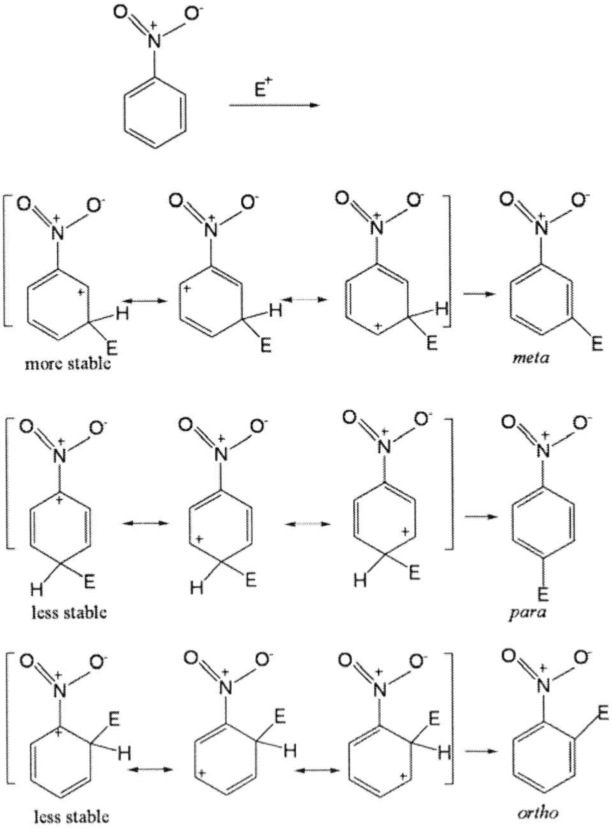

Figure 6.9. The resonance structures of intermediates leading to the *meta*, *para*, and *ortho* aromatic electrophilic substitutions, respectively.

The directing effect of the nitro group depends on the stability of intermediates formed during electrophilic substitution. From Figure 6.9, it is easy to see that destabilization is greatest for the intermediate arenium ions that lead to *ortho* and *para* products because one of the three resonance structures for the arenium ion has positive charge on the carbon atom bearing the positively charged nitrogen atom of the nitro group.

The intermediate leading to *meta* substitution is less destabilized because none of its three resonance structures has this proximity of positive charges. Thus the nitro group deactivates the ring and directs electrophilic substitution at the *meta* position. It is clear that $sp^2$ hybridization of carbon cannot explain the resonance structures of benzene, which are crucial in predicting the reactions. The resonance structures are based on alternate single and double bonds around the aromatic ring. With these, quaternity explanation of the chemical property of aniline and nitrobenzene agrees with established organic chemistry and provides a fresh insight into chemical reactions.

## 6.2. Molecular Chirality

The discovery of anisotropic 2p orbitals naturally leads to the idea that they may form a handed structure in a tetravalent carbon atom. Here we put forward that molecular chirality results from the orbital arrangement within the atomic sphere of carbon. Although it is generally correct to deduce the existence of molecular chirality based on whether a carbon atom is bonded to four different substituents, the four substituents of different electrophilic powers just fix and amplify the internal chirality of electron configuration of the central carbon, giving its stable optical activity. Inspired by the dynamic transformation of electrons within the atom, we further suggest that selective electron transport through a stereocenter follows the electronic circulation pattern of $2s \rightarrow 2p_x \rightarrow 2p_y \rightarrow 2p_z \rightarrow 2s$. Given the handedness of a chiral molecule, the sign of its optical rotatory effect and the pathway of selective electron transfer through the carbon center can be predicted theoretically to a large extent. For the first time, we discover the biological role of chiral carbons in directing electron flow in DNA.

### 6.2.1. Density Gradient of 2s2p Orbitals

It has been established that $2s2p_x2p_y2p_z$ electrons within neon shell take different shapes as well as orthogonal orientations. As was shown in Figure 2.14, a $2p_x$ orbital is a point displaced from the nucleus; a $2p_y$ orbital is a two-dimensional semicircular arc; a $2p_z$ orbital is a three-dimensional hemispherical surface; and a 2s orbital is a four-dimensional solid sphere. Because all electrons are in dynamic motion, these geometries are only typical illustrations of electrons at a certain moment. The geometries and dynamic motion of the orbitals determine that a carbon atom with outer electron configuration of $2s^1 2p_x^1 2p_y^1 2p_z^1$ is a potential chiral structure.

Figure 2.14 is only a snapshot of 2s2p orbital geometries in their exact dimensions. Figure 2.12 shows these orbitals in dynamic motion, which we shall illustrate in more detail here. First, orbital transformation from 2s to $2p_x$ is the gradual displacement of the whole sphere away from the nucleus and the condensation of the electron in the meantime (Figure 6.10). As point E traces along the hypothetical dashed semicircle OEB, the center of the sphere moves to point A and the radius of it decreases from OB to AB where point A is the vertical projection of point E onto line OB. The electronic orbital can be regarded as a volatile medium in spacetime, which we may call electron cloud. The time series images (Figure 6.10b) clearly indicate density gradient of the dynamic electron. The tiny dislocated point B represents a full $2p_x$ orbital at the rotation terminal.

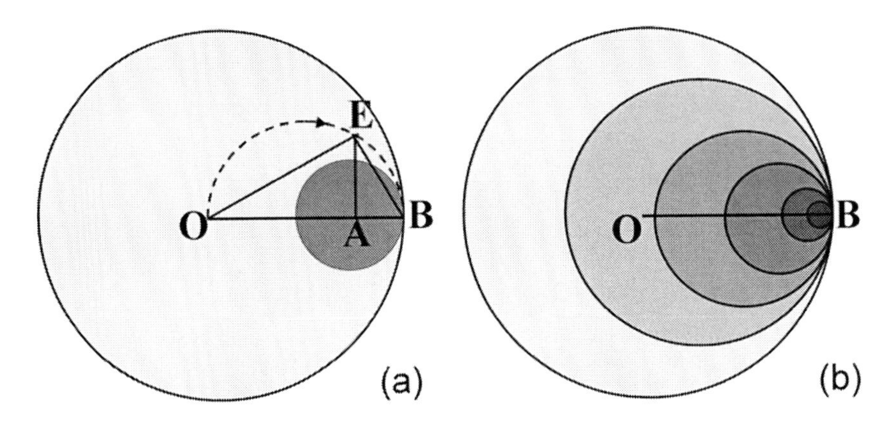


Figure 6.10. Polar orbital transformation from 2s to $2p_x$ as the whole sphere dislocates off the origin and contracts in the meanwhile.

Second, let's examine various phases of electronic transformation from both $2p_x$ to both $2p_y$ orbitals. As shown in Figure 6.11, electronic transformation from a pair of dense points (b) towards a pair of semicircular arcs (f) experiences middle phases (c)(d)(e) during the continuous extension process. If we capture this transformation by overlaying several of its middle stage images together into (a), it is easy to see that the dynamic process renders each $2p_y$ orbital gradient in density. The density gradient is schematically represented by the relative thickness of the arcs and their corresponding lengths. The longer the arc is, the thinner the line is. The density gradient is directional along the orbital. Directionality refers to dynamic geometries whose head and tail are distinguished in time series.

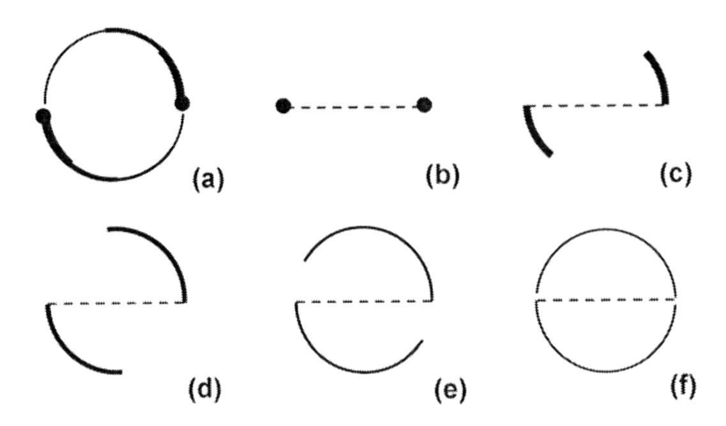


Figure 6.11. Electronic transformation from a pair of $2p_x$ orbitals (b) to a pair of $2p_y$ orbitals (f) through various middle phases (c)(d)(e) and their images in time series (a).

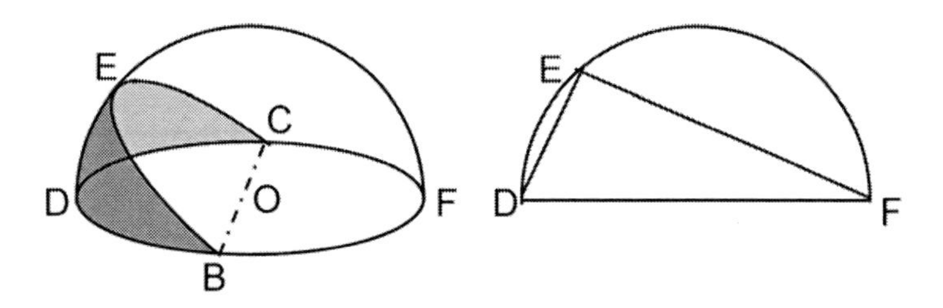


Figure 6.12. Electronic transformation from $2p_y$ to $2p_z$ orbitals as semicircle BDC sweeps through the hemispherical surface around BC axis.

Third, electronic transformation from $2p_y$ to $2p_z$ also reveals density gradient of the orbital. As shown in Figure 6.12, a $2p_z$ orbital experiences geometrical transformation of the shaded ribbon BDCE widening from arc BDC to the entire hemispherical surface BDCEF as point E traces from point D towards F along the semicircle DEF. During the spreading process, the density of electron decreases continuously as its covering area increases. Like a $2p_y$ orbital, the expanding process in space produces orbital directionality in time series for the $2p_z$ electron.

Finally, orbital transformation from $2p_z$ to 2s can be characterized by the waxing of a crescent volume. As shown in Figure 6.13, a full $2p_z$ orbital is a hemispherical surface represented by DF; and a full 2s orbital is whole sphere O. The transformation from a full $2p_z$ orbital to a 2s orbital involves the gradual inward filling of electron cloud in the sphere. The electron cloud is diffusing inwards over the atomic sphere during the process, which naturally leads to an orbital of ever sparser electron density with time. Thus the dynamic 2s orbital displays density gradient in time series.

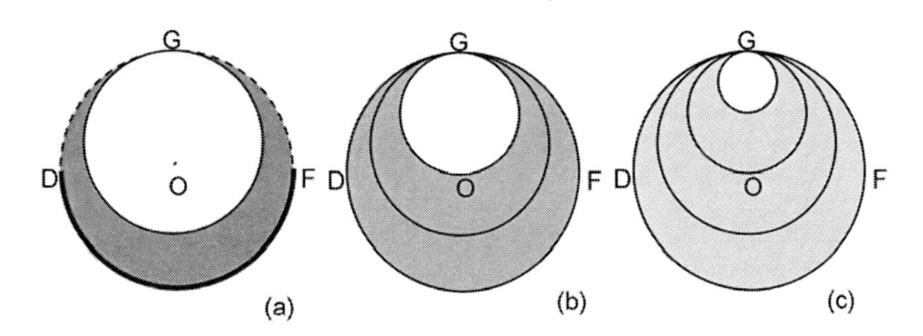


Figure 6.13. Density gradient of directional orbital implosion from $2p_z$ to 2s in three snapshots over time.

The foregoing description indicates that four harmonic oscillation processes of 2s→ $2p_x$ → $2p_y$ → $2p_z$ → 2s constitute a continuous geometrical evolution loop. Each process produces a gradient orbital in time series. The geometrical shapes of four orbital types are different. In their dynamic motion, a 2s electron is transforming into a $2p_x$ electron, and a $2p_x$ electron is transforming into a $2p_y$ type while a $2p_y$ electron is transforming into a $2p_z$ type, which in turn is evolving into a 2s type electron. The geometries and dynamic motion of the electrons determine that a carbon atom with four outer electrons of $2p_x^1 2p_y^1 2p_z^1 2s^1$ possesses non-uniform electronic density within the atomic sphere, i.e., the electronic density of $2p_x$ is higher than that of $2p_y$, and the electronic density of $2p_y$ is higher than that of $2p_z$, which in turn is higher than that of 2s. The continuous gradient of electronic density in the atomic sphere formed by various orbitals is the physical reason for carbon chirality. Light bends due to the existence of continuous electronic density gradient within the atomic sphere.

### 6.2.2. The Origin of Molecular Chirality

A molecule that is nonsuperimposable on its mirror image is chiral. A chiral molecule rotates the plane of incident polarized light and is said to be optically active. A carbon atom bonded by four different substituent groups constitutes a chiral center or stereocenter. In spite of the fact that thousands of compounds with one chiral center have been found to be optically active with no exceptions, the notion of chirality according to the existence of a dissymmetric atom in a molecule is empirical. There must be a fundamental reason underlying the optical phenomenon.

Here we investigate the cause of chirality at a deeper level than molecular configuration. We propose that chirality of a molecule originates from $2p_x^1 2p_y^1 2p_z^1 2s^1$ electron configuration of the outer shell of a carbon atom.

As was described in section 6.1.2, in a methane molecule, the carbon atom uses a $2p_x$ electron, a $2p_y$ electron, a $2p_z$ electron, and a 2s electron to form four covalent bonds with four hydrogen atoms respectively. Thanks to the spatial flexibility of the latter three orbital types, each bond is restricted to a corner of a tetrahedron under electrical repulsion between the covalent bonds.

However, due to dynamic transformation of the electrons, the nature of each covalent bond normally switches among four types of carbon-hydrogen bonds periodically from $2p_x$-1s bond to $2p_y$-1s bond, to $2p_z$-1s bond, then to a 2s-1s bond, and return to $2p_x$-1s covalent bond type.

The rapid dynamic cycling renders a methane molecule achiral or ambi-chiral because the molecule is switching instantly between two contrary kinds of chiral configurations, which are characterized as follows.

As shown in Figure 6.14, because three types of 2p orbitals are mutually perpendicular in the absence of electrical repulsion, if we set up Cartesian coordinate system in such a way that its origin sits at the nucleus, X-axis points to the orientation of $2p_x$ and Y-axis points to the orientation of $2p_y$, then $2p_z$ would be along Z-axis protruding either towards the negative direction (Figure 6.14a) or towards the positive direction (Figure 6.14b). Therefore, three orbitals of $2p_x$, $2p_y$, and $2p_z$ have two kinds of handedness corresponding to left-hand rule and right-hand rule of X, Y, and Z-axes in Cartesian coordinate system. Under electrical repulsion, the angles between the three 2p orbitals would be larger than $90^0$ without changing their relative handedness, and 2s orbital would be oriented to the position farthest away from the three due to its high spatial flexibility.

Hence anisotropy of 2p orbitals leads to anisotropy of 2s2p orbitals. We use both terms interchangeably. A carbon atom with outer electrons of $2p_x^1 2p_y^1 2p_z^1 2s^1$ has two possible chiral configurations. Because of the dynamic transformation of orbitals in the directions of 2s → $2p_x$ → $2p_y$ → $2p_z$ → 2s, a methane molecule exhibits both chiral properties with time. Many $CH_4$ molecules amount to a racemic mixture at any instant and appear achiral in current knowledge.

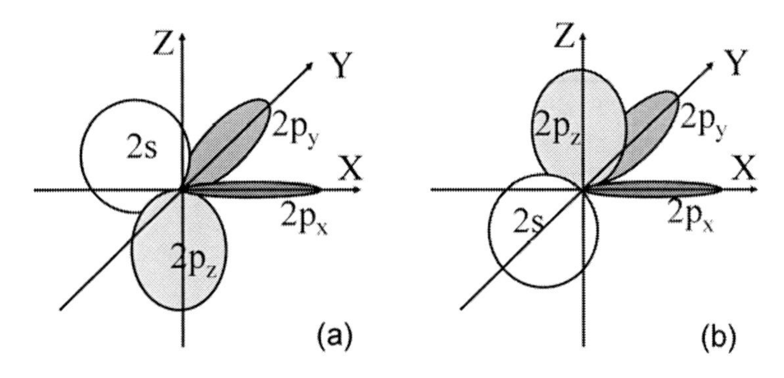


Figure 6.14. Two kinds of handed orientations due to anisotropic 2s2p orbitals in Cartesian coordinates.

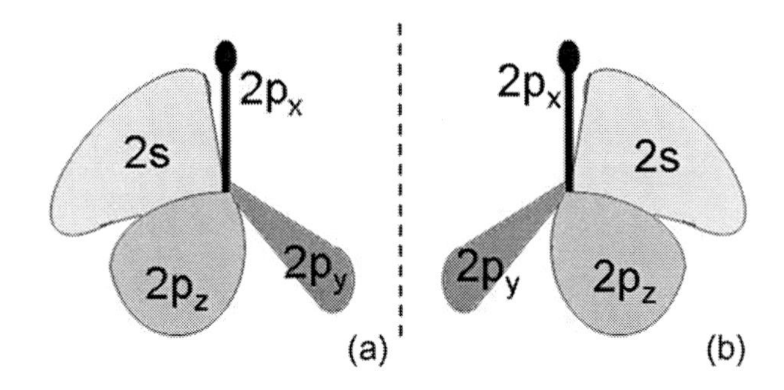


Figure 6.15. Schematic diagram of enantiomers of chiral carbon centers owing to $2p_x{}^1 2p_y{}^1 2p_z{}^1 2s^1$ orbitals of different volumes (size) and densities (shade).

A methane molecule does not show stable chiral property because four hydrogen atoms that are bonded to $2p_x{}^1 2p_y{}^1 2p_z{}^1 2s^1$ have similar actions on them. Only when the four types of electrons within a carbon atom are bonded to four different substituent groups, the compound shows stable chiral property because the four substituents, having different electronegativities and electrical polarizabilities, not only inhibit the free switch of $2p_x{}^1 2p_y{}^1 2p_z{}^1 2s^1$ orbitals among themselves within the carbon, but also fix and reinforce the handedness of the four orbital types. Thus, even though the judgement of a chiral compound based on the handedness of the entire molecule is apparently correct, the cause of chirality indeed rests with the anisotropy of 2s2p electronic orbitals. Corresponding to Figure 6.14, we depict both kinds of orbital handedness in Figure 6.15 for clarity, but bear in mind that the atomic sphere is a continuous spacetime medium of four dynamic orbitals. It is that chirality at the orbital level extending to the molecular level that accounts for its optical activity. Had not been caused by orbital gradient, chirality merely at the molecular scale might be too macro to produce any effects on light refraction.

### 6.2.3. Physical Principle of Optical Rotation

Two enantiomers shown in Figure 6.15 depict four types of electronic orbitals in discrete shapes for clarity, but in fact the four types of orbitals form a gradient medium for chiral characteristics. The smaller the volume is, the denser the electron cloud is. Light bends as it travels through a medium of density gradient by the principle of refraction. For example, as shown in Figure 6.16, when a beam of light enters a transforming $2p_x$ orbital (Figure 6.10b), it will be refracted downwards at the density inclines due to refractive index increase. Since the density gradient is in a continuous wave, light travels along a smooth curve within the atom as a result.

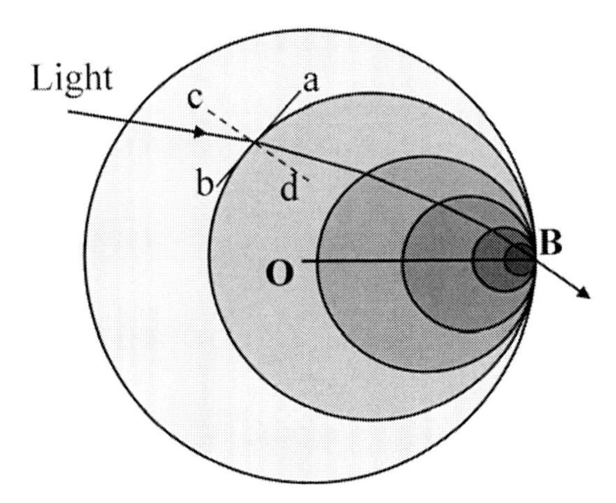


Figure 6.16. Light refraction at density gradients where ab denotes a plane tangential to a refractive index contour and cd indicates the line normal to the plane.

How does a chiral center rotate plane-polarized light? Plane-polarized light may be regarded as being made up of two kinds of circularly polarized light, which has the appearance of a helix propagating around the axis of light motion. One kind is left-handed while the other a right-handed helix. As plane light enters an uneven medium, it turns to bend towards the direction of denser region because of its relatively higher refractive index. The higher the index is, the slower the speed of light travels through the medium. This would seem to mean that the left- and right-handed circularly polarized components travel at different velocities because each is slowed down to a different extent. However, since both components are inseparable, the consequence is the curving of light's pathway and the rotation of the plane.

This phenomenon is analogous to a car running on the road. When the right wheels are faster than the left wheels, the car turns left and the driver's body rotates a certain degree counterclockwise accordingly. Thus the rotation of plane light is closely associated with light bending.

To explain chiral optical effect visually, suppose that the inner region of a chiral structure has a higher refractive index than the outer region, then as light enters the chiral structure, it will be refracted towards right direction all the way through the medium (Figure 6.17). Since a chiral carbon is composed of four orbitals in transformation pattern of $2s \rightarrow 2p_x \rightarrow 2p_y \rightarrow 2p_z \rightarrow 2s$, we assume that electronic orbitals that form a chiral center are arranged in a manner somehow like a screw pattern or a snail shell in spacetime. Light approaching the center will be routed forward in such a way as if a screw rod is guided forward by the screw thread complying with refraction. The rotation of plane light is in a coherent thread direction of clockwise or counterclockwise even though light may enter the medium at any directions.

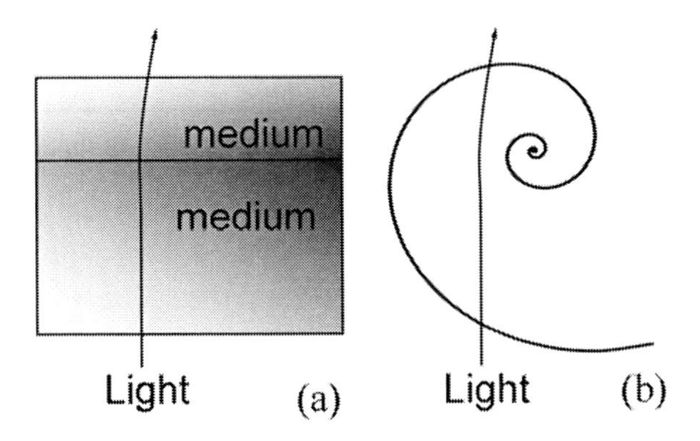


Figure 6.17. Planar illustration of light following (b) the spiral curve of a chiral center to a certain extent due to (a) the existence of density gradient.

Moreover, a coiled spring is right-handed forward when perceived from front and rear; and the spiral line of a field snail remains right-handed forward when perceived from both head and tail. One remarkable feature of a chiral carbon must be that its helix remains the same pattern when perceived from any directions. There is not a head and tail distinction among the orbitals that form the "snail" but the fixed spacetime direction of circular orbital transformation (2s → $2p_x$ → $2p_y$ → $2p_z$ → 2s) for a specific chirality, like a traffic circle at a junction of thoroughfares that routes the cars in a one-way direction. No matter which direction plane light approaches a chiral center, it will follow the spiral curve to rotate a certain degree at that fixed direction, thereby producing a measurable optical rotation magnitude proportional to the chiral compound concentration in the solution.

## 6.2.4. Chiral Disposition by Electrophilic Groups

The relationship between a chiral carbon center and its specific optical rotatory effect is complicated and defies theoretical generalization. By tradition, one can only know the direction and magnitude of rotation, $[\alpha]_D$, of an asymmetric carbon by experimentation. After discovering the essence of chirality at the orbital level, we wish to shed light on this important issue. First, since the property of a chiral carbon comes of orbital anisotropy, we may derive the relationship between the handedness of 2s2p orbitals and the rotatory direction of plane-polarized light. As was shown in Figure 6.3, electrons transform from $2p_x$ to $2p_y$, from $2p_y$ to $2p_z$, and from $2p_z$ to 2s following the rule of dynamic calculus. It has been demonstrated that electronic transformation is realized by transferring virtual photon (Figure 3.13). Virtual photons travel in the contrary direction to electrons. Therefore plane-polarized light entering a chiral carbon will be routed in a direction contrary to electronic transformation direction. Viewing from the opposite side of 2s orbital, we can tell the direction of plane light rotation based on the handedness of three 2p orbitals. The plane light rotation is along $2p_z$→$2p_y$→$2p_x$ direction. For examples, Figure 6.15(a) is levorotatory having a negative $[\alpha]_D$ value whereas Figure 6.15(b) is dextrorotatory having a positive $[\alpha]_D$ value.

Secondly, for four different substituents bonded to a carbon atom, we need to define their priority in forming covalent bonds to various orbitals. Because $2p_x$ orbital has the highest electron density in 2s2p shell, the most electrophilic substituent would form a σ-bond with it, the second electrophilic substituent would bind to $2p_y$ orbital, the third would attach to $2p_z$ orbital, and the least electrophilic substituent (which is –H more often than not) attaches to 2s orbital. Hydrogen atom usually bonds to 2s orbital of an asymmetric carbon also because 1s and 2s orbitals are similar orbital types that match perfectly in forming a covalent bond. In terms of attracting dense electron cloud, the electrophilic property must include two aspects: high electronegativity and low polarizability. For example, –F ligand bonded to a carbon atom would surely arrest its $2p_x$ electron to form a σ-bond. Thus the handedness of 2s2p orbitals is contingent on the relative electrophilic power of four substituent groups attached to them. This establishes the connection between chirality owing to intrinsic $2p_x^1 2p_y^1 2p_z^1 2s^1$ orbitals and chirality induced by extrinsic substituents bonded to them. It is the different substituents that fix and amplify the chirality of orbitals within the carbon atom, giving its stable chiral property. This fresh interpretation is in line with established facts.

The tetravalent nature of a carbon atom naturally leads to molecular handedness in any case where four different substituent groups attach to it. The early work by Brewster suggested an empirical method for the prediction of the sign and amount of optical rotation based on bond refractions and polarizabilities of groups in a molecule [1, 2]. The method predicted the direction of optical rotation using an asymmetric screw pattern of electron polarizability, the magnitude of rotation being related to the refractions of the atoms making up the pattern. Under Fischer projection of Figure 6.18(a), the tetrahedral system of xABCD could be described as dextrorotatory when A>B>C>D in polarizability. D is assumed to be hydrogen atom in most cases. This result agrees with our suggestion that substituent groups from low to high polarizabilities would attach to electrons in the priority order of $2p_x$, $2p_y$, $2p_z$, and 2s within the chiral carbon except the case of –H substituent that always attaches to 2s orbital in our suggestion (Figure 6.18b).

Recently, Zhu et al. devised a matrix model to predict specific optical rotations of acyclic chiral molecules [3]. The matrix elements included the substituents' comprehensive masses and radii, asymmetric factors, and electronegativities of the atoms or groups that are bound to the stereogenic center. Pauling electronegativities were directly used in the matrix and their contribution to the specific rotation was large. The effect of electronegative value on the rotatory direction of plane light in the matrix agrees with our statement that substituent groups from high to low electronegativities would attach to electrons in the priority order of $2p_x$, $2p_y$, $2p_z$, and 2s within the chiral carbon.

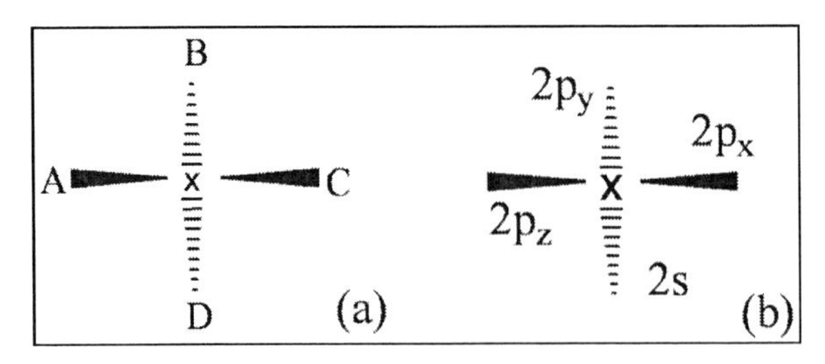


Figure 6.18. Fischer projection of the absolute configuration of xABCD and their bonding orbital types within the stereocenter.

The electrophilic power of substituent groups in terms of high electronegativity and low electric polarizability is a chemical property indicative of both electron transfer and stability.

If we can arrange various kinds of substituents in the order of electrophilic power, we would be able to predict whether a specific chiral molecule is dextrorotatory or levorotatory, which turns out to be exceedingly useful in the next section. Qualitatively, the electrophilic power is related to the relative rate constants for reactions of different electrophiles towards a common substrate, usually involving attack at a carbon atom in organic chemistry. It has something to do with Lewis acidity. Quantitatively, Parr et al. proposed that electrophilicity index of a specific chemical species should be defined as the square of its electronegativity divided by its chemical hardness [4]. This concept measured the propensity of a substituent group to attract electrons in a covalent bond, which seems to agree with our considerations of both electronegativity and polarizability. But it tends to assign the index value for hydrogen –H on the high side. How well the electrophilicity index fits into the framework of preferential selection of tetrahedral $2p_x^1 2p_y^1 2p_z^1 2s^1$ orbitals in a chiral carbon remains to be investigated. Obviously, the suggested order of substituents based on the electrophilic power is different from that based on the Cahn-Ingold-Prelog rules. The latter is useful in the nomenclature of R or S chiral molecular configuration but helpless in predicting optical rotatory effects, at which the former aims.

In spite of its profound level, the determination of specific molecular chirality based on the relative electrophilic powers of four substituent groups attached to a central carbon must prepare to encounter some exceptions for several reasons. First, the chiral configuration of an asymmetric molecule has something to do with its chemical synthetic route. Once a certain handedness has been preserved by four various substituents during synthesis, it takes a certain amount of activation energy to overturn its shape later on. Hence the effect of relative electrophilic powers does not always prevail and manifest in chiral compounds. Second, disorder always competes with order in molecular arrangement. Many organic compounds synthesized by chemical methods naturally result in a racemic mixture due to two accessible sites on the central carbon during the reaction, which shows no apparent optical activity. They do not comply with the rule of preferential orbital selection by relatively strong electrophile either. Even some biologically produced chiral compounds racemize in time under the driving force of entropy increment. Such racemizations have been exploited as an amino acid dating method [5]. Third, the relative electrophilic powers of four different substituents must be considered in a comprehensive manner. Some substituents are more likely to have $2p_x$ orbitals available for forming $2p_x$-$2p_x$ σ-bond with carbon atom, the others tend to expose their $2p_y$ orbitals to form $2p_y$-$2p_y$ π-bond with carbon. $2p_y$ orbitals are more polarizable electrically than $2p_x$ orbitals. For example, as was mentioned in section 6.1.5, nitrogen atom of the amino group has $2p_x$ orbital available for forming a covalent bond whereas that of the nitro group only has 2s orbital available. The electrophilic order of various substituents is not invariable. It depends on the specific role combination of four different substituents in forming a relatively low-energy chiral configuration. While the general rule is set, case by case analysis still deems necessary.

## 6.2.5. Selective Electron Flow Through Chiral Centers

As was shown in Figure 6.3, electrons transform in the direction of $2s \rightarrow 2p_x \rightarrow 2p_y \rightarrow 2p_z \rightarrow 2s$ within neon shell. In a chiral carbon atom with four different substituents fixing the orbital types, light hitting the chiral center would follow the reverse direction of electron transformation and rotate a certain degree due to the gradient of electron density within the atomic sphere. This results in the measurable optical rotation of plane-polarized light passing through the chiral compound medium, $[\alpha]_D$. In this section, we shall first apply this principle to a deoxyribonucleoside molecule and then examine electron transport behavior through various chiral centers in it.

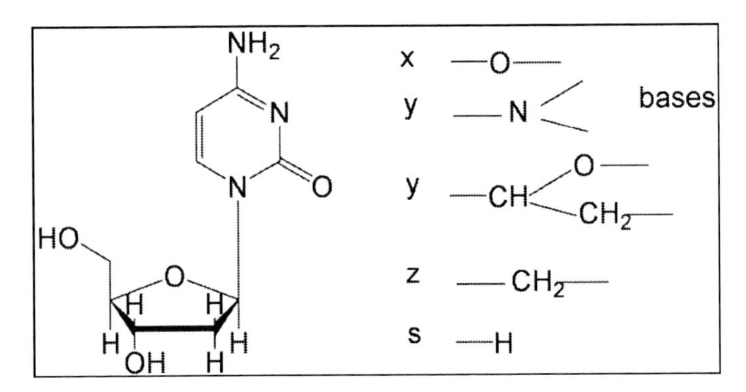


Figure 6.19. Three chiral carbon centers C1', C3' and C4' in a deoxyribose with their substituent groups in the order of relative electrophilic power x>y>z>s.

Consider deoxycytidine with three chiral carbon centers in its deoxyribose, C1', C3' and C4'. Four substituents bonded to each chiral carbon are labeled x, y, z, and s in Figure 6.19. Assuming the electrophilic powers of the substituents are in the order of x > y > z > s, we may predict the tendency of optical rotatory effect of each chiral center according to the above chiral disposition principle by electrophilic groups. For example, to C1' atom, the oxygen atom of the ether bond would bind to $2p_x$ orbital of the chiral carbon; the cytosine would bind to $2p_y$ orbital of the chiral carbon; the aliphatic chain $-CH_2-$ of C2' would bind to $2p_z$ orbital of the chiral carbon; and the hydrogen atom would bind to 2s orbital of it. A substituent of relatively higher electrophilic power binds to an orbital type of relatively higher electron density. Thus the C1' chiral configuration would produce levorotatory effect on plane-polarized light according to Figure 6.15(a).

Likewise, to chiral center C3', the hydroxyl group would bind to its $2p_x$ orbital; C4' would bind to its $2p_y$ orbital; C2' would bind to its $2p_z$ orbital; and the hydrogen atom binds to its 2s orbital. Hence C3' chiral center is dextrorotatory. The four substituents bonded to C4' are in a very similar way to those to C3', but they are arranged in a converse pattern for chiral C4' to be levorotatory. Although the optical activity of the entire nucleoside is difficult to predict, we can derive the sign of optical effect of each chiral carbon based on the relative electrophilic strength of the substituents.

If instead of light refraction, there are electrons flowing from C5' to C4', in which direction should the electrons move at the C4' junction? We suggest that electrons flowing through a chiral center would stick to the transformation pathway of $2s \rightarrow 2p_x \rightarrow 2p_y \rightarrow 2p_z \rightarrow 2s$ around the central carbon. Because C5' is bonded to the $2p_z$ orbital type of C4' chiral center, electrons entering C4' are directed towards 2s direction and then towards $2p_x$ orbital bonded to oxygen atom. Hence the negative charges would be routed from C5' to the ether bridge to C1'; and positive holes would travel in the reverse direction (Figure 6.20). Electric current from C4' to C3' is choked by C4' chiral center. Thus the chiral carbon center transfers electrons through itself along a selective pathway.

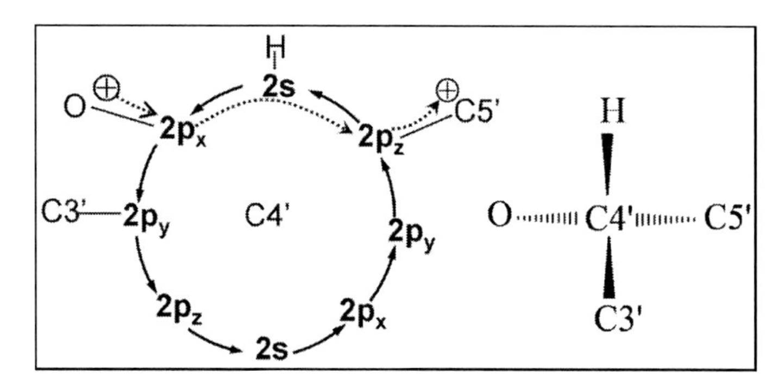


Figure 6.20. Electron circulation direction (solid arrows) within C4' chiral atom and selective positive hole migration (dotted arrows) through the center shown along with its Fischer projection (right).

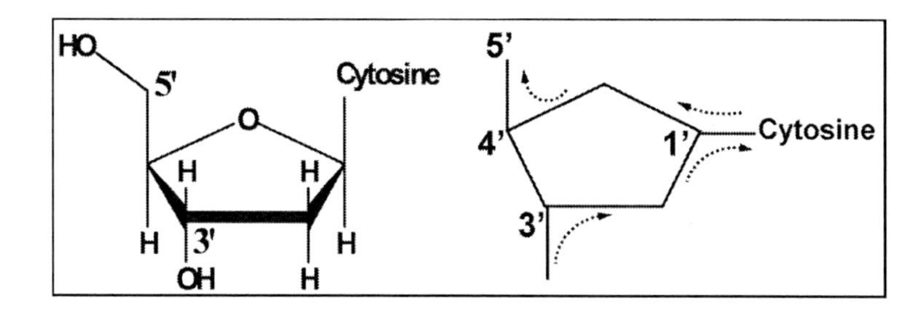


Figure 6.21. Electric currents directed by chiral carbons through a nucleoside. The dashed arrows represent electric current directions.

As electrons arrive at C1' from the ether bridge bonded to the $2p_x$ orbital of chiral C1', they are directed to the N-glycosyl bond that is connected to the $2p_y$ orbital of the chiral center, hence charging the nitrogenous base. Furthermore, suppose that electrons are overflowing from the nitrogenous base towards C1', they would be routed towards C2' direction by the chirality of C1', i.e., from $2p_y$ to $2p_z$ direction. By the same rule, electrons entering C3' from C2' would be routed towards the hydroxyl group rather than C4', i.e., in $2p_z \rightarrow 2s \rightarrow 2p_x$ direction. Electric current from C3' to C4' is choked by the chirality of C3'.

On the contrary, positive charge holes, representing electric current, travel in the opposite direction to electrons. We expect that electric current would be routed by the chiral centers in the direction from 3'-hydroxyl group → C3' → C2' → C1' → nitrogenous base, and then from nitrogenous base → C1' → ether bridge → C4' → C5' (Figure 6.21). Direct electric current between C3' and C4' is prohibited by the chiral carbons, detour being mandatory. Besides traveling direction reversal, another significant difference between light refraction and selective electron flow at a chiral center is that light hitting the atomic sphere immediately swerves out of the atom whereas electrons flowing into the chiral center from any connecting branch must follow the next covalent bond to another connecting branch to exit the center. By the nucleoside example, we have explained the stereo-selectivity of chiral centers in directing electric current.

Living organisms consist almost exclusively of chiral molecules. Specifically, the amino acids, the building blocks of proteins, are found in L-configuration with few exceptions whereas sugars are only found in D-configuration [6]. In explaining why living organisms discriminate one type of enantiomer from the other, it is traditionally stated that biochemical processes in living cells are normally catalyzed by enzymes, which are proteins that fold up into complex structures with an active site for reactants. Only chiral compounds of specific handedness can be locked into the site and participate in enzyme catalyzed reactions. Although such a stereo-specific explanation sounds logical, the more critical reason probably resides in the mechanism of chiral compounds in selective electron transfer. As was described, a chiral center would direct electron flow from a substituent branch to another in accord with 2s2p electronic circulation direction within the chiral atom. A pair of enantiomers, for example, would channel electron flow in contrary directions and produce different biological consequences. Hence one type of enantiomer cannot substitute another, their electrical properties being different. Living organisms must maintain the homochirality of biochemical molecules to ensure their proper functions in charge transfer, which might contain physiological signals. Racemization of homochiral biological substances leads to stray signals and death.

The discovery of chiral carbon in selective electron transfer in conformity with optical activity undoubtedly breaks new ground for biochemical research. Early in 1970s, Garay made a hypothesis that biological chiral molecules, such as chorophylls, during excitation might possess magnetic moment that caused a partial polarization in electron transport [7], but it has received not any attention later on. Modern scientists probing the electron transfer pathways within proteins and nucleic acids are totally unaware of the electron directing role of chiral carbon centers. In a consistent manner, we have set forth the anisotropic orbital theory for delving into the underlying secret of molecular chirality.

## 6.3. SUMMARY

Quaternity spacetime theory indicates that 2s2p orbitals have four various spacetime dimensions as well as orthogonal orientations and these electrons are circulating by the rule of dynamic calculus. After a brief review of quaternity theory, we proposed that a carbon atom has outer electrons of $2p_x^1 2p_y^1 2p_z^1 2s^1$. A $2p_x$ orbital is a linear segment; a $2p_y$ orbital is a semicircle; a $2p_z$ orbital is a hemispherical surface; and a 2s orbital occupies the entire sphere. The anisotropic orbitals of a carbon atom possess great geometrical flexibility to accommodate for the molecular structure of methane, ethane, ethyne, and benzene. They are also responsible for the reactivity of alkyl halides in $S_N2$ mechanism and the substituent effect of mono-substituted benzenes towards further electrophilic substitution.

The anisotropy of 2s2p orbitals is the origin of molecular chirality. When a carbon atom is bonded to four different substituents, it becomes a chiral center because the four bonded substituents would differentiate four outer electrons of the carbon atom, thereby preserving its orbital chirality. The handedness of a chiral carbon could be predicted to a certain extent according to the relative electrophilic powers of the four substituents, i.e., the substituents from the strongest electrophilic power to the weakest bind to outer electrons of carbon from the highest electronic density to the lowest in the order of $2p_x^1 2p_y^1 2p_z^1 2s^1$. The optical activity of a chiral carbon actually results from light refraction in the uneven spacetime medium. Based on the handedness of a chiral center, one could tell whether a chiral compound is levorotary or dextrorotatory because plane light rotates in a reverse direction to electronic circulation of $2s \rightarrow 2p_x \rightarrow 2p_y \rightarrow 2p_z \rightarrow 2s$. In contrast, electric current entering a chiral center from one covalent bond will be directed to another covalent branch along the electronic circulation direction. A chiral center, like a special automatic switch valve, is capable of directing electron flow along macromolecules. We have characterized the important role of chiral carbons in selective electron transfer in a deoxyribonucleoside. This fresh discovery is the application of quaternity spacetime theory, alternative to quantum mechanics.

## REFERENCES

[1] Brewster, J. H. A useful model of optical activity I. Open chain compounds, *J. Am. Chem. Soc.*, 1959, 81, 5475-5483..

[2] Brewster, J. H. The optical activity of saturated cyclic compounds, *J. Am. Chem. Soc*, 1959, 81, 5483-5493.

[3] Zhu, H-J; Ren, J.; Pittman Jr., C. U. Matrix model to predict specific optical rotations of acyclic chiral molecules, *Tetrahedron*, 2007, 63, 2292-2314.

[4] Parr, G. P.; Szentplay, L. V; Liu, S. Electrophilicity index, *J. Am. Chem. Soc.* 1999, 121, 1922-1924.

[5] Bada, J.L.; Luyendyk, B.P.; Maynard, J.B. Marine sediments: Dating by racemization of amino acids, *Science*, 1970, 170, 730–732.

[6] Bailey, J. Chirality and the origin of life, *Acta Astronautica*, 2000, 46, 627-631.

[7] Garay, A. S. On the role of molecular chirality in biological electronic transport and luminescence, *Life Sciences*, 1971, 10 (24), 1393-1398.

*Chapter 7*

# DNA CIRCUITS AND WAVE FUNCTIONS

We shall shift our focus from simple organic molecules to complex ones. One of the most conspicuous macromolecules is DNA that is composed of two chains of nucleotides wound around each other. Following the discovery of selective electron transfer through chiral carbons in a nucleoside in section 6.2.5, this chapter further explores the electrical property and wave function of a DNA molecule. There have been many experimental reports on the electrical conductivity of DNA fibers, providing a mixture of results. Here we sift through the results and present a DNA molecule as stepwise *LC* oscillatory circuits. Based on the sinusoidal signals of *LC* oscillatory circuits, we formulate the base pairs by trigonometric functions and the phosphate bridges of a DNA molecule by spherical quantities in dynamic calculus. Oscillatory circuits precisely characterize a DNA molecule on a fundamental level and provide sharp insight into its biological significance.

## 7.1. DNA ELECTRICAL CONDUCTIVITY

Long after the chemical structure of DNA has been elucidated, the electrical property of the double helix remains a puzzle. Previous studies by direct measurements of electric current in DNA fibers have produced various conclusions from no conductivity [1,2], semi-conductivity [3], good conductivity [4], to superconductivity of a DNA molecule [5]. By appending a photo-oxidant to one end of a DNA duplex and measuring the oxidative damage of guanine doublet site at distances, recent biochemical studies of DNA charge transfer have provided more convincing results. It has been established that very fast charge transport can take place over a short distance (~37 Å) and a slower transport may propagate over a long distance (~200 Å) [6-14]. The results have led to two hypotheses of charge transport mechanism in DNA, one indicating super-exchange, or tunnel, through the sugar phosphate bridge between the bound charge donor and acceptor for the short distance, and the other suggesting charge hopping between discrete bases through the DNA $\pi$-stack over long distance. Although both mechanisms are widely regarded, they are incompatible and rather ill-defined, giving the expression of loose or uncertain charge transfer along a DNA molecule. As genetic substance, DNA must possess well-defined electrical behavior. Here we show a DNA double helix as a stepwise *LC* oscillatory circuitry, in line with the experimental evidence. We demonstrate that a DNA molecule maintains an electron transport mechanism through both strands that matches the fidelity and reliability of its chemical structure. For the first time, we discover the physical meaning of the chemical structure of DNA. The original publication of this section is available at www.springerlink.com (Xu, K. Stepwise oscillatory circuits of a DNA molecule, Journal of Biological Physics, 35, 223-230, 2009, DOI 10.1007 /s10867-009-9149-9).

## 7.1.1. DNA Electric Elements and Circuitry

The circuit model regards each base pair as an electric capacitor because the base pair is composed of two heterocyclic amines placed at a semi-conductive distance, capable of storing opposite charges. The hydrogen bonds between the base pair deliver electric pulse but prohibit direct current between them, which is a typical characteristic of electric capacitors. In the double helix, the phosphate bridges are twisted physically like both strands of a rope under torsion. The wound phosphate bridge can be treated as an electric inductor, capable of storing energy as well. The rigid structures of the Watson-Crick parities and of the pentose sugars help in forcing the torsion onto the phosphate bridges mechanically. Charge transport through the strands, like electric current through a coiled wire, produces electric inductance. Hence DNA structure is a circuitry composed of multiple oscillatory $LC$ circuits (Figure 7.1).

The deoxyribose is a key electric element in the circuit, which serves as a switch at the juncture (Figure 7.2). It has been stated that a chiral carbon center may transfer electrons towards a selective pathway due to its asymmetric polarizations [15]. Three chiral carbons in a deoxyribose direct electric charges in a selective course (see section 6.2.5). Specifically, as electrons enter C4' from C5', it will be routed through the ether bond by the chirality of C4' because an oxygen atom is more electronegative than carbon in drawing the electron; and thereafter the electrons must build up to a threshold potential to escape oxygen attraction. In this way, the ether bond delivers electric pulses through itself. When the electrons arrive at C1', they will be directed to the N-glycosyl linkage, which is more electronegative than carbon, hence charging the nitrogenous base; and after electrons overflow out of the base towards C1', they will be transferred via C2' to C3', which in turn directs the electrons towards the oxygen atom connecting the phosphate, and then through the phosphate bridge. Direct charge flow between C3' and C4' is choked by both chiral carbons.

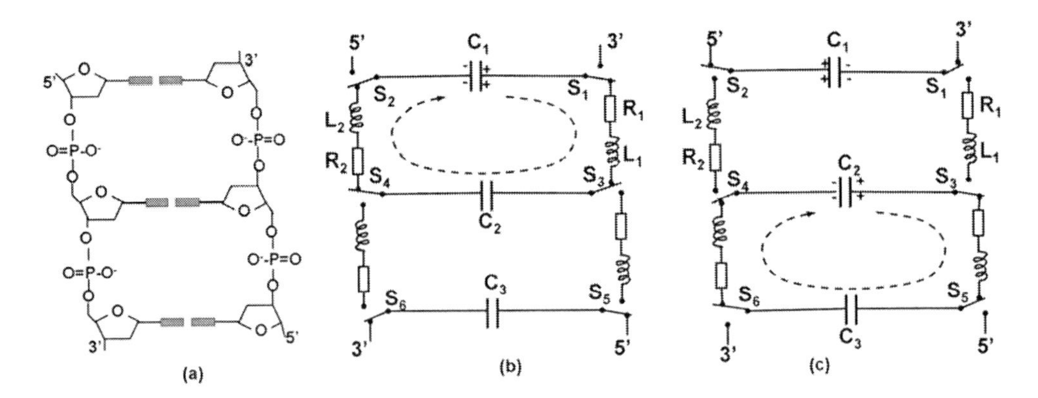


Figure 7.1. DNA double strands (a) along with stepwise oscillatory circuits (b) and (c), where the dashed arrows indicate electric current directions and the deoxyriboses are represented by electric switches.

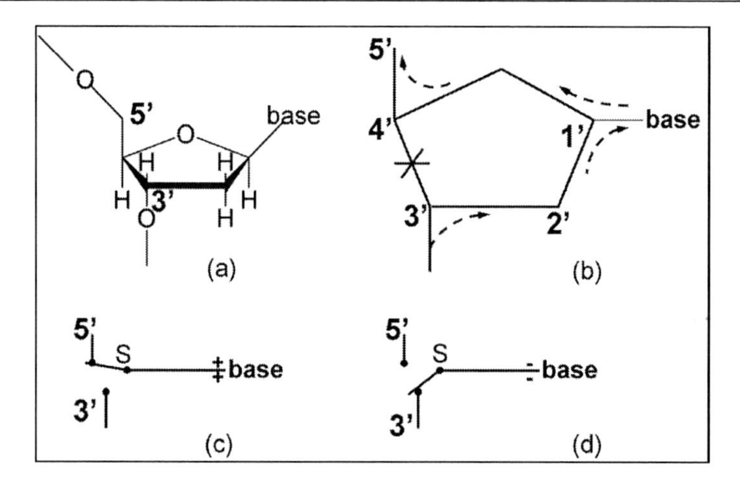


Figure 7.2. Chemical structure of a deoxyribose (a) and its electric switch property (b) where dashed arrows indicate electric current directions. Positive holes at the nucleotide base are routed to C5' (c) and electrons at the nucleotide base to C3' (d).

Positive holes transport in the reverse direction as indicated by the dashed arrows. Thus the role of a deoxyribose is an electric switch. Depending on whether the nucleotide base is positively charged or negatively charged, the switch connects C1' to C5' or otherwise connects C1' to C3' respectively. The anti-parallel alignment of the pentose sugars on both strands determines that electric current forms a stepwise closed loop between every two adjacent base pairs.

Suppose in the closed circuit (Figure 7.1b), charges stored in capacitor $C_1$ transport through the phosphate inductors $L_1$ and $L_2$ to reach capacitor $C_2$ under the routing of the deoxyriboses. The anti-parallel alignment of the pentose sugars on both strands determines that positive charges transfer along the strand in the direction of 3'→ 5' while negative charges transfer along the strand in the direction of 5'→ 3'. Electric current forms a closed circuit so that by Kirchhoff's loop rule, we have voltage drop relationship of

$$\frac{Q_1}{C_1} = IR_1 + L_1\frac{dI}{dt} + \frac{Q_2}{C_2} + IR_2 + L_2\frac{dI}{dt}, \tag{7.1}$$

where $I$ is electric current around the circuit unit, $Q_1$ and $Q_2$ are charges stored in base pair capacitances $C_1$ and $C_2$ respectively, and $R_1$ and $R_2$ are representative resistors along the path. Let

$$R_1 + R_2 = R, \tag{7.2}$$

$$L_1 + L_2 = L, \tag{7.3}$$

$$\frac{1}{C_1} + \frac{1}{C_2} = \frac{1}{C}, \tag{7.4}$$

$$Q_1 + Q_2 = Q_0 , \tag{7.5}$$

then upon differentiation on both sides, equation (7.1) becomes

$$L\frac{d^2I}{dt^2} + R\frac{dI}{dt} + \frac{I}{C} = 0 . \tag{7.6}$$

This second-order differential equation is a damped harmonic oscillator. Depending on the relative values of *L*, *R* and *C*, the system may be over-damped, critically damped, under-damped, or simple harmonic. When $R^2 < 4L / C$ , the circuit property falls into the two latter categories with a current function of

$$I = I_0 e^{-\frac{R}{2L}t} \cos(\omega t), \tag{7.7}$$

where $I_0$ is the initial current and $\omega$ is angular velocity of the oscillation with a value of

$$\omega = \sqrt{\frac{1}{LC} - \frac{R^2}{4L^2}} . \tag{7.8}$$

The circuit system transfers electric charges from a base pair to another in stepwise oscillatory processes, i.e., from capacitor $C_1$ to $C_2$ (Figure 7.1b) and then from capacitor $C_2$ to $C_3$ (Figure 7.1c) along the double helix. Simple as it is, this model is in line with experimental evidence [6-13] so far established concerning DNA conductivity and reconciles both super-exchange and multi-step hopping mechanisms.

### 7.1.2. Model Discussion and Prediction

It has been reported that the rate of electron transfer within a short distance decreases exponentially with increasing distance [7-10]. Such a phenomenon corresponds to the under-damped condition of the circuit system due to the relatively low ( $R^2 < 4L / C$ ) resistance response of DNA backbones to the artificial introduction of voltage drop between base pairs in the experiments. However, when *R* value is considerable, the exponential signal of electric current prevails over the sinusoidal cycle in equation (7.7) and vanishes within a few oscillatory cycles. Since the distance between the base pairs of B-DNA is a fixed value of 3.4Å and each stepwise oscillatory cycle takes a certain interval of time for that distance, we may replace the time variable in equation (7.7) with a distance variable $\Delta r$. In considering that electric current is a measure of electron transfer rate *k*, equation (7.7) is equivalent to the Marcus correlation [16, 17]

$$k \propto e^{-\beta \Delta r} \tag{7.9}$$

where $\beta$ values between 0.1 and 1.4 $Å^{-1}$ have been estimated for the double helix [7-10]. The dramatic difference can be ascribed to the uncertainty of resistance $R$ that is sensitive to various experimental conditions. The presence of considerable $R$ value along the circuit is because the phosphate bridges are under persistent high voltage drop induced artificially in experiments so that inductances partially manifest as resistances in the circuit.

Based on frequency value of $10^{-10}$ $s^{-1}$ measured by Fukui et al. [7], we simulate the under-damped oscillation with parameters of $C$=0.02pF, $L$=0.01μH, and $R$=100Ω in equation (7.8). Assuming three radical cations are initially generated to trigger charge migration from a base pair to another through the stepwise oscillatory cycles, the current function of equation (7.7) is calculated to be

$$I = 5.2e^{-0.005t}\cos(0.068t)\,, \tag{7.10}$$

where electric current is in the unit of nA and time in ps (Figure 7.3). Suppose at t = 0, capacitor $C_1$ carries charges in the polarity as shown in Figure 1(b), the deoxyribose switches ($S_1$ to $S_4$) will route the charges in the dashed arrow direction. After one oscillatory cycle at t = 92ps, the charges reach capacitor $C_2$ so that switches $S_3$ and $S_4$ change their connections. The charges at capacitor $C_2$ will then be transferred to capacitor $C_3$ in the next step (Figure 7.1c). The amplitude of the electric current along the double helix decreases exponentially with each oscillatory cycle in the under-damped situation (Figure 7.3). Because each oscillatory cycle takes 92ps for charges to migrate 3.4Å along DNA strands, the β value in equation (7.9) is 0.14$Å^{-1}$ in this case.

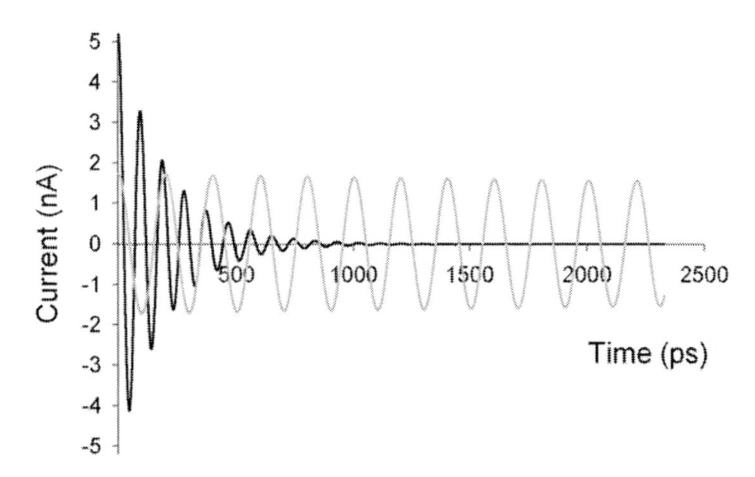


Figure 7.3. Charge transfer of DNA circuit by fast stepwise under-damped oscillations in vitro (black curve) in contrast to slower stepwise simple harmonic oscillations in vivo (gray curve) where each cycle spans 3.4Å in distance. Charges undergoing under-damped oscillations have a hasty speed but cannot reach a long distance.

In vivo, we believe that natural electron transport from a base pair to another should incur trivial electric resistance. Even if there exists certain electric resistance along the strands, the thermal energy produced by resistors would immediately be absorbed by both energy storage components of the base pair capacitor and the phosphate inductor. Hence electric resistance can be neglected. Let $C$=0.02pF, $L$=0.05μH, and $R$=5Ω, the circuit declines into a series of $LC$ oscillators that transfer charges step by step along the strands harmonically at a slower pace. The frequency of the harmonic oscillation is slower than that of under-damped oscillation as can be predicted from $\omega$ formula under the relatively high value of inductance and trivial resistance. It takes about 200ps for electrons to transfer from a base pair to the next. But the amplitude of the electric current remains almost the same in each oscillatory cycle. A comparison of under-damped oscillation and simple harmonic oscillation can be found in Figure 7.3.

In the stepwise $LC$ circuits, electric current is defined as positive when a base pair capacitor is being charged and negative when discharged in the next cycle down the chain. The stepwise oscillations agree with the evidence that has supported hopping mechanism through the base π-stack [6-14], but electric current through the sugar phosphate bridges is more reliable than the haphazard migration by hopping across the base rungs. Charges stored in capacitor $C_1$ transport through both strands to capacitor $C_2$, and will continue to move towards $C_3$ in the similar sinusoidal manner but at a lagging phase of $\pi$ in the cycle. During the processes, positive holes move in the direction of 3'→5' on one strand while electrons flow in the direction of 5'→3' on the other strand in good synchronization. The mechanisms for charge transport through long and short distances are the same. It takes more cycles of oscillations for charges to transport through longer sequences of base pairs. Without considering the effect of sequence dependence, traveling time is proportional to distance.

The stepwise oscillatory $LC$ circuits determine that a DNA fiber is a special AC wire. The oscillatory frequencies in the range $10^7$ to $10^{11}$ $s^{-1}$ have been reported [18-20]. Single strand turned out to be less conductive than double strands [2], which agrees with the circuitry model prediction because only double strands constitute closed circuits in each grid for proper conduction. It has been pointed out that the intensity of the torsion may influence the charge transport through the double helix [7, 21]. But we may attribute this effect to inductance change in the circuit rather than the variation of π-stack overlap for hopping.

We shall carry a further step to discuss the biological implication of the circuitry. From equation (7.8), we know that $\omega$ is codetermined by inductance $L$ and capacitance $C$ when $R$ is negligible. Assuming constant inductance for all nucleotides, the harmonic frequency would be determined by two neighboring capacitance values in the closed circuit, such as $C_1$ and $C_2$, which are specific to the base pairs. And this is perhaps the most subtle part of the story for it indicates that charge transfer rate is sequence dependent [8-10, 22]. Because nucleotide bases have ionization potentials in the order of G<A<C<T and electron affinities in the order of C<T<G<A (disregarding negative sign) [23], it takes the least amount of energy to charge G:C base pair in the polarity of +G:C- and requires the highest amount of energy to charge +T:A- capacitor. This means that the capacitances of the base pairs are in the order of +G:C- > +A:T- > +C:G- > +T:A- polarities. Thus there are four distinct capacitance values depending on the base pair and polarity. Since $\omega$ is determined by two neighboring capacitances in series, it may take eight possible values. It is predicted that charge transport along the strands will produce various frequencies reflecting the identity of the bases. In other words, the pattern of charge transport is precisely controlled by the gene sequence. The +G:C- capacitor has a higher capacitance than the other base pairs and carries higher amount of charges in experiments whereas +T:A- has a lower capacitance than the other base pairs and is the limiting step or bottleneck in charge transfer along the DNA strands [11, 17].

However, at long $(+T{:}A-)_n$ sequences the size of the bottleneck remains the same so that the sequence distance dependence vanishes [17]. Because +G:C- has a relatively high capacitance, it is easy to trap charges, so it is likely to become the end point of a charge transport [11, 17]. In this regard, the circuit model prediction agrees with experimental results completely.

Oscillatory current through the double helix is likely to have physiological significance. For example, if a base pair is mismatched at a certain position, then the oscillatory rhythms would be broken. Proofreading enzyme that scans the DNA sequence continually might easily locate the trouble point by the abnormal electric signal [24]. Furthermore, biomolecules are inherently unstable. Only constant flow of energy prevents them from being disorganized. It stands to reason that the incessant charge vibrations in the genetic substance are vital for living organisms to maintain the integrity of the gene sequence.

From organic chemistry perspective, the base pair is capable of storing considerable amount of charges in either polarity by at least two conceivable mechanisms. First, both pyrimidine and purine are heterocyclic rings composed of carbon and nitrogen atoms. On the one hand, nitrogen is more electronegative than carbon for attracting higher electron density in covalent bonds with carbon. On the other hand, nitrogen atom has lone pair electrons to share with carbon under electron deficiency. The combination imparts great flexibility to the nucleotide bases for either holding or releasing electrons. Second, the heterocyclic rings are aromatic that possess diamagnetic ring current. In the circuit, aromatic ring current may reduce the charge saturation of the nitrogenous base as if wind induces low pressure in the air, and as a result increase the electric capacitance of the base pair. Both properties enable a base pair to be a good bipolar capacitor. The recognition of a base pair as a valid capacitor provides a sharp insight into the electrical property of DNA.

In the stepwise oscillatory circuit, a capacitor must be charged up to a threshold potential for positive charges to overflow through the ether bond of the deoxyribose. And this is mediated by the ether bond shown in Figure 7.4 as an electric switch. The oxygen atom would preferably use a $2p_x$ and a $2p_y$ orbitals to form covalent bonds with carbon atoms in the ether bridge. A full electron octet is a stable configuration of the oxygen atom because eight outer electrons are in a complete circulating cycle. Electric current through the oxygen atom must comply with the circulation direction (Figure 7.4). Like a gas-filled tube, once a path is channeled by a positive threshold potential, the ether switch will remain on until the base pair is over-discharged to apply a negative potential of the same magnitude to the ether bond, thereby flipping the oxygen state from Figure 3(ON) to 3(OFF). To open the channel again requires the base pair to be recharged up to the initial threshold potential to flip the state over again. In the ON state, C1'—O covalent bond is $2p_y$-$2p_y$ overlap while C4'—O covalent bond is $2p_x$-$2p_x$ overlap; in the OFF state, the nature of the covalent bonds is reverse (Figure 7.4). The chirality of C3', C1' and C4' determines that electric current is unidirectional through the ether bond and the phosphodiester bonds while the base pair is being charged and discharged cyclically. Such a mechanism ensures that the oscillatory current is in strict stepwise process. It has been found that charge migration in DNA is an ion-gated transport depending on the hydrated counter-ions and configurations [25]. This phenomenon indirectly supports our model because any factors that influence the properties of the circuit would modify the efficiency of electron transport in DNA strands, which are especially true in vitro. But we believe that in vivo the circuit is gated by the dependable ether bond instead of by random ions coming from the environment.

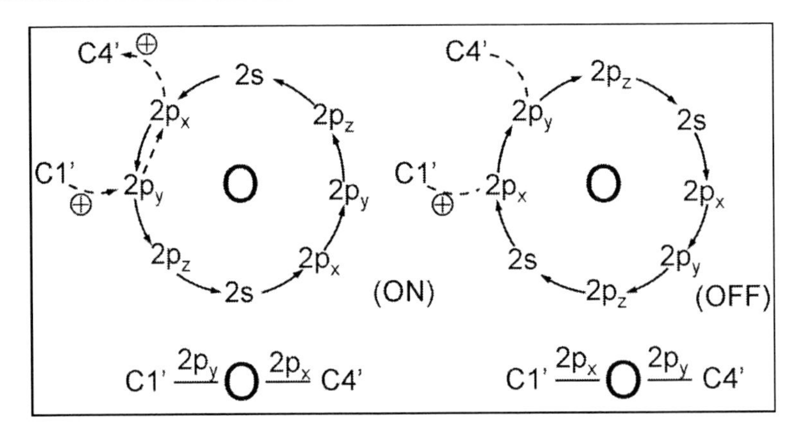


Figure 7.4. Atomic switch to control the flow of positive charges (dashed arrow) through the ether bond by flipping between two interconvertible states of contrary electron circulation (solid arrows) alignments.

Finally, we wish to present a general molecular model for phosphate in DNA backbone based on the electron circulations within oxygen atoms. The central phosphorus atom is bound by four oxygen atoms whose electron circulations are linked together forming four loops like a coiled wire, each loop involving an electron octet (Figure 7.5). Electric current passing through the molecule inevitably incurs considerable induction due to the convoluted electron flow, which explains why phosphate is an inductor and perhaps why ATP can serve as a physiological energy source when broken into ADP $+P_i$ even though the energy of the phosphoanhydride bond is not so large. The convolution collapse is the key factor in releasing a large amount of inductive energy. In DNA the electrical tension of phosphate chains must equilibrate with the mechanical torsion of the strands. It has been pointed out that the intensity of the torsion may influence the charge transport through the double helix [14, 21]. But we attribute this effect to inductance change in the circuit rather than the variation of $\pi$-stack overlap for hopping.

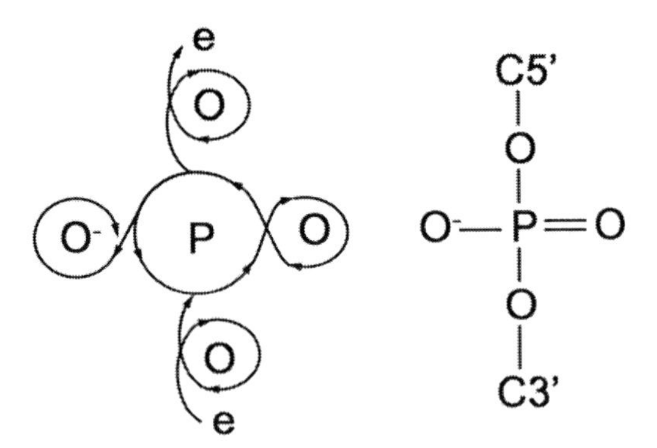


Figure 7.5. Schematic diagram of electron flow via the concatenated electron circulations within a phosphate bridge for electric induction corresponding to its chemical structure.

To summarize, electron transport mechanism in DNA is of paramount important to understand the biological function involving the genetic substance. Based on the accurate identification of chemical structures as electric elements, we describe electron transport mechanism in DNA by a reliable and well-structured oscillatory circuitry that distinguishes a DNA molecule from messy condensed matter. This alternative view is supported by the same established experimental evidence that has supported super-exchange and hopping mechanisms so far. The *LC* oscillatory circuit is robust and classical in physics, yet revolutionary and wonderful in chemistry and biology. This is an interdisciplinary analysis that produces a result of general interest in biological physics and must have potential influence in molecular electronics.

## 7.2. DNA Wave Functions

The stepwise oscillatory circuits of DNA indicate that the nucleotide sequence is linear in geometric alignment but spherical in electric function. From one nucleotide to another, the electrical signals follow harmonic oscillation principle, which transforms the wave function of each nucleotide by dynamic calculus instead of linear algebra. Upon close examination, we define the mathematical form of a DNA molecule in this section.

### 7.2.1. Harmonic Oscillations between the Base Pairs

The double helix DNA consists of two complementary polynucleotide strands wound around each other and crossed by hydrogen bonds. The nucleotides are joined into polynucleotide strands by phosphodiester bond that connects the 3' carbon atom of one deoxyribose sugar to the 5' carbon of the next. Though the polynucleotide seems to be linearly arranged and look monotonous in structure, DNA has a spherical elegance and beauty unsurpassed by any other macromolecules.

There are only four kinds of nitrogenous bases in DNA forming two pairs, i.e., guanine (G) matches with cytosine (C), and adenine (A) matches with thymine (T) through hydrogen bonds as were discovered by Watson and Crick. In each pair, a purine base with two rings matches with a pyrimidine base with one ring. If we assign two waveforms to the former, and one waveform to the latter mathematically, then these three waveforms must be coupled together harmonically in certain way because their mating is so unique and unassailable. We interpret the relationship of these three waveforms as a rotation pathway because it allows alternate current to pass through hydrogen bonds. Let spherical quantities $\Gamma_{00}$ and $\Gamma_{01}$ represent two aromatic rings in adenine, and $\Gamma_{10}$ represent the ring of thymine tentatively:

$$\begin{pmatrix} \Gamma_{00} \\ \Gamma_{01} \\ \Gamma_{10} \end{pmatrix} = D \begin{pmatrix} \cos\Theta\cos\theta \\ -\dot{\Theta}\sin\Theta\cos\theta \\ -w\sin\Theta\sin\theta \end{pmatrix}, \tag{7.11}$$

where $D$ is a dimension factor, $w$ is a velocity dimension, and functions $\Theta$ and $\theta$ are two radian angles as were analogously described in Chapter 5. There are differential and integral relationships between the three waveforms:

$$-\frac{\partial \Gamma_{00}}{\partial t} = \Gamma_{01}, \tag{7.12}$$

$$\int \Gamma_{01} dl = \Gamma_{10}. \tag{7.13}$$

when describing electronic orbitals, these two equations imply rotation transformation of one electron to another. Here these similar formulae are used to explain the interaction between a Watson-Crick base pair. Waveforms $\Gamma_{00}$, $\Gamma_{01}$, and $\Gamma_{10}$ constitute a rotation pathway that satisfies Faraday's equation, meaning that there is a harmonic resonance between the base pair. This wave function interpretation complies with DNA oscillatory circuitry.

In dimension diagram, a spherical quantity undergoes a rotatory operation (Figure 7.6). As radian angle $\Theta$ rotates from 0 to $\pi/2$, wave function $D_1 \cos\Theta$ transforms into $-D_1\dot{\Theta}\sin\Theta$; and as $\theta$ rotates from $\pi/2$ to 0, wave function $D_2 \cos\theta$ transforms into $-D_2 \frac{1}{\theta'}\sin\theta$. We may express this rotation by three of the four chords concisely as: $\cos\Theta - \dot{\Theta}\sin\Theta .... -\frac{1}{\theta'}\sin\theta$ with the dotted line indicating hydrogen bonds. If we designate this rotation in DNA base pair as A....T, then T....A base pair would be $-\dot{\Theta}\sin\Theta .... -\frac{1}{\theta'}\sin\theta + \cos\theta$.

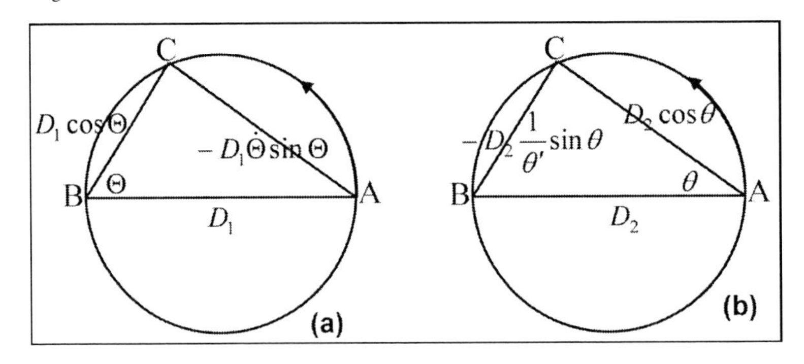


Figure 7.6. Dimension diagram of a rotatory operation to represent the electric oscillation between A....T base pair. $D_1$ and $D_2$ are dimension factors.

By the same manner, as radian angle $\Theta$ rotates from $\pi/2$ to 0, wave function $D_3 \sin\Theta$ transforms into $D_3\dot{\Theta}\cos\Theta$; and as radian angle $\theta$ rotates from 0 to $\pi/2$, wave function $D_4 \sin\theta$ transforms into $D_4 \frac{1}{\theta'}\cos\theta$ (Figure 7.7). We may express this rotation by three of the four chords simply as: $\sin\Theta + \dot{\Theta}\cos\Theta .... \frac{1}{\theta'}\cos\theta$. If we designate this rotation in DNA base pair as G....C, then C....G base pair would be $\dot{\Theta}\cos\Theta....\frac{1}{\theta'}\cos\theta+\sin\theta$ . In this manner, we associate the waveforms with the rings of the nitrogenous bases (Figure 7.8).

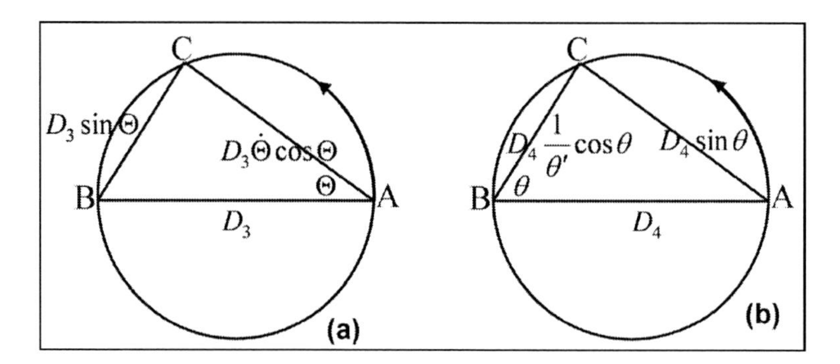


Figure 7.7. Dimension diagram of a rotatory operation to represent the electric oscillation between G....C base pair. $D_3$ and $D_4$ are dimension factors.

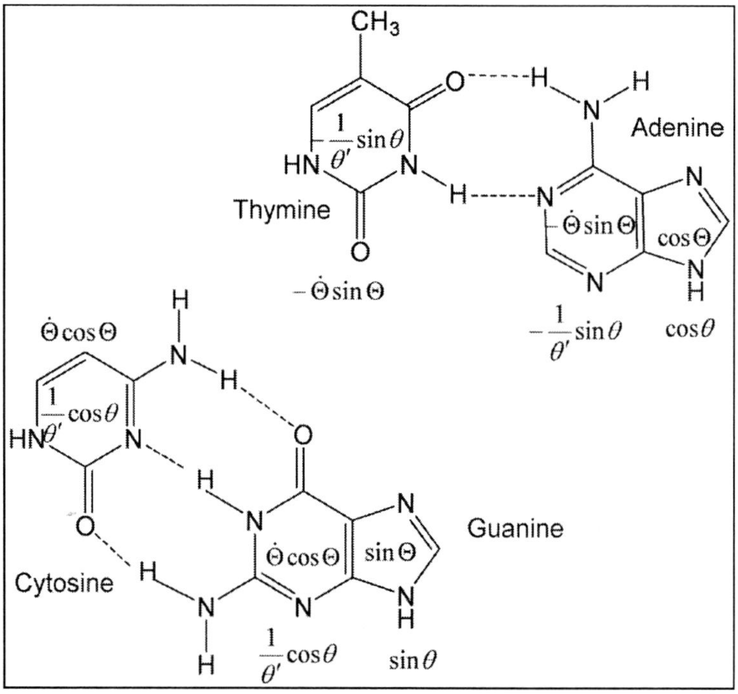

Figure 7.8. Watson-Crick base pairs A....T and G....C with waveforms in each rings for proper rotatory operations. The waveforms labeled outside the rings are for T....A and C....G pairs.

The communication of wave signals between each base pair is through hydrogen bonds, which are implemented by differential chain rule:

$$
\begin{aligned}
w &= -\frac{1}{\theta'} \cdot \dot{\Theta} = \frac{dl}{d\theta} \cdot \left(-\frac{d\Theta}{dt}\right) \\
&= \int A_2 \cos\psi \; dl \cdot \left(-\frac{dA_1 \cos\Psi}{dt}\right) \\
&= \Phi^0 \\
&= u\Phi \\
&= \Omega^0\Phi,
\end{aligned}
\tag{7.14}
$$

where $\Phi$ denotes an 2s2p electron wthin the nitrogen or oxygen atom involved in each hydrogen bond, and $\Omega$ denotes an 1s electron passing through each adjoining proton of the hydrogen bond. Thus the wave signal between the base pair is delivered by differential and integral chain rules. The wave signal becomes alternate current passing through the hydrogen bonds. A base pair is an electric capacitor. There are oscillating waves between each pair of nitrogenous bases within DNA nucleotides. Hydrogen bonding is a dynamic electronic action that delivers messages between the two bases, helping to balance the strain in a DNA double helix. Hydrogen bonds serve as the gateways to deliver messages. In this way, we describe the hydrogen bonds as a portal between the pair of bases. This dynamic interpretation is contrary to traditional static model of hydrogen bonds. Since the expression of any forces requires message exchanges, it is the wave information conveyed between pairs of bases that makes hydrogen bonds strong enough to hold the double helix of DNA together. Without the wave message, the atoms on both sides of hydrogen bonds would be detached and the organization of DNA would start to unravel. It is interesting that the formulae of electronic orbitals can be customized to describe molecular wave functions. Such mathematical uniformity for diverse phenomena reflects the universality of spherical view that has rich connotative meaning as well as general validity.

### 7.2.2. Mathematical Form of DNA

We have described the base pairs as electric capacitors and assigned waveforms to the rings in the bases. We have also described the phosphate bridges as electric inductors. In order to express the entire DNA molecule mathematically, we need to express the phosphate bridges in dynamic calculus as well. As shown in Figure 7.9, we treat both strands of the double helix as time and space components of a DNA molecule. Variables of Z series represent time while variables of $\zeta$ series represent space. Both are intertwined together to form double helix DNA. Here we emphasize the electric elements of base pairs and phosphate bridges disregarding deoxyribose because the sugar is an electric router only.

DNA molecule is a stepwise oscillatory circuitry. After alternate current passes through a base pair via hydrogen bonds, the electric wave reflects back into the phosphate strands to initiate the next cycle in the stepwise oscillatory circuits. The phosphate bridges in both strands undergo a rotatory operation. The time strand transforms by dynamic differentiation; and the space strand by dynamic integration. The rotatory operation produces electric tension on both strands. From a base pair to another, the electric signal proceeds along the closed circuit sinusoidally. The mathematical form of DNA complies with DNA circuitry.

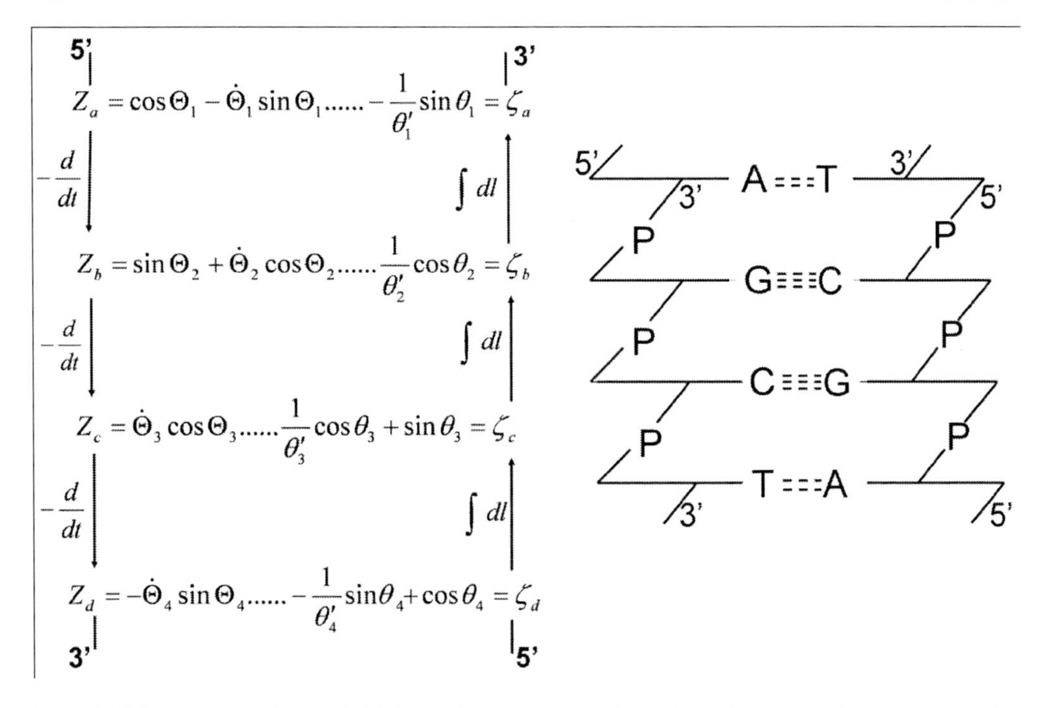


Figure 7.9. Mathematical form of a DNA molecule along with its abbreviated chemical structure. The arrows indicate electron flow in rotatory operations, contrary to electric current direction.

## 7.3. SUMMARY

A DNA molecule is insightfully characterized by a stepwise oscillatory circuitry where every base pair is a capacitor, every phosphate bridge is an inductor, and every deoxyribose is a charge router. The circuitry accounts for DNA conductivity through both short and long distances in good agreement with experimental evidence that has led to the so-called super-exchange and multiple-step hopping mechanisms to date. However, in contrast to the haphazard hopping and super-exchanging events, the circuitry is a well-defined charge transport mechanism reflecting the great reliability of the genetic substance in delivering electrons. Stepwise oscillatory charge transport through the nucleotide sequence that regulates its frequency along must contain great biological significance.

Stepwise oscillatory circuits of a DNA molecule indicate that the nucleotide sequence is functioning spherically. This permits us to define a DNA molecule by wave functions as well as spherical quantities in dynamic calculus. Three trigonometric waveforms are assigned to each base pair. They constitute a rotatory operation pathway with hydrogen bonds serving as the gateways to deliver electric pulses. Spherical quantities in dynamic differentiation and integration are assigned to phosphate bridges in both time and space strands respectively. Time and space strands are bonded together by hydrogen bonds. The time and space distinction of DNA double strands might contain biological significance that has not yet been discovered.

# REFERENCES


[1] Snart, R. S. Photoelectric effects of DNA. *Biopolymers*, 1968, 6, 293-297.

[2] Klein, H.; Wilke, R.; Pelargus, C.; Rott, K.; Pühler, A.; Reiss, G.; Ros, R.; Anselmetti, D. Absence of intrinsic electric conductivity in single dsDNA molecules. *Journal of Biotechnology*, 2004, 112, 91-95.

[3] Porath, D.; Bezryadin, A.; de Vries, S.; Dekker, C. Direction measurement of electrical transport through DNA molecules. *Nature,* 2000, 403, 635-638.

[4] Okahata, Y.; Kobayashi, T.; Tanaka, K.; Shimomura, M. Anisotropic electric conductivity in an aligned DNA cast film. *J. Am. Chem. Soc*., 1998, 120, 6165-6166.

[5] Kasumov, A. Y.; Kociak, M.; Gueron, S.; Reulet, B.; Volkov, V. T.; Klinov, D. V.; Bouchiat, H. Proximity-induced superconductivity in DNA. *Science,* 2001, 291, 280-282.

[6] Hall, D. B.; Holmlin, R. E.; Barton, J. K. Oxidative DNA damage through long-range electron transfer. *Nature*, 1996, 382, 731-735.

[7] Fukui, K.; Tanaka, K.; Fujitsuka, M.; Watanabe, A.; Ito, O. Distance dependence of electron transfer in acridine-intercalated DNA. *Journal of photochemistry and photobiology B: Biology*. 1999, 50, 18-27.

[8] Wan, C.; Fiebig, T.; Schiemann, O.; Barton, J. K.; Zewail, A. H. Femtosecond direct observation of charge transfer between bases in DNA. *Proc Natl. Acad. Sci. USA* 97, 14052-14055 (2000).

[9] Giese, B. Long-distance charge transport in DNA: The hopping mechanism. *Acc. Chem. Res.,* 2000, 33, 631-636 .

[10] Giese, B.; Amaudrut, J.; Köhler, A-K.; Spormann, M.; Wessely, S. Direct observation of hole transfer through DNA by hopping between adenine bases and by tunneling. *Nature*, 2001, 412, 318-320.

[11] Nunez, M. E.; Hall, D. B.; Barton, J. K. Long-range oxidative damage to DNA: effects of distance and sequence. *Chem. Biol.* 1999, 6, 85-97.

[12] Bixon, M.; Giese, B.; Wessely, S.; Langenbacher, T.; Michel-Beyerle, M. E.; Jortner, J. Long-range charge hopping in DNA. *Proc. Natl. Acad. Sci. USA*, 1999, 96, 11713-11716.

[13] Bixon, M.; Jortner, J. Long-range and very long-range charge transport in DNA. *Chem. Phys.* 2002, 281, 393-408.

[14] Schuster, G. B. Long-range charge transfer in DNA: transient structural distortions control the distance dependence. *Acc. Chem. Res.* 2000, 33, 253-260.

[15] Garay, A. S. On the role of molecular chirality in biological electronic transport and luminescence. *Life Sciences*, 1971, 10, 1393-1398.

[16] Marcus, R. A. Electron transfer reactions in chemistry theory and experiment. *Journal of Electroanalytical Chemistry* 1997, 438, 251-259.

[17] Giese, B. Electron transfer in DNA. *Current Opinion in Chemical Biology*. 2002, 6(5), 612-618.

[18] Murphy, C. J.; Arkin, M. R.; Jenkins, Y.; Ghatlia, N. D.; Bossman, S.; Turro, N. J.; Barton, J. K. Long-range photoinduced electron-transfer through a DNA helix. *Science,* 1993, 262, 1025–1029.

[19] Lewis, F. D.; Liu, X.; Liu, J.; Miller, S. E.; Hayes, R. T.; Wasielewski, M. R. Direct measurement of hole transport dynamics in DNA. *Nature*, 2000, 406, 51-53.

[20] Ladik, J.; Ye, Y-J.; Shen, L. The a.c. conductivity of aperiodic DNA revisited. *Solid state communications,* 2004, 131, 207-210.

[21] Zhang, W.; Ulloa, S. E. Structural and dynamical disorder and charge transport in DNA. *Microelectronics Journal* 2004, 35, 23-26.

[22] Nogues, C.; Cohen, S. R.; Daube, S.; Apter, N.; Naaman, R. Sequence dependence of charge transport properties of DNA. *Journal of Physical Chemistry* B, 2006, 110, 8910-8913.

[23] Wetmore, S. D.; Boyd, R. J.; Eriksson, L. A. Electron affinities and ionization potentials of nucleotide bases. *Chemical Physics Letters*, 2000, 322, 129-135.

[24] Rajski, S. R.; Jackson, B. A.; Barton, J. K. DNA repair: models for damage and mismatch recognition. *Mutation Research*, 2000, 447, 49-72.

[25] Barnett, R. N; Cleveland C. L.; Joy, A.; Landman, U.; Schuster G. B. Charge migration in DNA: Ion-gated transport. *Science* 2001, 294, 567-574.

*Chapter 8*

# THE PRINCIPLE OF CELLS AND LIFE

We have described the organization of various electronic orbitals within atoms in Part I. The order of electrons within the atoms provides the pattern for the order in which a biological cell should arrange its organelles. This is viable because the law of nature is single. In other words, the law that governs the cells is the same as the law that governs the atoms. It is of utmost important to examine the structure of the cells carefully if we are to characterize the various aspects of life phenomena incisively. In this chapter, we shall demonstrate that mathematical models for atomic shells may be applied to the characterization of biological cells. Helium shell is an electronic rope that may serve as a pattern for other duality systems in nature. Configuration of 1s2s2p electrons of a neon atom may serve as a pattern for various biochemical cycles within a cell; and the complementary relationship between $3d^{1s}$ and 1s electrons due to orientation shift within a zinc atom provides a clue for the relationship of endosymbiotic organelles within a cell. DNA double helix is a stepwise oscillatory circuitry, so is the human lineage. More fundamentally, if dynamic calculus inherent in electronic motion is followed by life phenomena, then the density distribution of biological variables may be characterized by trigonometric functions. The mind and body problem can be resolved mathematically by synchronized differential and integral operations. A human body can be regarded as a multiple ring gyroscope created and governed by the law of nature. And the evolutionary steps leading from low to advanced life forms should observe the same principle of spherical quantity in dynamic calculus.

## 8.1. HELIUM MODEL FOR ROPE STRUCTURE

Helium shell contains both 1s electrons in mutual rotation relationship: $\Omega_0^0 = \Omega_2$ ; $\Omega_2^0 = \Omega_0$ . Such a relationship was characterized by rope structure in section 5.1.4. Helium shell provides a model for many other systems with dual elements (Table 1). For example, light is one of the simplest ropes in nature. Light consists of electric field and magnetic field that interact with each other forming electromagnetic wave. As was explained in Chapter 3, the physical interaction between electrons within an atom is through emitting and receiving virtual photons that observe dynamic calculus. Thus electronic motion is closely associated with light.

**Table 8.1. Rope structure in life phenomena**

| Scope and entity | Two strands | | Interaction or specialization |
|---|---|---|---|
| General sense | Space | Time | Two symmetric dimensions |
| Light | Magnetic field | Electric Field | Electromagnetic wave |
| Helium shell | $\Omega_2$ | $\Omega_0$ | Rotatory operation |
| DNA | A | T | Hydrogen bonds between nitrogenous bases |
| | G | C | |
| Zygote | Ovum | Sperm | Fertilization |
| Family | Wife | Husband | Marriage and sex |
| Life | Plant | Animal | Aerobic respiration versus Photosynthesis |
| The earth's outer stratum | Abiotic | Biotic | Order process versus disorder process |

Atoms are the building blocks of various organic molecules such as nucleotides that form DNA. If we regard each atom such as nitrogen and carbon as possessing a space dimension, then macromolecules have multiple dimensions of space, but they are less stable than neon atoms. Thus from atoms to nucleotides to DNA, space dimension increases while time dimension decreases in the biochemical constructions. A nucleotide is a larger spherical layer built upon atoms; and a DNA molecule is another outer spherical layer built with the basic unit of nucleotides. In a DNA molecule, both strands of nucleotides are twisted together and bound through hydrogen bonds between their nitrogenous bases: Adenine pairs with thymine and cytosine pairs with guanine. As was formulated in Chapter 7, there is a rotation relationship between each pair of bases.

From a DNA molecule to a cell and to an organism, space dimension increases while time dimension decreases. DNA is more stable than man in the sense that a man is short-lived while DNA, as inheritance substance, passes down from generation to generation, maintaining remarkable stability over long history. An organism develops from a single zygote that possesses all genetic information to produce the entire body. A zygote results from the fertilization of an ovum with a sperm. There is a symmetric relationship between an ovum and a sperm, which may be mathematically represented by $\Psi$ and $\psi$ respectively (see equations 1.2 and 1.5). A sperm denotes a time dimension while an ovum denotes a space dimension, both forming a rope. A zygote is a rope.

The union of a sperm and an ovum requires impregnation that involves human intercourse and other social behavior. Thus we ascend a layer from the human individual to the human society. Husband and wife play the roles of time and space in a family, respectively, like two electrons in helium shell. Families are the basic units of a human society.

A human society is only a part of an ecological community in an ecosystem. Evolution creates various lives to colonize various niches on the earth. These include animals and plants that interact with each other. On the surface of the earth, solar energy is captured in organic matter by plants and free oxygen is released as a result of photosynthesis. Animals derive their energy by digesting organic compounds using oxygen as the ultimate oxidant in respiration (equation 8.1). There are a diversity of animal species besides human and a variety of plants including crops. The interactions between these two kingdoms are critical to their survivals such as procuring foods, fertilization, and pollination. Thus, we may likewise regard animals and plants as forming two functional strands of a twisted rope.

$$CO_2 + H_2O = CH_2O + O_2 \quad (8.1)$$

Furthermore, the earth outer stratum, featuring biological diversity and abundance, is the most sophisticated layer on the planet. Here biosphere and abiotic sphere also constitute two strands of a rope. Biosphere refers to the full array of life, including animals, plants, fungi, prokaryotes, and protoctists whereas abiotic sphere includes litheosphere, hydrosphere, atmosphere, etc. While biotic forces transform elements into ordered patterns actively, abiotic processes go on sponteneously leading the environment towards disorder characterized by entropy increment. Entropy is an indicator of disorder, which always increases according to the thermodynamic law. However, disorder has various criteria or meanings from different perspectives. For example, on the surface of the earth, living organisms are elements of disorder because they are creating parts of heterogeneity in the soil. But from the perspective of organisms, the biological bodies are in good order because the molecules are lined up in accordance with genetic information. As James Lovelock contemplated whether there is life on Mars and how life can be recognized, he proposed the Gaia hypothesis [1], which is, in brief, that the earth is itself a form of life in the sense that it reduces or reverses the entropy in a planet's system. He claimed that the earth's biosphere, tightly coupled with atmosphere, ocean, and soil, constitutes a feedback or cybernetic system that seeks an optimal physical and chemical environment for life on this planet. Thus the definition of disorder is not an absolute concept. The meaning is relative depending on the perspective and scope of view. Most common people don't really appreciate the content of relativity, but they think that nothing on earth is absolute. This plain impression turns out to be sounder than what relativity actually means. The globe is a rope of alternate order and disorder elements.

Finally, the earth rotates about itself, interlaces with the moon, and revolves around the sun that travels in the Milky Way. How stars form galaxies and move in the universe is still a mystery, but they obviously emit light and rays that travel a long distance and convey messages that could eventually change electronic behavior within atoms. As was described in Chapter 4, when an object travels at the speed of light, one of its space dimensions will transform into a time dimension. Hence light represents space to time dimension transformation.

On the emitting side, light is the infinite enlarging space, and on the receiving side, light provides a time dimension driving the undulation of the receiving particles. Light is the smallest time dimension and largest space dimension in the sequentially linked entities of light, electrons, atoms, nucleotides, DNA, cells, humans, societies, animals, biosphere, the earth outer stratum, the earth, the solar system, galaxies, and the universe brimming with light. This is a large cycle where life plays a key role in the middle rings. The world is continuous in spacetime. Figure 8.1 draws the structure of a large ring that contains multiple rings.

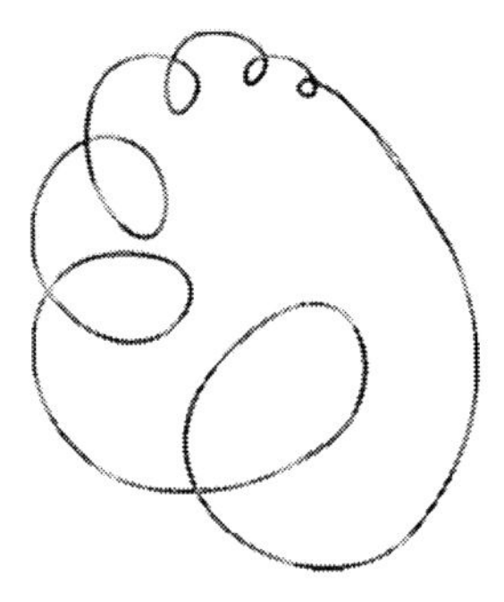

Figure 8.1. The continuity of the world. How many loops are shown in the sketch? Which loop represents properly the whole universe?

Table 8.1 summarizes a variety of ropes in terms of space and time strands. As we enlisted various structures as ropes, one may raise the question why the world pattern must be a rope? The secret lies in the unique property that a rope is able to endure more strength than its two separate strands combined, i.e., "one plus one is bigger than two". Like the win-win situations of a bilateral trade, the harmonic interplay of two strands of a rope is so wonderful. From the perspective of aesthetics, a rope is extremely beautiful besides feasibility. Rope is the theme of nature as well as culture. No wonder a pithy Chinese idiom runs: "Solidarity as a rope."

## 8.2. Neon Model for Biochemical Cycles

According to the expression of genetic information, biochemical activities within a cell include DNA replication, DNA transcription, RNA translation, enzymatic catalyses, and ATP cycle to provide energy for the entire cell. These processes correspond to electronic orbitals in a neon atom (Figure 8.2). DNA replication keeps the genetic substance as a double helix that resides at the nucleus of a cell, corresponding to the rotation of both 1s electrons in helium shell; the other four processes constitute four loops that manifest life phenomena surrounding the nucleus, corresponding to the rotations of $2p_x^2$, $2p_y^2$, $2p_z^2$, and $2s^2$ electrons in neon shell respectively. We shall discuss these similarities as follows.

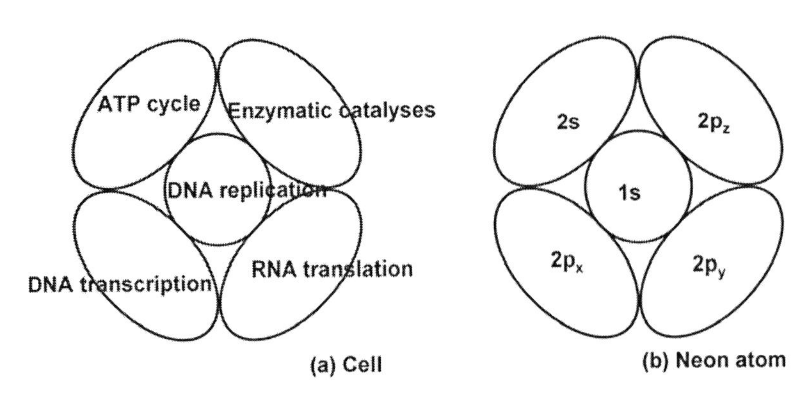


Figure 8.2. Biochemical cycles within a cell in comparison with electronic orbitals within a neon atom.

### 8.2.1. DNA Replication

All living organisms must possess rapid and accurate DNA synthesis and DNA repair mechanisms. DNA replication must occur before every cell division. The mechanism by which DNA copies are produced is similar in all living organisms. After the two strands of polynucleotide separate, each subsequently serves as a template for the synthesis of a complementary strand. A DNA double helix is composed of two complementary polynucleotide strands like helium shell composed of two complementary 1s electrons. Because a DNA molecule may be characterized by stepwise oscillatory circuits, the relationship between each base pair observes dynamic calculus.

DNA is at the core of life. The relationship between the DNA double helix and the whole cell is comparable with that between 1s electrons and 2s2p electrons in a neon atom. Like the inner electrons driving the rotations of the outer electrons, DNA is at the headquarter controlling the construction of the whole cell. DNA replication is not independent of other biochemical processes. On the one hand, DNA replication requires the input of four kinds of nucleoside triphosphate (ATP, TTP, GTP, and CTP) that come from metabolic activities. These required raw materials as well as ATP energy for the synthesis of new polynucleotide strands are represented in Figure 8.2 as ATP cycle. On the other hand, DNA replication produces a DNA double helix, which upon unwinding provides the necessary polynucleotide template for DNA transcription. DNA replication preserves the genetic information for the proper function of the entire cell. Thus, DNA replication is closely coupled with ATP cycle and gene expression processes.

### 8.2.2. DNA Transcription

The ability of living organisms to grow and function properly depends on the precise expression of the genetic information. This information is encoded in DNA, which consists of a linear sequence of deoxyribonucleotides. In the first step in gene expression, information present in one of the two strands of DNA (the template strand) is transferred into an RNA complement through the process of transcription. This RNA complement is called messenger RNA (mRNA) because it carries the message from DNA and associates with ribosomes for further translation later on.

Modern biochemistry has established that transcription from DNA template strand to mRNA is based on nitrogenous base complements, i.e., C to G, G to C, A to U, and T to A parities. DNA transcripts from the template strand to mRNA base by base. As was shown in Figure 7.9, although DNA looks like a linear stream of nucleotides, the nucleotides are functionally connected through dynamic calculus instead of simple addition mathematically. The coupling of DNA and mRNA during transcription is like the coupling of two cogwheels in motion (Figure 8.3). We have described such a relationship by differential chain rule in Figure 2.9. The unfolding of a deoxyribonucleotide in DNA involves a differential operation while the elongation of a ribonucleotide in mRNA follows the differential chain (see Figure 5.9). The chain rule of dynamic calculus is observed between both nucleotide sequences as long as the transcription proceeds. The intimate coupling by dynamic calculus ensures the fidelity of message delivery.

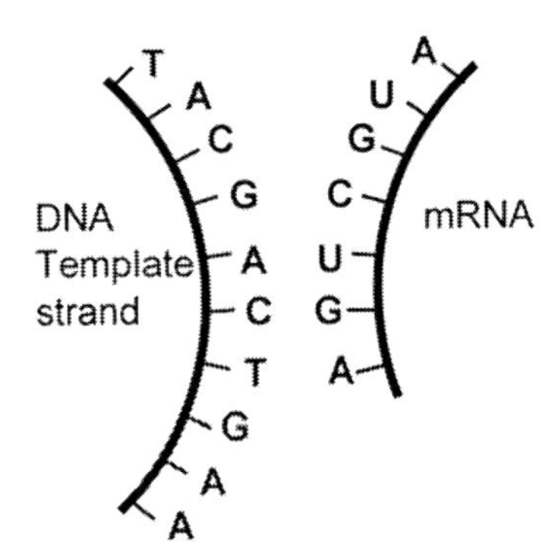


Figure 8.3. Transcription of DNA to mRNA based on the complementarity of nitrogenous bases. Note that both strands are coupled in a manner like two cogwheels in mesh.

It is also known that some RNA-containing viruses, called retroviruses, may replicate by reverse transcription. Their RNA serves as a template for the synthesis of the complementary DNA molecule. In Chapter 5, we demonstrated that the rotation of 1s electrons in helium shell drives the rotation of 2s2p electrons in neon shell, but the rotation of $3d^{2s2p}$ electrons drives the rotation of $3d^{1s}$ electrons, which are complementary to 1s electrons. It is interesting that DNA may transcript mRNA and some RNAs may reversely transcript DNA, the initiative depending on the circumstances.

## 8.2.3. RNA Translation

Once genetic information has been transferred from DNA to mRNA, it is ready to be translated into a polypeptide chain, the end product of most gene expression. Biochemical experiments indicate that a gene's nucleotide sequence specifies in a collinear manner the sequence of amino acids in a polypeptide chain. The ribonucleotide sequence in mRNA is divided into groups of three ribonucleotides, each group, called a codon, specifying one amino acid. The code is thus a triplet. During translation, transfer RNA (tRNA) serves as the adaptor molecule between an mRNA triplet and the appropriate amino acid. Under the direction of mRNA in association with ribosomes, the translation ultimately converts the sequence of codons in mRNA into corresponding sequence of amino acids making up the polypeptide.

Current knowledge of the characteristics of the genetic code is obtained through experiments. Even though we now know which codon corresponds to which amino acid, the mechanism of their correspondence is hidden in tRNA. The mechanism could be biochemical or electromagnetic, which remains largely unknown to date. However, this magic tRNA relates each codon to an amino acid, translating the mRNA sequence into amino acid sequence in great fidelity. Here we characterize the coupling of mRNA sequence and amino acid sequence during translation as two cogwheels in mesh. Mathematically, two cogwheels in mesh have been characterized as observing the chain rule of dynamic calculus. Codons are input by a differential operation while amino acids are output following the differential chain rule. The input of codons drives the production of a polypeptide chain. Like DNA transcription, RNA translation is a process featuring dynamic calculus.

### 8.2.4. Enzymatic Catalyses

The proteins produced by RNA translations may serves as enzymes to catalyze various biochemical reactions as well as structural materials of a cell. Life is inconceivable without enzymes. Enzymes are enormously powerful catalysts exhibiting high specificity. Each type of enzyme has a unique, intricately shaped binding surface called an active site. An enzyme-catalyzed biochemical reaction achieves significant reaction rate because the active site of the enzyme possesses a structure that is uniquely suited to promote the catalysis. Because of the specificity of an enzyme for a biochemical reaction, we may characterize the coupling of the enzyme and the biochemical reaction as two cogwheels in mesh. The catalysis of the enzyme drives the reaction of the biochemical process. Thus, we may treat the catalysis as a differential operation and the biochemical change as a differential chain. In this manner, enzymatic catalysis observes dynamic calculus as well.

### 8.2.5. Metabolism and ATP Cycle

Metabolism, the sum of all the enzyme-catalyzed reactions in a living organism, is a dynamic, coordinated activity. Many of these reactions are organized into pathways. There are two major types of biochemical pathways: anabolic and catabolic. In catabolic pathways, large complex molecules are degraded into smaller, simpler products. In the mean time, some of them release free energy by converting ADP into ATP and $NADP^+$ into NADPH. In anabolic pathways, large complex molecules such as polysaccharides and proteins are synthesized from smaller precursors such as sugars and amino acids. The energy and reducing power generated in catabolic pathways are consumed during the biosyntheses. For example, as mentioned above, DNA replication and transcription require the input of free energy. Like 2s orbitals permeating the entire atomic sphere, ADP-ATP cycle is omnipresent within the organism. Metabolism is closely coupled with ADP-ATP and $NADP^+$-NADPH cycles.

To summarize, we have identified five levels of interrelated processes in a biological cell that correspond to $1s2p_x2p_y2p_z2s$ electrons within a neon atom. DNA replication produces a double helix that resides in the nucleus; DNA transcription produces mRNA that carries the genetic information from the nucleus towards ribosomes; RNA translation produces proteins that scatter over the cell; enzymatic catalysis dominates the biochemical reactions; and ATP cycle maintains the balance and mobility of free energy currency within the whole cell. The scopes of these activities in a cell correspond to those of 1s, $2p_x$, $2p_y$, $2p_z$, and 2s electrons respectively within the sphere of a neon atom. They all proceed by strict dynamic calculus, the law of nature.

## 8.3. Zinc Model for Endosymbiotic Organelles

The degree of mathematical application is a reliable indicator of the maturity of a discipline. Although biology has been developed since antiquity, it remains a young discipline in terms of mathematical involvement. In spite of the widespread biological variable quantifications and analyses nowadays, neither biological statistics nor linear computational model has captured the fundamental aspect of the cells. After the discovery of electronic orbitals within a zinc atom, we have confidence to address the phenomenon of endosymbiosis in a fundamental way.

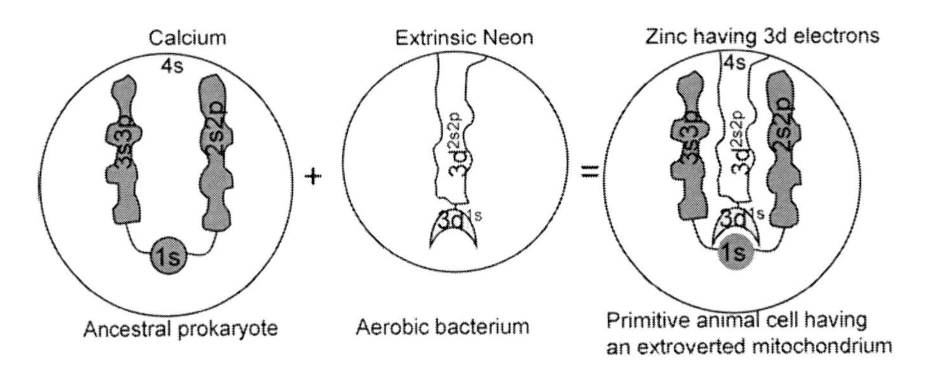


Figure 8.4. The endosymbiotic hypothesis of a mitochondrium evolved from an aerobic bacterium ingested by a larger anaerobic organism, analogous to the complementary relationship between extroverted $3d^{1s2s2p}$ electrons and 1s2s2p3s3p electrons in a zinc atom.

As was described in section 5.2.5, the relationship between $3d^{1s}$ and 1s electrons is complementary in geometry. We assume that the relationship between $3d^{2s2p}$ and 2s2p/3s3p electrons is complementary as well. As outer electrons of a calcium atom, 4s electrons constitute a membrane enclosing 2s2p/3s3p sphere. In a zinc atom, 3d electrons that attach to this membrane are extroverted as if they came from an extrinsic neon atom engulfed by a calcium atom. We shall not discuss whether it is possible for a calcium atom to fuse with a neon atom under any conditions. The point here is not about the genesis of atoms, but about complementary relationship between the extroverted 3d electrons and introverted 1s2s2p electrons (Figure 8.4). Thanks to the extroverted property of 3d electrons, multiple zinc atoms may form a piece of metal with high electrical conductivity.

If the law that governs the electronic orbitals is similar to that controls the functions of organelles within a cell, then we may attempt to establish a comparison between the structure of a zinc atom and a eukaryotic cell. According to the endosymbiotic hypothesis by Lynn Margulis [2], a mitochondrium evolves from an aerobic bacterium ingested by a larger anaerobic cell. In exchange for benefits such as protection and a constant nutrient supply, the smaller cell provided its host with energy generated by aerobic respiration. Thus the aerobic bacterium lived inside the larger cell in a harmonious way and gradually became an integral part of it. The striking evidence is that a mitochondrium performs complementary function to the functions of the other parts of the cell. Thanks to the extroverted property of mitochondria and other organelles, many eukaryotic cells may join together to form a multicellular organism. The point here is not about whether endosymbiotic hypothesis is correct or not, but about the similarity between a primitive animal cell having an extroverted mitochondrium and a zinc atom having $3d^{1s}$ electrons. The complementary relationship between $3d^{1s}$ and 1s electrons due to standpoint shift has been defined mathematically in equations (5.77) to (5.80); and the complementary relationship between an aerobic mitochondrium and the anaerobic nucleus due to organelle orientation shift follows the same logic. Both are similarly explained by spherical quantities in dynamic calculus. The natural law that governs the atomic structure is indeed the same as that that governs the cellular structure.

We may carry a further step to discuss the structure of a human body. Here the function of mitochondria corresponds to the function of the lungs that are responsible for aerobic respiration. There is a complementary relationship between the lungs and the heart in a human body. The heart drives blood circulation around the body while the lungs inhale oxygen into the blood and exhale carbon dioxides out of the body. The extroverted lungs and introverted heart are coupled together to realize gas exchanges between the entire body and the environment. We certainly don't suggest that the lungs should come from outside the body abruptly. However, if we regard a human body as a cell, then the origin of the human organs can be traced back to that of cellular organelles. Like a cell consisting of nucleus, cytoplasm, and various organelles surrounded by a semipermeable cell membrane, a human body consists of a heart and a brain, muscle and fluid, and various organs and structures enclosed and protected by the skin. The extroverted lungs endow humans with language capability, crucial to the social animals for making harmonious communities. The comparisons among the atomic structure, the cellular model, and the human body are informative because these three entities have certain similar structures. Demonstrating the similar structures affirms that the law of nature is single and universal.

## 8.4. DNA Circuitry for the Human Lineage

As was described in Chapter 7, a DNA molecule functions as stepwise *LC* oscillatory circuits. Charges transport from one base pair to another through the double helix constitutes harmonic oscillations. Like DNA double helix, the human lineage is a stepwise *LC* oscillatory circuitry where male and female form a capacitor and the individual development under living stress constitutes an inductor.

Each person is a multicellular diploid organism that begins life as a fertilized egg called zygote. A zygote consists of two poles originated from a sperm and an ovum. Like a charged electric capacitor driving electric current along a circuit, the dipolar zygote drives the growth of the individual. Genes contained in DNA within the zygote are the blueprint of our bodies, a blueprint that creates a variety of proteins essential to human functioning and survival. During the development process, the body is gaining mass while losing growth momentum. DNA is a spherical quantity with multiple time dimensions, and the entire body has multiple space dimensions in correspondence with DNA time dimensions. Body growth involves the expression of genes through transcription and translation, which observe dynamic calculus. Thus the growth of the zygote observes dynamic calculus as if charges would pass through an inductor. During the process, the energy flows from the capacitor to the inductor. Body growth is a process of time dimensions converting into space dimensions.

By the time an organism reaches the mature age, a male produces sperms while a female produces ova, both containing DNA substance that reflects the condition of the subjects. The seeds memorize enough details in the genes to replicate the next generation offspring. This process can be regarded as a space to time conversion. During the process, the energy transforms from the inductor to the capacitor. Thus the space and time oscillation carries on from a zygote to a body and from a body to a sperm or an ovum in a human lineage. A human lineage constitutes stepwise *LC* oscillatory circuits.

The division of humans into both sexes has profound biological significance. Male and female can be expressed mathematically by

$$M = \cos\Psi \cos\psi - \dot{\Psi}\sin\Psi\cos\psi \;, \qquad (8.2)$$

$$W = \cos\psi \cos\Psi - \psi'\sin\psi\cos\Psi \;, \qquad (8.3)$$

where $M$ denotes a boy rotating counterclockwise, $W$ denotes a girl rotating clockwise, and $\Psi$ and $\psi$ denote a sperm and an ovum respectively. Even though the male and the female are rotating in different clockwise directions, they are in a similar chirality (see Figure 5.5). Each human body is a chiral entity. Sex distinction allows males and females to rotate in different directions, maximizing human capabilities. He is thinking and driving while she is feeling and expecting. Male organ is basic and female organ is acidic. Both sexes form a capacitor so that electric potential may be built up between them. The division of both sexes normally prohibits electric current or chemical neutralization between them. Only during sexual intercourse do pulses of spermatic fluid pass through the reproductive organs. Thus we may compare this sex capacitor with the capacitor of DNA nitrogenous bases where hydrogen bonds deliver electric pulses between them.

A sperm and an ovum represent time and space dimensions respectively, the two poles of the sex capacitor. They are the haploid gametes resulted from the meiosis of diploid cells. The union of a sperm and an ovum produces a zygote that bears great momentum to grow due to the energy release between both sexes. This explains why organisms developed from the zygotes have stronger vitality than organisms developed by artificial cloning technology. After meiosis, a sperm and an ovum possess great mobility. Like the rotations of 1s electrons driving the rotations of 2s2p electrons (Figure 5.9), the mobility of a sperm and an ovum serves as an engine driving the growth of the zygote. An organism developed by artificial cloning technology, without going through the process of meiosis, lacks this innate driving power so that the rhythms of life are much weaker. As a result, the cloned organism has a relatively poor immune system. In a certain sense, every organism is a biochemical battery capable of doing certain kinds of work far more efficient than the finest artificial battery. Man is an energy-storing capacitor or inductor, like a battery, capable of working. The capacity of the energy is related to the capacitance of the initial capacitor derived from a zygote.

Just as two strands of polynucleotide form DNA, numerous couples of husbands and wives form the human lineages. Like hydrogen bonds between double helix DNA, the bond of marriage or love affair plays a key role in social structure (Figure 8.5). The structure of a human society is based on the unit of family. At least, the human lineage is like a multiple-strand rope or a web of couples. DNA double helix are a stepwise $LC$ oscillatory circuitry that transfers electric charges from one base pair to another; and the human lineages are a stepwise $LC$ oscillatory circuitry that passes down genetic information from one generation to the next. The gene-body oscillation is a wave, like an electromagnetic wave, traveling at spacetime where the rotation of genes indicates growth or body construction and the rotation of body produces sperms or ova that contain genes ( $Genes^0 = Body;\; Body^0 = Genes$ ). Thus the human lineage observes dynamic calculus in reproduction.

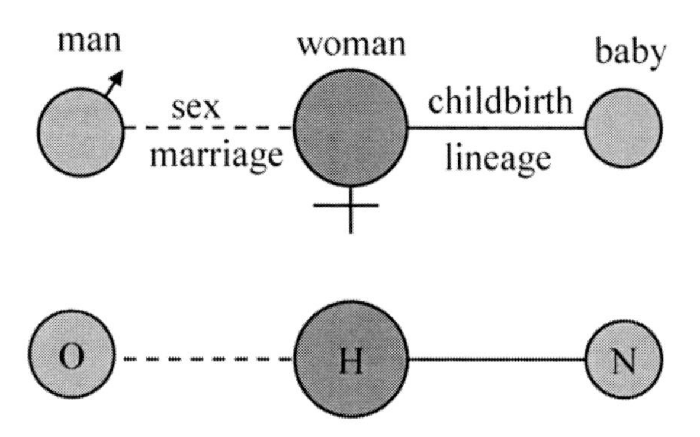


Figure 8.5. The relationship between man, woman, and baby through the reproductive canals (the dotted and solid lines) analogous to a hydrogen bond.

## 8.5. Wave Functions for Biological Variables

Biostatistics is a science of applying statistical methods to the solution of biological problems. As we defined the electronic orbitals within atoms, outer layer electrons are constructed upon inner layer electrons. The wave functions for electrons are invariantly expressed by sine and cosine functions that curl up multiple dimensions of a certain spherical layer. This provides a clue on how to model physical quantities of life. Should we use sine and cosine functions to account for biological data? Or should we always stick to data-oriented model such as the normal probability distribution?

To analyze biological data, it is customary to assume that the data obey the normal distribution. Nowadays, the normal distribution remains a cornerstone of biostatistics. But is this assumption based on sound biological knowledge? No, the normal distribution is a pure statistical theory based on central limit theorem. The conditions that tend to produce normal frequency distributions are all non-biological properties: a) that there be many influencing factors; b) that these factors be independent in occurrence; c) that the factors be independent in effect --- that is, that their effects be additive; and d) that they make equal contributions to the variance. There are surely many factors that affect the behavior of organisms, but it is hard to tell whether these factors meet the other three conditions. Many factors could be interrelated in occurrence and effect while some factors are doomed to be more important than the others.

Looking at the normal probability density function, there is a loophole in its definition of the variable scope:

$$f(x) = \frac{1}{\sigma\sqrt{2\pi}} e^{-\frac{(x-\mu)^2}{2\sigma^2}}, \ -\infty < x < \infty \tag{8.4}$$

Although $\mu \pm 3\sigma$ contains 99.73% of the items, the normal distribution predicts that it is possible for $x$ to have a negative number or a huge positive number. If human weight obeys the normal distribution, then can a man have negative mass? Or can he weight a hundred tons? These are impossible from the viewpoint of biology, but from the normal distribution these are possible even though the probabilities are very low. Thus the application of the normal probability to biological variables encounters insolvable difficulties. The variable must have an upper bound and a lower bound values.

In light of the fact that most biological entities exhibit certain sort of wave behavior, we hereby introduce trigonometric function as a probability density function:

$$f(x) = A\sin(ax + b), \tag{8.5}$$

where $A$, $a$, and $b$ are parametric constants and $x$ is the variable. The scope of $x$ may be defined within one or preferably half periodic cycle, but in rare circumstances it could cover two periodic cycles with bimodal peaks, which may be alternatively interpreted as belonging to two different characteristics. The suggestion of sine as a probability density function might not be new, but it deserves to be emphasized when most of contemporary biostatistics textbooks do not recognize the trigonometric distribution. Mathematically, the theorem for Fourier series indicates that every periodic function satisfying general conditions of continuous and piecewise very smooth can be represented as a trigonometric series. So trigonometric functions as probability density functions have widespread applications.

The trigonometric function agrees with the general style of wave functions. It has clear upper and lower limits for the variable. When one considers the ubiquitous phenomena of harmonic oscillations and waves, the use of trigonometric functions for describing biological data seem justified. However, it is not our intention to replace the normal probability distribution with trigonometric functions in all biological circumstances. We just provide a practical alternative method for analyzing data. The normal probability distribution still has a place in experimental biology when the conditions of collecting data fit the normal distribution conditions as mentioned above. Although mathematics tries to be exact and accurate, statistics is not an exact science. The choice of using which data analyzing techniques should be based on the specific case of concern, and as always readers are encouraged to analyze biological data using multiple approaches to derive useful information.

## 8.6. Rotatory Operation for Mind and Body Relationship

The mind and body problem is one of the oldest and most disputed topics in Western philosophy. The early Atomists, such as Democritus (about 460 to 360 BCE), suggested that the nature of things consists of an infinite number of particles called atoms and ascribed the human mind to the movement of those atoms. By contrast, Plato proposed that the world consists of Forms and objects. Forms are changeless, eternal, and nonmaterial essences or patterns; and objects are only concrete instances of Forms. Our souls (and thereby our rational minds) were distinct from our bodies and could not be reduced to material constituents. The dualism of mind and body was further developed by Descartes, who described the interaction of mind and body through the pathway of pineal gland within our brains. Dualism refers to the view that the world consists of or is explicable as two fundamental entities, such as mind and body. Even though dualism was widely accepted by later philosophers especially religious scholars, contemporary biologists tend to revert to the Atomist's view. According to materialism, our mental experience can be reduced to brain activities. But how are the activities of a physical brain related to the spiritual mind? Without clear definition of mind and body in mathematics, the question is open to dispute.

We have defined spherical quantity in dynamic calculus as the law of nature. Rotatory operations are the standard way of biological activities as well as electronic motion. The probability density functions of biological variables are considered to be trigonometric functions. A rotatory operation includes dynamic differentiation with respect to time and dynamic integration over space, both coupled together inherently. If we define spherical quantity in dynamic differentiation with respect to time as mind, the experience of time dimension change, then spherical quantity in dynamic integration over space is body, the construction or destruction of tissue due to space dimension change. To specify, given a spherical quantity with time component of $\Theta$ and space component of $\theta$, the mathematical expression $-\partial\Theta/\partial t$ refers to mind while $\int\theta\, dl$ indicates body, both processes synchronized automatically. In other words, mind and body (soul and heart) are represented by time and space quantities respectively, which are intertwined through rotatory operation. As point C traces along the semicircle in dimension diagram (Figure 2.11), the organism loses a time dimension $A_1$ and gains a space dimension $A_2$ simultaneously. We assume that each velocity dimension has specific physiological significance that was derived through evolution. The lapsing of a time dimension provides the organism with the experience of a spiritual meaning while the acquiring of a new space dimension has a corresponding physical consequence. Mind and body are coupled together as if time and space components were in a rotatory operation. Time reflects mind while space connotes body. Any kind of thinking, feeling, or sensing requires a period of time; and any kind of matter occupies a region of space. The dualism of mind and body can be precisely explained by the synchronized relationship of time and space components of a spherical quantity. Thus, we explain the relationship of mind and body by the mathematical mechanism of a rotatory operation.

## 8.7. The Poisson Distribution and Evolution

This section explores the biological significance of the Poisson distribution in statistics. Let $x$ be a constant dimension serving as the parametric mean of the Poisson distribution, we have:

$$1 = e^{-x} + xe^{-x} + \frac{x^2}{2!}e^{-x} + \frac{x^3}{3!}e^{-x} + \ldots + \frac{x^k}{k!}e^{-x} + \ldots \tag{8.6}$$

where each term on the right-hand side represents the expected frequencies corresponding to the $k$ rare event. Although the Poisson distribution is normally applied to discrete variables, we believe that its application may be extended to continuous spherical quantities because exponential function $e^x$ in the polynomial series has the property of a circular complex function (see equation 5.57). If we regard the expected frequency of $k$ rare event as the difference of counts ( $\Delta M$ ) between two time points delimiting a time interval ( $-\Delta t$ , the minus sign denotes the direction of time procession), then we may write the probability of $k$ rare event as $-\Delta M/\Delta t$ or in differential expression as:

$$-\frac{\partial M}{\partial t}=\frac{x^k}{k!}e^{-x}. \qquad (8.7)$$

Let $N$ denotes the probability of $k+1$ rare event:

$$N=\frac{x^{k+1}}{(k+1)!}e^{-x}, \qquad (8.8)$$

then performing a differentiation with respect to $x$ dimension on both sides of the equation gives a corollary to the Poisson distribution:

$$\frac{\partial N}{\partial x}=-N-\frac{\partial M}{\partial t}, \qquad (8.9)$$

or under dynamic differential rule that the differentiation is carried out with respect to the most immediate dimension in $x^{k+1}$ factor only, then instead of equation (8.9), we get

$$\frac{\partial N}{\partial x}=-\frac{\partial M}{\partial t}. \qquad (8.10)$$

This indicates that there is a rotation relationship between the probability of $k+1$ rare event and the accumulative number of counts of $k$ rare event. There are biological implications associated with this conclusion if we interpret $k$ or $k+1$ rare event as the emergence of organisms holding $k$ or $k+1$ dimensions.

A mutation shifting from $M$ to $N$ in equation (8.10) may be understood as in a smooth transformation manner where dynamic calculus is involved in the implementation of rotatory operation. Thus we may say that biological evolution is a process of gradual rotation of spherical quantities. The probability of evolving into a $k+1$ dimensional organism from a $k$ dimensional organism is related to the accumulative counts or population of $k$ dimensional organisms. If the intrinsic time signal favors the evolution of organisms towards more dimensions, then given enough time, it is certain that organisms will evolve from $k$ dimensions into $k+1$ dimensions. However, it is not certain whether an organism with higher numbers of dimensions is always more suitable for a given environment. This is where external space signal or natural selection comes into play. The space and time tides of ever-changing nature determine the flourish or recede of a certain species under sustained strain.

Aside from probability calculation, here we shall shed light on gene mutation and evolution briefly. Firstly, since DNA sequence contains genetic information, the biological evolution must ultimately reflect in the change of the sequence or in the elongation of the sequence, or affect the structure of DNA between or within chromosomes that may alter an organism's phenotype, giving it greater or lesser advantage in the process of natural selection. While the mechanism of genetic mutations remains largely unknown to biologists, the idea of treating genetic mutations as random and contingent events is never satisfactory. It is hard to imagine how chaotic changes in DNA nitrogenous bases without any orders and directions could lead to an upgrade in the genes for better fitness.

Secondly, the environment plays an important role in gene mutation. Organisms are not living within a vacuum. Man is embedded in nature. He needs to interact with the elements to obtain food, clothe, and shelter. Most importantly, he must gain a sense of space and time from the outside world, and pass this reference to the inside body so that genes may transcript accordingly as if a C++ program should pass parameters to initialize a class. The eyes, "the window to soul", serve as the major portal or gateway to human mind and emotion (time) from the environment (space), which marshal the appearance of the external world into the internal brain. Time is the humans' worldview or innate sense of the environment. Light, which carries images and messages, implements this space to time transformation. On the other hand, human's hands mold the environment according to their mind and feeling. Thus the message passings are bidirectional. A man works at daytime to gain the sense of his living environment while sleeps at nighttime to receive the feedback from the internal biological clock. This rhythm is going on day and night as a diurnal spacetime oscillation. Whether a man can live in harmony with nature determines his health condition to a great extent. This is a major foothold of Chinese medicine that has come a long history. Genes are the children of the struggle between space and time tides, the ongoning signatures of evolution. The environmental factors, boiled down to space and time effects, may change the genes. The genes so conditioned by the space and time climate determine an offspring that in turn may adapt better to the environment in the future.

Thirdly, as humans develop into societies, it is necessary for us to arrive at that degree of sophistication where natural environmental factors refer not only to the physical elements (such as earth, air, fire, and water) but also to the social structures. For examples, politics, economics, and customs have important influences on people's lives; languages, sciences, and technologies play key roles in shaping people's minds. This outer environment (space1) interacts with human mind and emotion (time1) arousing conscious experience that soon permeates the inner body fluid through mind and body oscillations. The body fluid (space2) then produces the seeds (sperms or ova) (time2) for the next generation, and the next generation is destined to change the outer environment as they grow up. Each generation has to meet the challenge of its days or recedes.

In short, evolution follows such a route that features interactions and oscillations. Evolution is directed by the law of nature towards the better harmony of the organisms and the environment, reflecting the space and time trends of the general oscillating cycle that encompasses them.

## 8.8. The Law of Nature and Gyroscope

The law of nature is defined as spherical quantity in dynamic calculus. A fundamental law must be simple. If it is complex, it is likely to be a composite of more fundamental ones. The law of nature is fundamental because it only sets the rule of operations for a spherical quantity. Throughout this book, the law of nature is repeatedly illustrated by dimension diagrams. The most direct derivation from the natural law is the principle of rotatory operation, which relates a spherical quantity in time to a spherical quantity in space. The principle of rotatory operation describes the motion of electrons in time and space simultaneously. Taking the example of equation 8.10, the spherical quantity that is differentiated with respect to a time dimension, M in the equation, can always be regarded as a time component whereas the spherical quantity that is differentiated with respect to a space dimension, N, is its space counterpart. Space and time components undergo harmonic oscillations in various dimensions (see Figure 2.12). The fundamental law of nature governing the oscillations is the single even though it was explained in various contexts and formats (see Appendix III). For example, as will be seen, the famous Pythagorean theorem is only a variant of the natural law (see equations 2.37, 9.2, and 9.3).

Harmonic oscillation is the theme of nature. Each spacetime oscillation is going on within a certain scope, and various oscillatory cycles may form a chain or more generally form a series of layered rings as in a gyroscope (Figure 8.6). A gyroscope originally refers to a device consisting of a spinning mass, typically a disk or wheel, mounted on a base so that its axis can turn freely in one or more directions and thereby maintain its orientation regardless of any movement of the base. Under spherical view, we replace each directions of movement with an oscillating entity or a ring. When one of the rings tilts in one direction, the other rings will adapt to it so that the overall orientation of the gyroscope maintains unchanged. The mechanism underlying the interactions between various rings could be Green's theorem, Stokes's theorem, and Gauss's theorem, etc. These theorems relate a spherical quantity in space to another spherical quantity in space. They are faithful interpretations of the law of nature (see section 3.4). While the principle of rotation operation defines harmonic oscillations, the general Stokes's theorem synchronizes them.

If we regard the human body as a multiple-ring gyroscope, each ring representing a physical, physiological, or psychological cycle in a certain manner, then the change in one ring will certainly affect the other rings, but the net orientation of the organism remains constant. Meanwhile, the change of a specific ring will incur counteractions from the other rings. Such a kind of capacity to adjust its various biological cycles in various layers to cope with a trouble ring in a body is known as the immune system, which enables the body to repair its injuries and fight against diseases to a great extent.

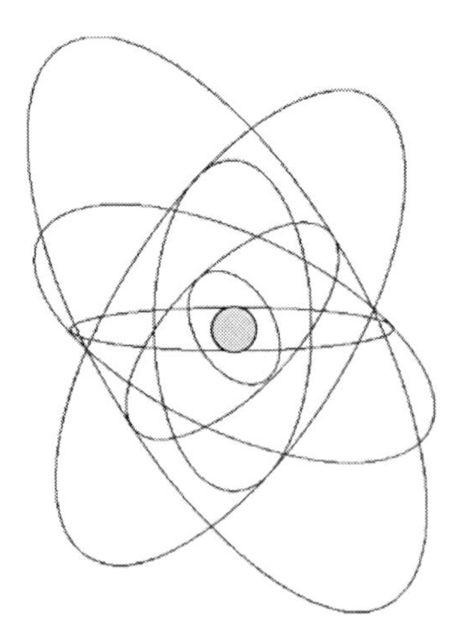

Figure 8.6. Schematic diagram of an imaginary multiple-ring gyroscope where each coupling ring represents a physical oscillation within a certain spacetime scope.

Furthermore, if we regard the whole universe as a super multiple-ring gyroscope where one ring of spherical layer is tightly coupled with another ring of spherical layer, then a change in the galaxies may affect the solar system. The effect may propagate down to the earth revolution cycle, the biospherical rhythm, and the cycle of a specific biological species in a chain rule. Time signal is passed down from light to electrons, atoms, nucleotides, DNA, cells, humans, societies, animals, biosphere, the earth outer stratum, the earth, the solar system, galaxies, and returns to light while space information is transmitted in the reverse order. This bi-directional exchange determines the direction of evolution.

However, the gyroscopic property must not be construed as fatalism. Although the orbit of biological evolution is mainly determined by prevailing spacetime tides, each individual organism has many degrees of freedom in his or her lifestyle as if in a gyroscope even when the net orientation is fixed, each individual ring in the gyroscope can still move actively, inducing the movements of others, or responsing passively to the actions of others as in a community of social animals. Moreover, as we mentioned before, everything is interactive, and none is absolute. The net orientation of a gyroscope is a relative concept, which is never fixed at a direction, so that human creatures can still change their destiny through their prudential collective efforts. We also insist that no any parts of the world are isolated from the others. The second law of thermodynamics that studies the isolation system must be taken with cautions when applied to the description of multiple-layer open system.

In short, space and time oscillations are harmonic and inter-coupled in various layers. Despite their diversity, the law of nature still governs their coupling interactions in certain physical manners. The universe is a natural gyroscope analogous to the atomic structure where each ring is inductively coupled with another by the process of dynamic calculus, so is a human body.

Naturally, one may raise the question why the human body or the whole universe must be a gyroscope, or why organisms must be evolved from entities of less dimensions to ones of more dimensions. This is not an easy question, but we shall explain it curtly from the origin and creation of the universe. Suppose before the advent of anything, there was nothing, which can be expressed by

$$0^{*}=0 \tag{8.11}$$

where the right-hand side means zero, and the left-hand side means nothing. The difference between nothing and zero is that nothing was soon taken over by space and time vibrations that change the equation into:

$$\frac{\partial W_1}{\partial t}+\frac{\partial W_2}{\partial l}=0, \tag{8.12}$$

where $W_1$ and $W_2$ are two orthogonal spherical quantities obeying the law of nature. Hence the universe became a system featuring two phases. These two phases were then replaced by more phases in a somewhat recursive way such as

$$\frac{\partial W_3}{\partial t}-\frac{\partial W_4}{\partial l}=W_1, \tag{8.13}$$

$$\frac{\partial W_5}{\partial l} - \frac{\partial W_6}{\partial t} = W_2, \tag{8.14}$$

where the right-hand sides of the equations can be regarded as quasi-zero or the equilibrium states between $W_3$ and $W_4$, and between $W_5$ and $W_6$ phases, as if we would treat mass as charge neutral. Hence the universe became enriched and more colorful with the emergence of more phases. Equations (8.13) and (8.14) could take the shape of equation (8.9) derived from the Poisson distribution. Bear in mind that equation (8.9) implies the rule of organism evolution from less dimensions to more dimensions. Combining equations (8.12), (8.13), and (8.14) yields

$$\frac{\partial^2 W_5}{\partial l^2} - \frac{\partial^2 W_6}{\partial l \partial t} + \frac{\partial^2 W_3}{\partial t^2} - \frac{\partial^2 W_4}{\partial t \partial l} = 0. \tag{8.15}$$

This equation in dynamic calculus may be interpreted as a transformation cycle. As shown in Figure 8.7, the four terms on the left-hand side of the equation constitute a two-dimensional oscillatory system, which may be instantiated by the relationship of both 1s electrons in counterclockwise and clockwise rotations. If the universe has been always created and evolved according to dynamic calculus, then no matter how many layers and spherical quantities it has now, it must be a natural gyroscope whose various rings are still interrelated by the law of nature. We have some reservations concerning the Big Bang theory even though our description of the origin of the universe is consistent with it in large.

By the way, it does not escape my observation that even social economy complies with the law of nature. If we tactfully define money as a time component and merchandise as its space counterpart, then the interactions between both components produce diverse economic phenomena. The faster money turns over in time, the longer merchandise travels in distance (see equation 8.10). With the introduction of stock markets and various firms and professions, economic activities become complicated with more dimensions. We briefly open this gate for economists to investigate the fine integration of social economics with natural science.

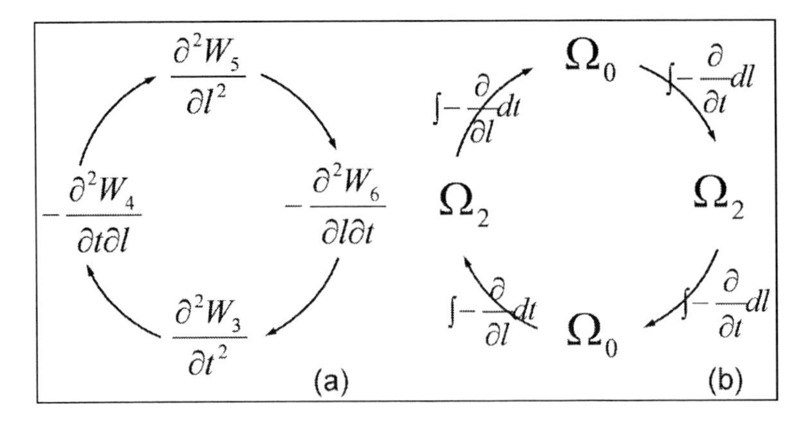


Figure 8.7. Transformation cycle of equation (8.15) in (a) dynamic calculus interpretation with (b) a concrete instance in wave functions.

We indicate the integration because we believe that social economy is a middle ring of the natural gyroscope, connecting the physical elements with the human behavior. After all, the human society is an inseverable part of nature.

## 8.9. Concluding Remarks

Life is a kaleidoscope, the most splendid part of nature. While many biological behaviors and metabolic activities have been investigated and discovered by biologists, the controlling mechanisms underneath the activities remain largely an untouched territory, awaiting the penetration and colonization by spherical view. This chapter has lightly tapped the great potentialities of spherical view in a passing review over biology, enough to whet the appetite of researchers for further exploring the structure and evolution of life. Indeed, until spherical quantity in dynamic calculus has been fully realized, the full array of life phenomena will remain a mystery.

## References

[1] Lovelock, J. E. *The Ages of Gaia: A Biography of our Living Earth*, Oxford University Press, Oxford, 1988, pp. 1-80.

[2] Sagan L. On the origin of mitosing cells, *J. Theor. Bio.* 1967, 14 (3), 255–274.

*Chapter 9*

# CIRCULAR MOTION AND CENTRAL FORCE

Uniform circular motion is the motion of an object in a circle at a steady speed. As an object moves in a circle, it is constantly changing its direction. At all instances, the velocity of the object is tangent to the circle while the acceleration of it is directed towards the center of the circle. In Chapter 1, we characterized harmonic oscillation by a rotatory vector with its tip on a circle. Electronic oscillation within a helium atom is simple harmonic and hence related to circular motion. A planet revolves around the sun in an elliptical orbit. An elliptical orbit is the projection of a circle onto a tilted plane. Thus both electronic motion and planetary movement can be unified in the framework of circular motion. Instead of an elliptical orbit of a planet around the sun, an electronic orbital takes the form of matter state transformation cycle. While the kinematic movement of a planet is governed by Kepler's first law, electronic transformation obeys Pythagorean theorem, both being equivalent in physical principle. This chapter tries to establish the correspondence between electronic orbitals and planetary orbits. From the primary wave function of a 1s electron, we shall also derive the mathematical expression of central force that guides the surrounding bodies along the orbits. The result is exciting and surprising that questions the exactness of the venerable Coulomb's law.

## 9.1. MANIFESTATIONS OF CIRCULAR MOTION

Electronic transformation and planetary revolution represent two extremes of circular motion. An electron revolves around the nucleus through changing matter state instead of changing position whereas a planet revolves around the sun through kinematic movement without changing state in itself. Because these two ideal extremes are orthogonal in their nature of circular motion, they can be distinguished by a hypothetical angle $\delta$ in their wave functions. When $\delta$ is equal to $\pi/2$, the orbital motion refers to the pathway of orthogonal state induction; and when $\delta$ is zero, the orbital motion declines into a kinematic circular track. We ordinarily call the former an electronic orbital and the latter a planetary orbit. As shown in Figure 9.1, both electronic orbital and planetary orbit are in two perpendicular planes with a right angle of $\delta$. Radian angle $\delta$ measures the degree of an oscillation deviating from kinematic movement towards matter state transformation.

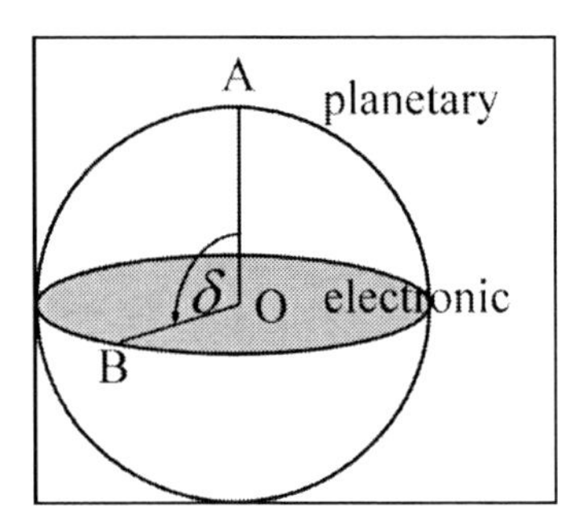


Figure 9.1. The orthogonal relationship of an ideal electronic orbital (shaded circle) and an ideal planetary orbit (blank circle) on two perpendicular planes.

Between electronic orbitals and planetary orbits, the motion of many entities follows circular motion on tilted flat planes, having $\delta$ values in the range of $(0, \pi/2)$. For example, during a year, a longan tree in front of my house experiences four seasons. The plant lives out a seasonal cycle similar to circular motion by changing matter state in response to the weather changes. During the same period, a flock of geese migrate from Canada to the United States in the fall and return to the north in the spring when the weather is getting warmer. The animals are able to lessen the effect of harsh temperature changes upon themselves through the strategy of migrations. While the plant has to withstand harsh weather changes by increasing its physical adaptability, the animals choose kinematic movement instead of physical adaptation. To a certain extent, the strategies of the plant and the animals in four seasons are analogous to the circular motion of electrons and planets, respectively. However, both plants and animals have physical adaptation and kinematic movements in the view scope from electrons to planets, their wave functions lie on the tilted planes of $0< \delta <\pi/2$.

For real entities whose circular orbit is within a tilted plane ($0< \delta <\pi/2$,), its orbit may manifest as an ellipse on the equatorial plane ( $\delta =0$) instead. Here the equatorial plane of the sun refers to the orbital plane of nine major planets including the earth. The equivalence of circular orbit on the tilted plane and elliptical orbit on the equatorial plane can be illustrated by the geometrical property of an ellipse as shown in Figure 9.2. In the first form (Figure 9.2a), a circular orbit with a radius r remains its circular shape (shaded) but its orbital plane rotates a certain angle $\delta$ away from the equatorial plane whereas in the second form (Figure 9.2b), an orbit keep in the equatorial plane but deforms into an ellipse whose focus shifts from central point O to $O_1$. When the vertical projection of the circular orbital onto the equatorial plane in the first form is in congruence with the ellipse in the second, we have a relationship concerning the eccentricity of the ellipse:

$$\cos\theta = \sin\delta\,, \tag{9.1}$$

where $\theta$ is the angle between $OO_1$ and $O_1C$ in right triangle $OO_1C$. In the right triangle, the distance of the focus from the geometric center of the ellipse ($OO_1$ = c) serves as a side, semi-minor axis (OC = b) serves as another side, and semi-major axis ($O_1C$ = a) serves as the hypotenuse. As $\delta$ angle increases, $\theta$ decreases, and the eccentricity (c/a) of the corresponding ellipse increases. If $\delta$ increases up to the maximum of right angle rendering the oval become a line with two foci at both ends, then the kinematic movement becomes matter state transformation as in the case of electronic orbital. The relationship is a well-known geometrical property of an ellipse, but it has a physical implication that goes unnoticed.

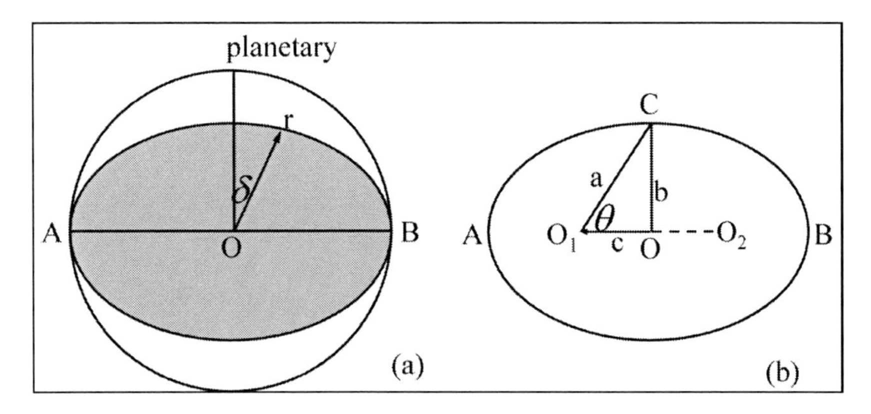


Figure 9.2. Schematic illustration of the equivalence of (a) a circle on a plane tilted $\delta$ angle from the equatorial plane with (b) an elliptical orbit on the equatorial plane.

The equivalence of an elliptical orbit in the equatorial plane and a tilted circular orbit reflects the space and time symmetry of nature. If we regard the focus shift $OO_1$ in the equatorial plane as in space, then the tilting departure of the shaded plane from the equatorial plane is in time direction, orthogonal to space. Equation (9.1) can be interpreted as a calculus relationship in a way similar to equation (1.19) that a movement in space is equivalent to a movement in time. The movements in space and in time are not only a mathematical trade-off but also have different physical manifestations. We believe that electrons travel in time primarily via state transformation whereas planets take the route of space movements along elliptical orbit during the early genesis. Thus, both electronic orbital and planetary orbit are governed by a unified natural principle, and as such Kepler's laws that govern the latter can be logically extended to the former.

## 9.2. Kepler's Law in the Atomic Spacetime

Kepler's first law states that a planetary orbit is an ellipse with the sun at its focus. What is an ellipse? An ellipse is the projection of a circle onto a plane tilted at a certain angle $\delta$ as was discussed previously. Mathematically, an ellipse is defined as the locus of points, the sum of whose distances from two fixed points, $O_1$ and $O_2$, known as the foci, is a constant (Figure 9.2b). We believe that such an elegant mathematical property must have its physical significance in the atomic spacetime. Under the atomic spacetime, if the movement of $OO_1$ corresponds to the departure of the orbital plane from the equatorial plane at an angle of $\delta$, then the movement of $OO_2$ must be equivalent to the rotation of the orbital plane at the opposite direction in symmetry to $\delta$. Since within helium shell both 1s electrons transform at opposite directions, they are symmetric about space and time, i.e., space component of one electron is equivalent to time component of the other, and time component of one electron matches space component of the other. Appling the property of the ellipse to the electronic orbitals, we draw a conclusion that the sum of time components of the two electrons is a constant, i.e., time conserves in helium sphere. In symmetry to this, space conserves too. However, the sum must be understood as the addition of two orthogonal quantities that observe Pythagorean theorem as was illustrated by dimension diagram (Figure 1.7).

$$(C_1 \cos\alpha)^2 + (-C_1\omega\sin\alpha)^2 = C_1^2, \tag{9.2}$$

$$(C_2 \cos\beta)^2 + \left|(C_2 r\sin\beta)^2\right| = C_2^2, \tag{9.3}$$

$$\cos\alpha = \sin\beta, \tag{9.4}$$

where the hypotenuse corresponds to $O_1O_2$ of an ellipse at its maximum eccentricity, which naturally degenerates into a line segment AB (Figure 1.7). Equation (9.2) indicates electronic time component conservation whereas equation (9.3) refers to electronic space component conservation, but they are indeed two aspects of the same process. Equation (9.4) synchronizes the space and time components of an electron in motion, which states that both radian angles $\alpha$ and $\beta$ are complementary at any moment. Hence Kepler's first law for planetary orbits takes the form of Pythagorean theorem for electronic orbitals.

Kepler's second law is closely associated with the first one. To illustrate its analogy in electronic orbital, we need to clarify the relationship between angular velocity $\omega$ and orbital radius $r$. As shown in quaternity coordinates (Figure 9.3), orbital radius $r$ is a vector rotating around the origin O from dimension meter to 1 while angular velocity $\omega$ is a vector tangential to arc ACB at point C, both orthogonal with their product $v = \omega r$. As point C moves along the arc, angular velocity rotates smoothly from dimension second to 1. For a planetary orbit, Kepler's second law states that the line connecting the planet and the sun sweeps equal area within equal time interval. If we interprets that area swept by that connecting line within a unit time interval in an elliptical orbit as corresponding to velocity $v$ in the atomic spacetime, then $v$ have a constant magnitude by Kepler's second law disregarding its direction. Since electronic transformation is an electromagnetic induction, Kepler's second law dictates constant speed of electromagnetic waves, which is a well known fact in physics.

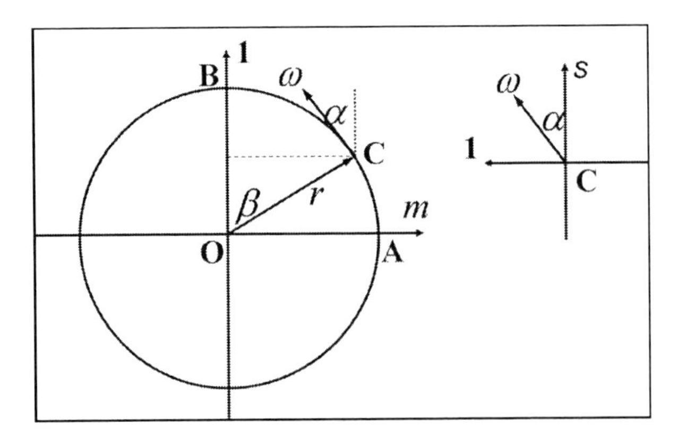


Figure 9.3. The orthogonal relationship between angular velocity and orbital radius in the atomic spacetime.

For a planetary orbit, Kepler's first law clarifies the elliptical orbit and the second law further specifies speed of movement. Both laws combined together to describe an elliptical motion in a certain way. For an electronic orbital, Kepler's first law governs the spherical quantities at the dimensional level of $-\partial\Omega_0/\partial t$ and $\int\Omega_1 dl$ while the second law governs spherical quantities at the dimensional level of $\omega$ and $r$. When an electronic orbital is specified at two consecutive orthogonal levels, it is well defined. This is the physical significance of Kepler's two laws in spherical view.

## 9.3. CENTRAL FORCE

Force is of paramount important in physics. For a planetary orbit, Newton deduced the famous gravitational law as follows:

$$F = G\frac{m_1 m_2}{r^2}, \tag{9.5}$$

where $m_1$ and $m_2$ are the masses of the two entities, $r$ is the distance between them, $G$ is a proportionality constant, and $F$ is the gravitational force on either object. For electrostatic interaction, Coulomb's law states:

$$F = k\frac{q_1 q_2}{r^2}, \tag{9.6}$$

where $q_1$ and $q_2$ are the charges of the two particles and $k$ is a proportionality constant. For centuries, physicists observing the similarity of gravitational law and Coulomb's law have tried to generalize them into a unified theory to cover other interactions. This approach turns out to be nearsighted. After all, Coulomb's law is obtained experimentally rather than theoretically. In addition to experimental evidences, the expression of a formula as complex as Coulomb's law must be a natural result of certain mathematical deduction. If theoretical derivation of its expression has not been made satisfactorily, the law is not completely established and should be regarded as an empirical formula only. In the context of electronic wave function in the atomic spacetime, we shall explore a probable theoretical background of Coulomb's law as follows.

From equation (1.14), we acquire the first characteristic root to duality equation as

$$\Omega_0 = C_1 C_2 \cos\alpha \cos\beta, \tag{9.7}$$

where $\alpha$ and $\beta$ are complementary so that by multiple angle formula

$$\Omega_0 = \frac{C_1 C_2}{2}\sin 2\beta, \tag{9.8}$$

where $\beta$ can be derived from integral operation of $\int 1/r dl$ by equation (1.6). Geometrically, this radian angle is a central angle subtended by a certain arc length $L$ of the radius $r$ of electronic revolution cycle (Figure 9.4) and can be expressed as

$$\beta = \frac{L}{r}. \tag{9.9}$$

Substituting $\beta$ value into equation (9.8) produces

$$\Omega_0 = \frac{C_1 C_2}{2} \sin \frac{2L}{r}. \tag{9.10}$$

Since $\Omega_0$ was an inherent characteristic of electronic orbital, it is best interpreted as the energy potential of the electron within helium sphere so that central force that the electron incurs at any moment can be derived from

$$F = -\frac{\partial \Omega_0}{\partial r}, \tag{9.11}$$

whence

$$F = L\frac{C_1 C_2}{r^2} \cos \frac{2L}{r}, \tag{9.12}$$

or in terms of infinitive progression

$$F = L\frac{C_1 C_2}{r^2}(1 - \frac{2^2 L^2}{2! r^2} + \frac{2^4 L^4}{4! r^4} - \frac{2^6 L^6}{6! r^6} + \ldots). \tag{9.13}$$

Because the infinite series of terms converge, each subsequent term is small relative to its preceding term. If we retain only the first term of the series and neglect the others, then the result would be

$$F \approx L\frac{C_1 C_2}{r^2}, \tag{9.14}$$

which conforms to equations (9.5) and (9.6). Constants $C_1$ and $C_2$ are inherent electronic dimensions in space and time respectively, and the force is the interaction between the nucleus and a 1s electron. If this deduction of force proves to be true and fundamental, then other electromagnetic interactions may follow suit. This indicates that Coulomb's law is not an exact law, but an approximation of equation (9.12). There have been some experimental reports of the inexactness of Coulomb's electrostatic law, but without theoretical guidance they are generally considered to be within measurement errors. Theoretical attempts for modifying the inverse square law in relativistic or quantum context are normally cumbersome and farther from the truth and hence have never been accepted either.

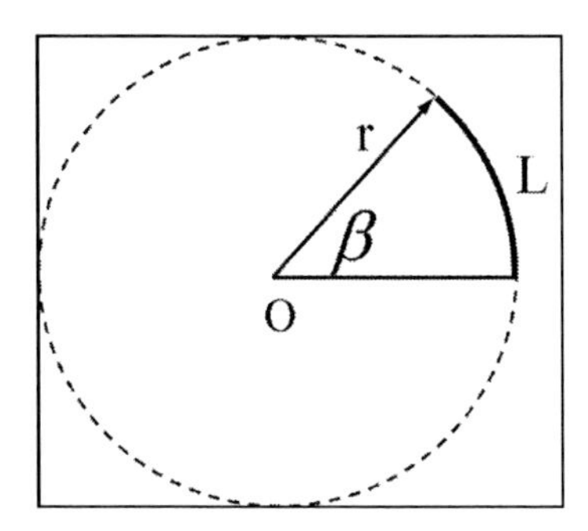


Figure 9.4. Geometrical relation of $\beta$ as a radian angle and its corresponding arc length $L$ in a circle.

Nowadays college physical experiments usually include the measurements of electrostatic forces for verifying Coulomb's law. Students are taught to use the inverse square relationship for curve regression on the data of forces versus distances. When they do not get a satisfactory goodness of fit, they normally blame experimental conditions. To think of the trigonometric term in equation (9.12) is out of the question. It is believed that a carefully designed experiment should be able to confirm our prediction and estimate the radian angle involved in the expression.

Because electronic motion and planetary movement are manifestations of the same physical principle in different directions, the expression of central force can be applied to gravitational force as well. However, the difference between equations (9.12) and (9.5) might be so trivially small for gravitational force that current experimental methods and observations are difficult to discern. While the modification of Newton's gravitational law and Coulomb's law stands to be verified experimentally, we are happy to see that the law of central force is a natural outcome of mathematical derivation from primary wave function beyond dirty data analysis.

## 9.4. SUMMARY

This chapter has provided an excellent example on the unification of physical entities. Even though planetary orbits and electronic orbitals are remarkably different, they obey the same physical principle in essence. While planets orbit around the sun through kinematic movement, electrons loop around the nucleus via matter state transformation. While the planets and orbits belong to different concepts, the electrons and electronic orbitals are indistinguishable. Kepler's first law for planetary orbits takes the form of Pythagorean theorem for electrons in the atomic spacetime; and Kepler's second law describes the constant speed of electromagnetic waves. This equivalence demonstrates that spherical view has grasped the fundamentals of nature rather than its diverse superficial manifestations. People have been searching for a grand unification theory in earnest for many centuries, but the Holy Grail of science lies on the more fundamental level than Euclidean space with Newtonian time. One can never manufacture a large ship on board a small boat. It is only on the atomic spacetime platform that various physical entities could be compared and forces and wave functions could be unified in a coherent manner.

*Chapter 10*

# THE ORDER OF PHYSICAL QUANTITIES

A spherical quantity assumes a fixed number with a variable dimension. When the dimension variable takes the value of a whole dimension in space and/or time, the spherical quantity corresponds to a physical quantity of that dimension; and when the dimension variable takes another whole dimension value, the spherical quantity corresponds to another physical quantity accordingly. The dimensions of physical quantities are discrete points on the continuous dimension curve of a spherical quantity as if integers were located at a real number axis. Thus a spherical quantity connects multiple physical quantities into a series of spacetime dimensions. This chapter tries to establish such a dimension series of physical quantities and order them into a periodic table. This approach provides sharp insight into the properties and relationships of various physical quantities.

## 10.1. DIMENSIONS OF SPHERICAL QUANTITIES

In Chapter 2, we provided an orbital interpretation of quaternity spacetime in neon shell. There are four kinds of dimensional combinations termed scalor, polor, metor, and vitor in a four-dimensional world, each being separated by a velocity dimension (m/s). We normally label the positive directions of four quaternity axes with $1$, $u$, $u^2$, and $u^3$ dimensions and the negative directions with $u^4$, $u^5$, $u^6$, and $u^7$ dimensions. Between $1$ and $u$ dimensions, there are also $\dot{\Psi}$ and $-1/\psi'$ dimensions, and other adjacent axes have similar dimensions between them. For spherical layer 2, it is necessary to use sixteen physical quantities for describing 2s2p electrons in neon shell and another sixteen physical quantities for describing 3s3p electrons in argon shell. While physical quantities of space or time dimensions higher than three cannot be easily understood in Euclidean geometry, they can be expressed concretely and conceivably in quaternity spacetime that provides a wider scope in interpreting physical phenomena.

As shown in Table 10.1, the first two rows correspond to dimensions of 3s3p electron octet and the last two rows correspond to dimensions of 2s2p electron octet. Each electron consists of two adjacent quantities in a vertical column as was defined by wave functions (2.18), (2.25), (5.5), and (5.11). The second row of the table has $n^i$ dimensions whereas the third row features $u^i$ dimensions where $i$ takes a value from 0 to 7. Within spherical layer 2, a spherical quantity completes a cycle after undergoing sixteen consecutive rotations so that

$$u^8 n^8 = 1 \tag{10.1}$$

which means that the unit of $s^8/m^8$ is equivalent to that of $m^8/s^8$ in an unperturbed argon atom. The atomic sphere is a spacetime continuum.

**Table 10.1. Space and time units of spherical quantities for characterizing electronic orbitals within an argon atom where s represents a time dimension and m represents a space dimension**

| s | $s^2/m$ | $s^3/m^2$ | $s^4/m^3$ | $s^5/m^4$ | $s^6/m^5$ | $s^7/m^6$ | $s^8/m^7$ |
|---|---|---|---|---|---|---|---|
| 1 | s/m | $s^2/m^2$ | $s^3/m^3$ | $s^4/m^4$ | $s^5/m^5$ | $s^6/m^6$ | $s^7/m^7$ |
| 1 | m/s | $m^2/s^2$ | $m^3/s^3$ | $m^4/s^4$ | $m^5/s^5$ | $m^6/s^6$ | $m^7/s^7$ |
| 1/s | $m/s^2$ | $m^2/s^3$ | $m^3/s^4$ | $m^4/s^5$ | $m^5/s^6$ | $m^6/s^7$ | $m^7/s^8$ |

Each cell in the table represents a dimension of a spherical quantity within an argon atom, which can be interpreted as a physical quantity as well. If spherical quantities reflect electronic properties correctly, then their whole dimensions must correspond to physical quantities. This means that there are arrays of physical quantities that are entirely composed of space and time dimensions so that we can arrange them according to their dimensions. Given that quaternity spacetime has a wider scope than Euclidean space, how do current existing physical quantities fit into quaternity framework? Basically, current existing physical quantities having only space and time dimensions (with units of meter and second only) must fit nicely into Table 10.1 with proper interpretation of their physical significance. Examples are acceleration, wavelength, frequency, etc. Physical quantities bearing other units such as mass, electric charge, and magnetic moment entail new definitions through rediscovery of their space and time dimensions. These integration processes could be long and painful. Can we tabulate all physical quantities in a pattern similar to Table 10.1? We shall approach the question through dimensional analyses of current existing physical quantities in the following sections.

## 10.2. DIMENSIONS OF PHYSICAL QUANTITIES

The foundation of physics rests upon physical quantities, in terms of which the laws of physics are expressed. A physical quantity in the general sense is a property ascribed to phenomena, bodies, or substances that can be quantified for, or assigned to, a particular phenomenon, body, or substance. It has been well accepted that the dimension of a physical quantity is the qualitative nature of the physical quantity. Dimensional analysis is often used to derive or check formulas by treating dimensions as algebraic quantities. For example, quantities can be added or subtracted only if they have the same dimensions, and quantities on both sides of an equation must have the same dimensions. We shall analyze the dimensions of physical quantities in status quo and then carry a farther step by organizing physical quantities in proper space and time orders as below.

Mechanical quantities include kinematic and dynamic quantities. Kinematics is the study of bodies without reference to mass or forces involved, in which bodies are considered particles so that we can ignore the object's size, orientation, and internal structure. Commonly used kinematic quantities include position, length, time, velocity, acceleration, period, frequency, etc. They serve to tell us how a particle moves in space with time. On the other hand, dynamics describes the motion of objects taking forces, mass, and shapes into considerations. Commonly used dynamic quantities include mass, force, momentum, energy, pressure, etc. Here we arrange some mechanical quantities into a table according to their space and time dimensions (Table 10.2). Three SI base units (second, meter, and mass) are used to characterize their dimensions.

It can be seen from Table 10.2 that the major difference between dynamic and kinematic quantities is the additional kilogram unit associated with dynamic quantities. No direct connection between kinematic and dynamic quantities is implied here. When considering kinematic and dynamic quantities separately, from left to right each subsequent column increases a velocity dimension, and from top down each subsequent row reduces a time dimension. A space or a time dimension is the smallest possible dimension, which is indivisible. By the way, we never ponder over the atomicity of mass, i.e., whether or not mass could be divided into space and time dimensions.

Electromagnetic quantities describe various phenomena in regard to electricity and magnetism. In SI unit system, they can be expressed in terms of four base quantities namely length, time, mass, and electric current. These four base quantities are assumed to be mutually independent, from which other physical quantities can be derived. Some commonly used electromagnetic quantities are tabulated in two groups reflecting their space and time relationships inside the groups (Table 10.3).

**Table 10.2. Units of mechanical quantities where s represents a time dimension, m represents a length dimension, and kg represents a mass dimension**

| Kinematic quantities | | | |
|---|---|---|---|
| | | | Area<br>$m^2$ |
| | Time<br>s | Length<br>m | Kinematic viscosity<br>$m^2/s$ |
| | Number<br>1 | Velocity<br>m/s | |
| Wave number<br>1/m | Frequency<br>1/s | Acceleration<br>$m/s^2$ | |
| Dynamic quantities | | | |
| | Mass<br>kg | Momentum<br>kg.m/s | Energy, work, heat<br>$kg.m^2/s^2$ |
| | | Force<br>$kg.m/s^2$ | Power<br>$kg.m^2/s^3$ |
| Dynamic viscosity<br>kg/ms | Surface tension<br>$kg/s^2$ | | |
| Pressure, stress<br>$kg/ms^2$ | | | |

**Table 10.3. Units of some electromagnetic quantities where A denotes an ampere dimension**

| Quantities with an [A] unit | | | |
|---|---|---|---|
| | | | Electric charge sA |
| | | | Electric current A |
| | Electric flux density sA/m$^2$ | Magnetic field strength A/m | |
| Electric charge density sA/m$^3$ | Electric current density A/m$^2$ | | |
| Quantities with a [kgA$^{-1}$] unit | | | |
| | | Magnetic flux kg A$^{-1}$ m$^2$s$^{-2}$ | |
| | | Electric potential kg A$^{-1}$ m$^2$s$^{-3}$ | |
| Magnetic flux density kg A$^{-1}$s$^{-2}$ | Electric field strength kg A$^{-1}$ms$^{-3}$ | | |

One group of quantities all bear an ampere unit in addition to meter and second, and the other group of quantities contain [kgA$^{-1}$] units. Other electromagnetic quantities are less regular in units for such a categorization. For each group in the table, from left to right, each subsequent column increases a velocity dimension whereas from top down, each subsequent row reduces a time dimension. No relations are implied here between both groups as shown in Table 10.3. After listing physical quantities in this way, it becomes obvious that there are many blank holes in the table, which indicates that many possible physical quantities remain undefined.

Ampere was defined in SI system as that constant current which, if maintained in two straight parallel conductors of infinite length, of negligible circular cross section, and placed 1 meter apart in vacuum, would produce between these conductors a force equal to $2\times10^{-7}$ newton per meter of length. The effect of this definition was to fix the magnetic constant (permeability of vacuum) at exactly $4\pi\times10^{-7}$ Hm$^{-1}$. However, if the permeability of vacuum were actually a non-unit constant, then the unit of ampere would be dependent upon the unit of newton! In that case, ampere would become a derived quantity. We shall explore the possibility of unifying the four base quantities as follows.

## 10.3. Interpretation of Electromagnetic Quantities

Compared with spherical quantities as were shown in Table 10.1, it is obvious that current existing physical quantities are a hodgepodge, not in unified units. They are severely fragmented because we have arbitrarily introduced the units of kilogram and ampere without realizing that both units might be the composites of some space and time dimensions. Since Chapter 1, we have decided that two fundamental spherical quantities, space and time, are enough to express all spherical quantities. Therefore one way to unify current physical quantities is to find spacetime dimensions for kilogram and ampere.

*Postulate 1. Every physical quantity is in a certain combination of space and time dimensions. All physical quantity units can be expressed in terms of two base units: meter (m) and second (s).*

This postulate assumes that the world is a rendition of space and time. If space refers to the whole universe, then time is only a snapshot of it. If space refers to a particle, then time is the complex motion of it. For a given entity, the smaller its space is, the denser its time is, and vice versa. In a four-dimensional sphere, it is reasonable to deduce that space and time quantities are enough to describe all phenomena, one serving as a reference while the other as a measuring rod.

We have defined four spherical quadrants in terms of various space and time combinations for describing electrons within neon shell. We have also used wave functions of various spacetime dimensions to account for electrons in larger atoms. Space and time are the ultimate dimensions of spherical quantities. It stands to reason that we may use two base units, meter and second, to express all physical quantities, among which mass and electric current are the most prominent ones.

To find the spacetime dimension of ampere, we note that there is a rotation relationship between electric field intensity $E$ and magnetic field strength $B$ as expressed by Maxwell's equations:

$$\nabla \cdot B = 0 , \tag{10.2}$$

$$\nabla \times E = -\frac{\partial B}{\partial t} , \tag{10.3}$$

$$\nabla \cdot E = \frac{\rho}{\varepsilon_0} , \tag{10.4}$$

$$\nabla \times B = \frac{1}{c^2}\frac{\partial E}{\partial t} + \mu_0 J , \tag{10.5}$$

where $\rho$ and $J$ are charge density and current density, respectively. The equations also involve two fundamental constants characterizing the electrical permittivity $\varepsilon_0$ and magnetic permeability $\mu_0$ of free space which are related by

$$\varepsilon_0 \mu_0 c^2 = 1 \tag{10.6}$$

Because operators $\nabla\cdot$ and $\nabla\times$ reduce their operands a space dimension but in different orientations distinguished by $l_0$ and $l_1$ as was discussed in section 2.1.2, the results of the operations on a operand have the same space and time dimension but must be distinguished by an complex number identifier *j*, which transforms a space dimension to its orthogonal orientation.

$$jl_1 = l_0 ; \tag{10.7}$$

$$\nabla\times = j\nabla\cdot \tag{10.8}$$

If we treat $\nabla\cdot$ as $\partial/\partial l$, and label $\nabla\times$ as $-\partial/j\partial l$ to reflect the sequence of space dimension shifting, then three Maxwell's equations can be rewritten as

$$\frac{\partial E}{j\partial l} = \frac{\partial B}{\partial t}, \tag{10.9}$$

$$\frac{\partial E}{\partial l} = \frac{\rho}{\varepsilon_0}, \tag{10.10}$$

$$-\frac{\partial B}{j\partial l} = \frac{1}{c^2}\frac{\partial E}{\partial t} + \mu_0 J, \tag{10.11}$$

where $l$ means the space dimension along vector *E* direction in equations (10.9) and (10.10) and along vector *B* direction in equation (10.11). In spherical quantity expression, we discarded dimension shifting indicator *j* because differentiation is performed with respect to the most immediate space dimension of the operand by default and the shifting of the space dimensions in consecutive differentiations, such as from $l_0$ and $l_1$, is implied automatically.

By the way, we summarize three different meanings of a complex number identifier. First, when placed before a trigonometric function, a complex number identifier indicates a space or a time dimension as was shown in equations (1.2) and (1.5). It changes the unit of the function. Second, when placed directly before a spherical quantity, a complex number identifier is an operator shifting the spherical quantity to its orthogonal function. This usage was exemplified in equation (1.46). Third, when placed before a dimension, a complex number identifier shifts the orientation of the dimension to its orthogonal direction such as equation (10.7) without changing its SI unit.

Due to their rotation relation, *E* and *B* must be a velocity dimension apart, i.e., they must be in adjacent positions in the same row if we merge two groups of physical quantities in Table 10.3 together. We label the second group by a negative complex number identifier *-j* so that Table 10.3 becomes Table 10.4 in continuous space and time dimensions. From equation (10.9), after considering the antiparallel relationship between $\partial l$ and $\partial t$, we deduce dimensional relationship between *E* and *B* by their units as

$$j(kg.A^{-1}.m.s^{-3}) = (m/s)(A.m^{-1}), \tag{10.12}$$

where *kg*, *A*, *m*, and *s* represent the units of kilogram, ampere, meter, and second, respectively. This equation can be simplified into

$$kg.m.A^{-2}.s^{-2} = -j \tag{10.13}$$

so that ampere is no longer independent of the other three base units. The left-hand side of equation (10.13) happens to be the unit of vacuum permeability $\mu_0$, which leads to the second postulate:

*Postulate 2. The dimension of the magnetic permeability of vacuum is a negative complex number identifier -j.*

**Table 10.4. Dimensions of some commonly used electromagnetic quantities**

| | | | | |
|---|---|---|---|---|
| | | | Electric charge<br>sA | Magnetic flux<br>-*j*.A m |
| | | | Electric current<br>A | Electric potential<br>-*j*.A m/s |
| | Electric flux density<br>sA/m$^2$ | Magnetic field strength<br>A/m,<br>Magnetic flux density<br>-*j*.A/m | Electric field strength<br>-*j*.A/s | |
| Electric charge density $\rho$<br>sA/m$^3$ | Electric current density<br>A/m$^2$ | $\frac{\rho}{\varepsilon_0}$, -*j*.A/ms | | |

It can be seen from Table 10.4 that dimensional interval between electric charge density $\rho$ and $\rho/\varepsilon_0$ is velocity square because they are in the same row and separated by a slot. The position of $\rho/\varepsilon_0$ is defined by equation (10.10) so that $\rho/\varepsilon_0$ is a space dimension less than electric field strength. If we adopt $c^2$ to indicate velocity square, then we have the following relationship concerning $\rho$ and $\rho/\varepsilon_0$:

$$j\frac{\rho}{\varepsilon_0} = \rho c^2, \tag{10.14}$$

whence

$$\varepsilon_0(-j)c^2 = 1. \tag{10.15}$$

Comparing equation (10.15) with (10.6), we draw a conclusion that the magnetic permeability of free space $\mu_0$ has a dimension of $-j$ and the electrical permittivity $\varepsilon_0$ has a unit of $s^2/m^2$, which agrees with the second postulate.

The above assignments of $\mu_0$ and $\varepsilon_0$ dimensions clearly honor SI system and abandon CGS system of units because we may easily derive from Postulate 2 that electric field strength and magnetic field strength have a velocity dimension interval in SI system and hence are orthogonal, obeying the principle of rotation operation and Maxwell's equations, whereas in CGS system, both quantities have a same spacetime dimension, leading to the difficulty of explaining their orthogonal relationship (10.3). Indeed, Postulate 1 implies that the dimension of a physical quantity must reflect its spacetime reality and is not a matter of arbitrary definition. The junction between quaternity spacetime and SI system is coherent with the principle of rotatory operation. The dimensional interpretation of electromagnetic induction brings two groups of physical quantities in Table 10.3 together into one group in Table 10.4.

## 10.4. On the Unification of Physical Quantities

A space or a time dimension is the smallest quantized dimension in physics. Below the quantum, Postulate 2 suggests that the property of a vacuum such as the unit of vacuum permeability [$m.kg.s^{-2}.A^{-2}$] is a negative complex number identifier without unit. This means that the unit of ampere is a derived unit rather than a base unit.

$$A = \sqrt{j\frac{kg \times m}{s^2}} . \tag{10.16}$$

Hence our question becomes how to express mass in terms of space and time factors to ensure that ampere and other electromagnetic quantities have meaningful expressions given equation (10.16). At least, the right-hand side of the equation must be able to get rid of the square root sign.

A significant breakthrough is made as we ponder over the concept of central force. What kind of spacetime dimension should force be assigned abstractly? Suppose an entity, A, is orbiting around a massive object at point O, which is exerting a centripetal force on the entity. Since time and space are symmetry, we may assume that there is a point $\infty$ at the very far end imposing a centrifugal force on the entity (Figure 10.1). Because point $\infty$ is a hypothetical point at the very far end, it is good to treat it as a vacuum property. If we use a dimensionless quantity $-j$ to represent the centrifugal property of the vacuum, then spacetime dimension of the centripetal force from point O must be in symmetry to it. In a four-dimensional sphere such as an argon atom, an axis rotated sixteen times consecutively returns to its original place. We assign the dimension of $ju^8$ to the centripetal force because the quantity of $j.m^8/s^8$ dimension is in symmetry to a negative complex number identifier in quaternity coordinates. By the way, since vacuum permeability is postulated to be of $-j$ dimension, it can of course be treated as a special kind of spacetime force, or centrifugal force due to spacetime curvature.

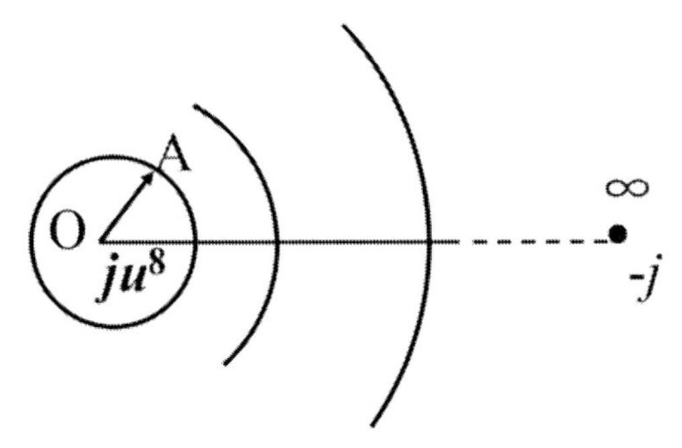


Figure 10.1. Schematic diagram of entity A orbiting around a massive object O under centripetal and centrifugal forces from points O and ∞, respectively.

*Postulate 3.* Force has a unit of $j.m^8/s^8$ in symmetry to a negative complex number identifier.

Based on the above three postulates, it is easy to derive that kilogram has a dimension of $j.m^7/s^6$ according to Newton's second law, and ampere has a dimension of $j.m^4/s^4$ as calculated from equation (10.16). The dimension of other physical quantities can be deduced accordingly. For example, inductance has a dimension unit of (henry = $m^2kgs^{-2}A^{-2} = -j.m$), confirming equation (1.37). Some commonly used physical quantities are ordered into Table 10.5. Such a unification of physical quantities results from an original exploration of physical quantity dimensions. In this periodic table, each column has a same space dimension. From top down, each subsequent row reduces a time dimension vertically. From left to right, each subsequent column increases a velocity dimension horizontally. Obviously, the periodic table is not perfect in its first version. Like a word puzzle, there are many blank holes to be filled and many dimensional relations to be verified in the table. Many existing physical quantities have not been framed in the table yet. Even some mistakes might exist in the table. We broach the methodology for organizing physical quantities with the hope that readers may study it carefully and improve it in the end.

After organizing various physical quantities into a unified space and time dimension table, we may deduce dimensional relationships between physical quantities. Although the conditions that govern those relationships remain unknown, the periodic table provides a clue to the possibility of certain relationships between physical quantities. It also indicates that there are a lot of physical quantities and relationships still untouched by physicists. A good periodic arrangement must be able to predict and retrodict important physical relationships. For example, Table 10.5 predicts that there is a rotation relation between acceleration and electric flux density, and there is also a rotation relation between $\rho/\varepsilon_0$ and mass density. These indicate strong coupling relationships between mechanical properties and electromagnetic properties even though the conditions for producing such inductions remain unspecified. We therefore suggest that careful experiments should be designed and conducted to verify the validity of the table.

Table 10.5 provides a clue to dimensional relationship between two physical quantities. For example, although subject to verification, any two adjacent quantities in the same row obey the principle of rotatory operation, i.e., Faraday's equation; any two adjacent quantities in the same column have differential relation with respect to a time dimension; any two adjacent quantities in the top-right diagonal direction have integral relation over a space dimension; and any two quantities having a dimensional gap of $m/s^2$ obey Newton's second law in the form of $F = Ma$ (Figure 10.2). If the architecture of the periodic table of physical quantities turns out to be correct, then it is a very useful guide to discover new physical quantities and the laws of physics.

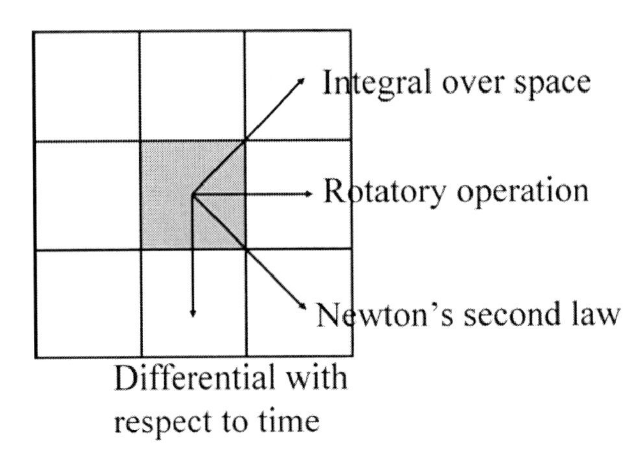


Figure 10.2. Physical laws that govern various adjacent quantities in the periodic table of physical quantities.

The significance of Table 10.5 goes far beyond what have been expected from traditional dimensional analyses. The periodic table of physical quantities is comparable to the Mendeleev periodic table of elements in the nineteen's century. The table effectively lays the foundation for the grand unification of physical quantities and physical laws. There are still many physical quantities to be defined. Each physical quantity must have its physical correspondence and reflect the reality of the world. No matter which physical quantity is discovered and filled in the table in the future, it must observe dimensional rule of the periodic table, obey the theory of quaternity, show its familial characteristics, and reflect physical reality by its unique property and behavior.

## 10.5. Thermodynamic Quantities

There are still some difficulties in finding all the connections between the dimensions of spherical quantities and physical quantities. It might be easier to define a new set of physical quantities to account for a specific phenomenon than to define them with existing physical quantities. Because the laws of physics have been constructed upon physical quantities, the crucial point in understanding physical laws lies in the correct definition and rediscovery of the physical quantities. Take the unit of joule as an example. It may refer to quantities of energy, work, torque, heat, enthalpy, etc., which share the same spacetime dimension. Energy is an important quantity that can be expressed as the product of other quantities with various space factors such as:

$$W = Fl, \tag{10.17}$$

$$W = \tau A, \tag{10.18}$$

$$W = PV , \tag{10.19}$$

where $W$ represents energy or work; $F$, $\tau$, and $P$ are force, surface tension, and pressure, respectively; and $l$, $A$, and $V$ represent one, two, and three space dimensions, respectively. In quaternity spacetime, these three equations can be written as:

$$\frac{\partial W}{\partial l} = F , \tag{10.20}$$

$$\frac{\partial^2 W}{\partial l^2} = \tau , \tag{10.21}$$

$$\frac{\partial^3 W}{\partial l^3} = P , \tag{10.22}$$

which mean that the gradient of energy is force, the flux of energy is surface tension, and the density of energy is pressure where $\partial / \partial l$, $\partial^2 / \partial l^2$, and $\partial^3 / \partial l^3$ are defined as gradient, flux, and density respectively in spherical view. Since quaternity spacetime consists of four basic space dimensions, we need to search for the four-dimensional space quantity beyond three-dimensional cubage.

One probable candidate for a four-dimensional space is entropy because it is an extensive state variable that measures the disorder of a system. We suspect that entropy has a space dimension more than three-dimensional volume, the fourth space dimension being the order of elements that gives entropy a statistical significance associated with the number of possible microstates. The connection between order and volume is not far-fetched as it might seem at first glance. To illustrate, we all know that a disordered office gives us the feeling of crowding. After properly organizing the items in the office, more working space becomes available. Thus the order of elements occupying a certain volume possesses space property because the order arrangement directly affects the apparent space. For another example, as shown in Figure 10.3, when four streaky squares are bound properly, they make five squares, the fifth one being composed of the four.

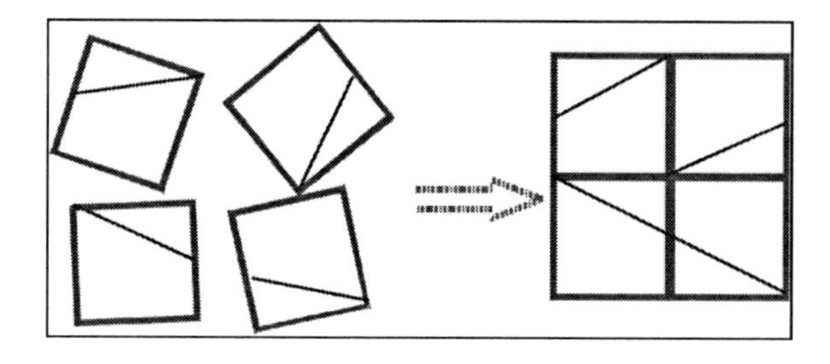

Figure 10.3. When four scattered squares are ordered properly, they form a bigger square unit with a similar streak.

**Table 10.5. Periodic table of commonly used physical quantities and their units in space and time dimensions**

| $s^3/m$ | $s^2$ | sm | $m^2$ area | $m^3/s$ flow | $m^4/s^2$ | $j.m^5/s^3$ electric dipole moment | $j.m^6/s^4$ magnetic dipole moment | $m^7/s^5$ | $m^8/s^6$ | $j.m^9/s^7$ angular momentum |
|---|---|---|---|---|---|---|---|---|---|---|
| $j.s^2/m$ capacitance (farad) | s time | m, length, $-j.m$ inductance (henry) | $m^2/s$ kinematic viscosity | $m^3/s^2$ | $j.m^4/s^3$ electric charge (coulomb) | $m^5/s^4$ magnetic flux (weber) | $m^6/s^5$ electric flux | $j.m^7/s^6$ mass (kg) | $j.m^8/s^7$ momentum | $j.m^9/s^8$ energy, heat (joule) |
| $j.s/m$ electric conductance (siemens) | $-j$ permea-bility | m/s velocity, $-j.m/s$ electric resistance, (ohm) | $m^2/s^2$ specific energy, absorbed dose(gray) | $m^3/s^3$ | $j.m^4/s^4$ electric current (ampere) | $m^5/s^5$ electric potential, electromotive force (volt) | $m^6/s^6$ | $m^7/s^7$ | $j.m^8/s^8$ force (newton) | $j.m^9/s^9$ power (watt) |
| 1/m wave number | 1/s frequency (hertz) | $m/s^2$ acceleration | $j.m^2/s^3$ electric flux density | $m^3/s^4$ magnetic flux density, $j.m^3/s^4$ magnetic field strength | $m^4/s^5$ electric field strength | $m^5/s^6$ | $j.m^6/s^7$ dynamic viscosity | $j.m^7/s^8$ surface tension | $m^8/s^9$ | $m^9/s^{10}$ |
| 1/sm | $1/s^2$ | $j.m/s^3$ electric charge density $\rho$ | $j.m^2/s^4$ current density | $m^3/s^5$ $\rho/\varepsilon_o$ | $j.m^4/s^6$ mass density | $m^5/s^7$ | $j.m^6/s^8$ energy density, pressure (pascal) | $j.m^7/s^9$ heat flux density, irradiance | $m^9/s^{10}$ | $m^9/s^{11}$ |

The entropy of the fifth square has to do with the permutation of the four squares without breaking the function of the fifth one as a streaky square. To further illustrate with a real entity, within a human body, many cells form a liver, an organ that can be regarded as a large cell capable of independent functioning. The fourth dimension of the liver, besides length, width, and height (as measured by the number of liver cells in certain lineups), is the alignment of liver cells to form the liver ensuring the proper functions of it.

*Postulate 4. The functional order is a space dimension like length, and entropy is a four-dimensional space quantity*

If this guess proves to be a correct rediscovery, then the differential change of $dS$ in the entropy of a system when differential heat transfer $dQ$ accrues to the system at temperature $T$ can be expressed as:

$$dS = \frac{dQ}{T}, \tag{10.23}$$

$$S = Vl\ , \tag{10.24}$$

so that

$$\frac{\partial^4 Q}{\partial l^4} = T\ , \tag{10.25}$$

In this way, we have identified four space dimensions as physical quantities of length, area, volume, and entropy and effectively integrated thermodynamic quantities into the periodic table of physical quantities where entropy bears a spacetime unit of $m^4$ and temperature has a dimension of $j.m^5/s^8$. The relationship between temperature and entropy is parallel to that of pressure and volume in the sense that pressure is acting upon three dimensions of the volume as in the case of compressed gas whereas temperature influences four dimensions of the entropy. Both pressure and temperature are intensive state variables, whose values are independent of the size of the system, while volume and entropy are extensive quantities, whose values are proportional to the size of the system.

## 10.6. SUMMARY

The urge to discover a fundamental theory underlying all natural phenomena has been expressed since the beginning of civilization. However, in seeking the Holy Grail of physics, it was a temptation for physicists to jump to the prescription of certain physical laws without first unifying physical quantities. After all, physical laws are based on physical quantities that describe nature in its crude shape. A successful unification of physical laws should be conducted at the low level physical quantities first. In this chapter, we have approached physics from ground up by delving into very basic concepts of space and time and analyzing dimensions of current physical quantities. Through a periodic organization of physical quantities, we have been paving the way to the unification of physical laws. We believe that physical quantities have definite patterns like those of chemical elements and tabular relationships between any two or more physical quantities constitute the so-called physical laws.

We are confident that a grand unification theory that physicists have long dreamed of does exist and have tried to set up the methodology to accomplish the discovery. However, to discover the grand unification theory, we still need substantial amount of detailed researches on the junctions of spherical quantity and physical quantity. How to translate physical quantities in Cartesian coordinates into spherical quantities in quaternity coordinates is one major issue; and how to interpret spherical quantities in terms of well-known concepts is another. Considering the fragmentation of current physics in various fields, there is not such a readily available grand unification theory to account for everything. A programmer who wants to develop an all-embracing version of a software application based on existing scattered codes must design unified interfaces and integrate the old and new codes. The grand unification theory will require a consistent manner in describing the world, and quaternity spacetime has provided this new platform for defining spherical quantities in dynamic calculus that bridge the dimensional gaps between physical quantities. Although great barriers exist for many physical theories remain only isolated successes in specific areas, we have shown the shining light of the grand unification theory on the other end of the tunnel.

Although this chapter has put forward a theoretical unification based on mathematical and logical deductions, it is quite possible that one might probe a certain physical quantity and discover an oversight or conflict in this chapter under conventional paradigm. On the one hand, one must be prepared to change his or her mindset as we migrate from Euclidean geometry to quaternity spacetime, which might have unexpected properties. On the other hand, this chapter is an initial attempt to suggest a probable way of unifying physical laws. It is a promising step towards the grand unification theory but it should not be credited as the grand unification theory itself. To get there, there is still a long way to go. Mendeleev did not provide the periodic table of elements as it is nowadays, but his original arrangement of chemical elements gave us a sound approach for studying elemental behaviors. By the same manner, this chapter furnishes the clue and methodology towards a grand unification theory.

*Chapter 11*

# RATIONAL CORE OF CHINESE MEDICINE

After writing the first ten chapters of this book, I was trying to expound the implication of the theory of quaternity in philosophy, but ended up with the finding that a good part of what I wanted to say had been stated in traditional Chinese medicine. This chapter is a bidirectional exploration, extending the theory of quaternity and the law of nature to traditional Chinese medicine on the one hand, and explaining the implication of quaternity by the philosophy of traditional Chinese medicine on the other hand. The merging of the new science with ancient philosophy is a process of mutual validations. The relationship between both electrons in helium shell can be explained in terms of yin and yang interaction; five element theory can be illustrated by the geometrical characteristics of electronic orbitals $2s2p_x2p_y2p_z2s$; and eight trigrams express consecutive rotatory operations, coinciding with 2s2p electron octet within neon shell. Our rational faculties are capable of conceiving the law of nature for it is by that law that we have evolved from and been an integral part of nature. Rational thinking to reflect truth is universal regardless of modern or ancient era and of Western or Eastern origin. Here we try to integrate our spherical view with traditional Chinese medicine.

## 11.1. YIN AND YANG INTERPRETATION OF SPACE AND TIME

Since Chapter 8, we have found that spherical quantities of time and space are so broad in their meanings that they may not refer to the same properties as their physical quantity counterparts. Spherical quantity is a dimensional concept, not a numerical concept. In dimension diagram, a dimension is a diameter for an electron to traverse via its subtended semicircular arc during dynamic differential or integral transformation. For a spherical quantity in rotatory operation, time dimension is the one, with respect to which a dynamic differentiation is performed upon the spherical quantity; and space dimension is the one, over which a dynamic integral operation is performed upon the spherical quantity. Spherical quantities of time and space change over the dimensions sinusoidally rather than uniformly. They transform by dynamic calculus instead of linear algebra. Nevertheless, the abstraction of spherical quantities from diverse entities is understandable in terms of harmonic oscillation. Time and space as spherical quantities in harmonic oscillation equations agree with those as physical quantities. The mathematical forms of both dynamic and infinitesimal calculi are similar as well.

In a two-dimensional world, space and time refer to the two outstanding intertwined dimensions respectively such as $\Omega_0$ and $\Omega_2$, male and female, animal and plant, biotic and abiotic, etc. as were shown in Table 8.1. Space means medium while time denotes propensity. They must always be associated with physical entities as if two sides of a coin cannot be divorced from the coin itself. In a four-dimensional world, space and time are blended into four spherical quadrants: scalor (space end), vitor, metor, polor, and scalor (time end). Space is the whole sphere while time is a black box unit or the core. If time refers to a seed, then space is the whole plant developing from it. Space and time are not absolute concepts. They are meaningful only within a certain spherical layer, like the relationship between a mother and a daughter in a specific family. Beyond that scope, the mother is a daughter of the grandmother and the daughter may become a mother in the future, too. Whether a spherical quantity is a space or a time quantity depends on its role in the spherical layer of concern. In a system of more than four dimensions, we may still use the four-dimensional framework to account for most of it by insulating its inner dimensions. This structural paradigm is scalable. The abstraction of space and time characterizes the main structure of the world and is a step towards better understanding of the world order.

Spherical quantities of space and time correspond to yin and yang concepts in traditional Chinese medicine. Yin and yang are opposite in meaning. Yin refers to feminine, passive, and negative principle in nature; and yang refers to masculine, active, and positive principle in nature. Yin refers to the moon, north or shady side of a hill, and south of a river; and yang refers to the sun, south or sunny side of a hill, and north of a river. The yin and yang pair is one of the most fundamental concepts in traditional Chinese medical theory, as it is the foundation of diagnosis and treatment.

Like space and time concepts, the concepts of yin and yang are relative. They can only be spoken of in relationships. If yin refers to cool, down, front, and static; then yang refers to warm, up, back, and dynamic. For example, water is yin relative to steam but yang relative to ice. They are a pair of opposites, such as the opposite ends of a cycle like the seasons of the year. If yin refers to spring, then yang refers to autumn. If yin refers to night, then yang refers to day.

Space and time are complementary and constantly transforming between each other, so are yin and yang. We have characterized both 1s electrons within a helium atom in space and time wave functions. Both electrons are complementary within the shell. The two-dimensional system is analogous to Taiji diagram, which features a pair of yin and yang fishes shown in black and white respectively (Figure 11.1). Both fishes are complementary inside a circle. The most interesting point is that the black yin fish has a white eye whereas the white yang fish has a dark point, the meaning of which often escapes public notice. As a matter of fact, the ancient Chinese already knew that yin and yang are transforming between each other. The white eye of the black fish must have indicated its connection with the tail of the white fish; and the tail of the black fish is the eye of the white fish. Nothing remains totally yin or totally yang. Just as a state of total yin is reached, yang begins to grow. Yin contains the seed of yang and vise versa. We have described the geometrical transformation of both 1s electrons in a sphere (Figure 1.15). As an alternative, Taiji diagram is the best way that I can think of to portray the transformation of both 1s electrons on a plane.

Figure 11.1. The Taiji (supreme ultimate) diagram showing the interdependent relationship of yin and yang.

However, as we merge our space and time theory with yin and yang theory where appropriate, please bear in mind that the transformation between space and time is by strict dynamic calculus while the transformation between yin and yang is undefined mathematically. A pair of space and time spherical quantities can be regarded as a pair of yin and yang safely; but a pair of yin and yang is not necessarily a pair of space and time. Therefore, the extension of space and time theory to cover yin and yang must be taken with caution.

Relative levels of yin and yang are continuously changing. Normally this is a harmonious change, but when yin and yang are out of balance they affect each other. Too much of one can eventually weaken or consume the other; and too little of one may be taken over or surpassed by the other. It is suggested that the human body should always maintain the equilibrium between yin and yang. The imbalance of yin and yang elements might result in illness. This is a basic principle of traditional Chinese medicine.

## 11.2. INTRODUCTION TO "*YI JING*"

Yin and yang theory originated from an ancient Chinese scripture, "*Yi Jing*" (or "*I Ching*"). Written more than three thousand years ago, "*Yi Jing*" is considered to be one of the oldest extant scriptures in the world, China's unique heritage, and a great treasure for all mankind. Some diagrams in it were created five thousand years ago, preceding Chinese characters. By the time of Zhou Dynasty (about 1100 BCE), the whole scripture of "*Yi Jing*" had been completed after experiencing three dynasties and over the hands of three generations of saints. The classic text describes an ancient system of philosophy that is intrinsic to Chinese cultural beliefs. Its influence on Chinese people is far-reaching and persistent.

When the ancient clan Bao-Xi ruled the country, they observed upwards the images of sky, watched downwards the methods of earth, paid attentions to the characteristics of beasts and birds and their native habitats, abstracted bodies nearby and multifarious objects far away, and started to draw eight trigrams so as to understand the revelation of the divine power and categorize the circumstances of all lives. It is evident that the quest for truth is a noble cause of intellectuals both ancient and modern. We don't know exactly how much the ancient Chinese learnt about the law of nature, but their unmistakable diagrams indicated that they had mastered the law in reasonable details, about which even modern scientists do not know. "*Yi Jing*" is the very source of traditional Chinese medicine. However, since

Confucius's humanistic interpretation of "*Yi Jing*" in ten commentaries, known as ten wings, the study of "*Yi Jing*" has been focused on the ethics and getting farther and farther irrelevant to science over time. Thus the best way for me to do here is to borrow the terms and diagrams from "*Yi Jing*" and vest them with my own views, ignoring their original denotations. What I am trying to accomplish is to establish the proper connections between my fresh discovery in mathematical physics and the ancient philosophy to avoid reinventing the wheel.

To bridge the gap between the East and the West is one of my life dreams. While taking biochemistry course at the University of Delaware in 1996, I discussed with Professor Waite about my initiative of integrating traditional Chinese medicine with western sciences, and his opinion was that that isn't easy though he believes in the pharmaceutical effects of Chinese herbs. Humbled by the tremendous task ahead of me, I devoted my time to it assiduously. The Occident developed linear mind of thinking for it seemed that there were inexhaustible resources across the Atlantic Ocean and farther in the West to support economic expansion whereas the Orient were inclined to spherical mind of thinking by the evidence that limited resources had been recycled once and again over thousands of years of civilizations. Apart from the different mindsets, English is a phonetic language, its fluidity allowing it to become a global vehicle for communication, whereas Chinese began as a pictographic language, the solidity of its compact square characters making it efficient and durable in recording history. When we consider the contrary frames of mind and dissimilar phyla of languages, the barrier to the integration is as formidable as the ten thousand miles long Great Wall, which had insulated the Chinese from Hsiung-Nu nomads for peace in various times over long history. And I could not expect to make a great progress in the integration but start with a trivial step as follows.

Although not any part of my spherical view has been proposed based on the knowledge of "*Yi Jing*", some rudimentary thoughts of "*Yi Jing*" coincide with my ideas. This is where the East meets the West. The very East is the West. So I find myself actually coding the program prototyped by my ancestors. This section accounts for three correspondences between spherical view and "*Yi Jing*" philosophy, leaving more for later sections to discuss. First, the law of nature is simple to govern the world robustly; the law of nature is invariable no matter how weird some things may appear to be; and the law of nature characterizes continual changing of everything in the universe. The title of book "*Yi Jing*" expresses these three properties of the law. In Chinese, *jing* means scripture, and *yi* has three basic meanings. *Yi* means simple and invariable as an adjective, and to change or exchange as a verb. Thus *yi* accurately characterizes the properties of the natural law, which are simple, invariable, and about changing. "*Yi Jing*" may be literally translated as "The book of changes".

Second, as was discussed in section 11.1, our definitions of space and time dimensions correspond to the core concepts of yin and yang in "*Yi Jing*". Spherical view deals with space and time dimensions whereas the entire scripture of "*Yi Jing*" is about yin and yang interaction. *Yi* has Taiji, Taiji produces two poles of yin and yang, two poles produce four quadrants, four quadrants produce eight trigrams, and eight trigrams determine good or ill luck, which in turn determines the prosper or decline of an enterprise. The interaction of yin and yang is called *yi*. If yin and yang refer to both 1s electrons, then Taiji can be understood as helium shell. When one electron occupies the entire helium sphere representing a space dimension, the other is an infinitesimal point at the center representing a time dimension. If yin and yang refer to an ovum and a sperm respectively, then Taiji is the zygote at the moment when the sperm enters the ovum, a fertilized egg ready to multiply.

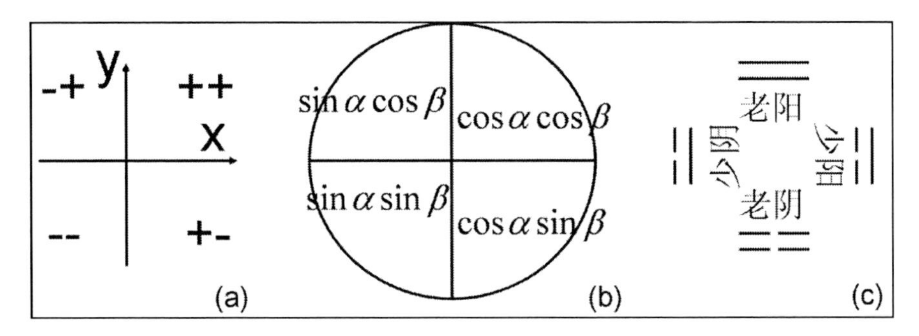


Figure 11.2. Comparison of signs, sines, and lines in four quadrants.

Third, four quadrants produced by both yin and yang poles are analogous to four quadrants on Cartesian plane. Yin and yang were drawn as a divided line and a whole line to symbolize female and male respectively, and two pairs of yin and yang produced four quadrants. As shown in Figure 11.2, the left plot indicates positive and negative signs of X- and Y-coordinates in four quadrants in Cartesian coordinate system; the middle plot lists sine and cosine functions of matrix (1.14), four roots to duality equation; and the right one shows four quadrants from "*Yi Jing*", termed old yang, young yin, old yin, and young yang counting counterclockwise from the top. The plus sign corresponds to cosine and to yang; and the minus sign corresponds to sine and to yin. The plots speak for themselves the unity of three systems even though Cartesian coordinates are related to linear algebra, trigonometric functions observe dynamic calculus, and yin-yang theory is not mathematically implemented. A truth can be approached in diverse but unified manners.

## 11.3. Four Spherical Quadrants in "*Yi Jing*"

In analytic geometry, four quadrants normally refers to the four areas into which a plane is divided by X and Y axes in Cartesian coordinate system, designated first, second, third, and fourth, counting counterclockwise from the area in which both coordinates are positive (see Figure 11.2a). After the introduction of quaternity theory, four spherical quadrants refers to the four regions into which a sphere is divided by four spacetime axes, designated first, second, third, and fourth, counting counterclockwise from the region bounded by *1* and *u* axes (see Figure 2.4). Four spherical quadrants are termed variously: a polor is a spherical quantity of the first spherical quadrant, a metor of the second, a vitor of the third, and a scalor of the fourth. Geometrically, a polor is a point displacing along a line segment; a metor is an arc along a semicircle; a vitor is a curved surface along a hemisphere; and a scalor is a volume region within a sphere. Four spherical quantities may be vividly expressed by various arrows as shown in Figure 11.3.

For translation, the proper Chinese characters matching four spherical quantities are readily available from "*Yi Jing*", namely, Yuan for polor, Heng for metor, Li for vitor, and Zhen for scalor, respectively. Regarded as the very classic of Chinese culture, "*Yi Jing*" expresses our ancestor's good understanding of the natural law in pithy vocabulary and vivid diagrams. Besides defining four spherical quadrants in physics and geometry, it is noteworthy to dig out their literal meanings from both English words and Chinese characters.

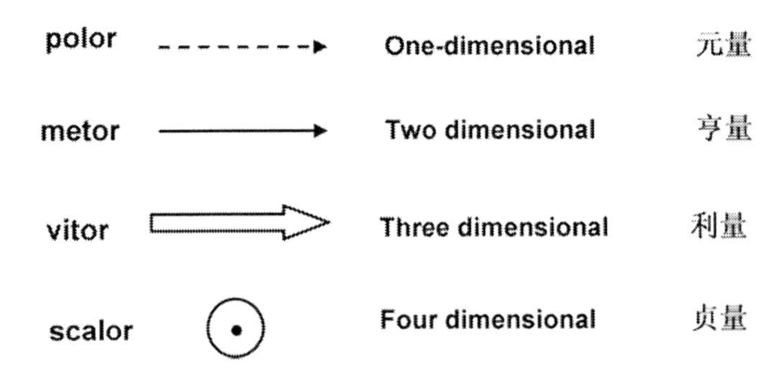


Figure 11.3. Symbols of four spherical quantities.

Yuan is in the first spherical quadrant, referring to an entity that is polarizing as "polor" implies. Yuan means inchoation, agglomeration, origination, growth, debut, and walking out of fog with great ambition. Heng is in the second spherical quadrant that connotes metamorphosis after polor as prefix "meta" implies. As the second developmental stage after Yuan, Heng means flow, penetration, traverse, expedition, delivery, and marching circuitously without hesitation. Li is the third developmental stage of the affair that manifests vitality and maturity, so is the significance of the coined word "vitor" from prefix "vita". Li indicates surrounding, enclosure, sincerity, well-being, production, and steering around aptly and harmonically. Zhen is in the last spherical quadrant that concludes a development cycle fairly. With a certain scale that delimits a spherical layer, by the meaning of "scalor", Zhen indicates richness, soundness, pureness, chastity, perseverance, and awaiting orders humbly and distinctly for the next cycle. Four spherical quadrants also refer to four seasons in a year, forming a full cycle (Table 11.1).

In addition, according to "*Wenyan Zhuan*" that literally means "*Classic commentaries on texts and characters*" by Confucius, Yuan is the growth of kindness because man is innately benevolent at the outset; Heng is the gather of goodness or politeness after education; Li is the sum of justice in conduct; and Zhen is the undertaking of work firmly on principle. Such an ethical interpretation identifies the four characters as gentlemen's morality in four developmental stages of virtue cultivation and reflects the philosophy and value of ancient Chinese culture.

**Table 11.1. Bidirectional translations of four spherical quantities**

| Spherical quadrant | Spherical quantity | English meaning | Chinese character | Chinese meaning | Geometric meaning | Season |
|---|---|---|---|---|---|---|
| I | Polor | Polarizing | Yuan | Start | Displacing point | Spring |
| II | Metor | Meta-morphosis | Heng | Flow | Extending arc | Summer |
| III | Vitor | Vitality | Li | Mature | Surrounding surface | Fall |
| IV | Scalor | Scale | Zhen | Chastity | Waxing sphere | Winter |
| Polor initiates after full scalor to begin another cycle. | | | | | | |

## 11.4. Geometrical Interpretation of Five Element Theory

Based on ancient Chinese philosophy, there are five elements in the natural environment: metal, water, wood, fire, and earth. Everything is made of these five basic elements. The relationships between five elements include generation and overcoming cycles. Each elemental force generates or creates the next element in a creative cycle. Metal generates water because water usually condenses on metal; water generates wood because rain nourishes a tree; wood generates fire because wood is easy to be burnt; fire generates earth because ash is created from fire; and earth generates metal because metal is mined from earth (Figure 11.4).

According to five element theory, each elemental force also controls or overcomes another element in an overcoming cycle. Metal overcomes wood because an ax cuts wood; wood overcomes earth because tree roots into the earth; earth overcomes water because earth absorbs and conceals water; water overcomes fire because water puts fire out; and fire overcomes metal because fire melts metal (Figure 11.5).

In traditional Chinese medicine, five elements are abstracted as five properties or five characteristics that support or control one another. Metal behaves like leather. Those having properties of clearing away, quelling down, and convergence belong to metal. Water slicks down. Those having properties of coldness, lubrication, moving downwards belong to water. Wood is straight and resists bending.

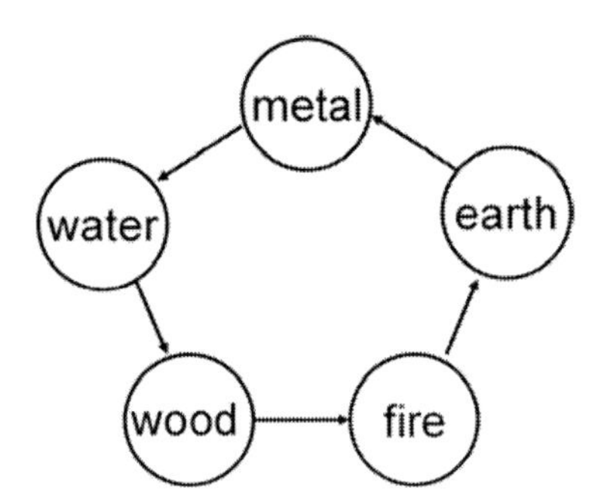


Figure 11.4. Cycle of five element generation.

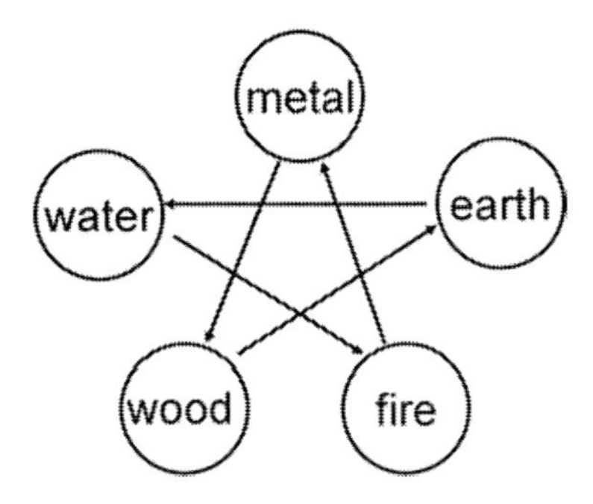


Figure 11.5. Cycle of five element overcoming.

Those having properties of growth, development, extending, and stretching belongs to wood. Fire is hot and oriented upward. Those having properties of warmness and rising up belong to fire. Earth supports crops. Those having properties of production, supporting, and reception belong to earth.

In a human body, every organ may be categorized according to its roles and properties. For example, the lungs have the functions of removing carbon dioxide from the blood and providing it with oxygen so that the lungs belong to metal; the kidneys have the functions of maintaining proper water and electrolyte balance, regulating acid-base concentration, and filtering the blood of metabolic wastes, which are then excreted as urine, so that the kidneys belong to water; the liver has the function of discharging bile and tends to be elastic and not to be depressed so that the liver belongs to wood; the heart has the function of warmness so that the heart belongs to fire; and the spleen has the functions of storing blood, disintegrating old blood cells, filtering foreign substances from the blood, producing lymphocytes, and nourishing various organs so that the spleen belongs to earth. Because of their belongings, these five organs have the relationships of generation and overcoming cycles as illustrated in Figures 11.4 and 11.5. For instance, the liver (wood) is said to be the mother of the heart (fire), and the kidneys (water) the mother of the liver. One observation is that kidney deficiency affects the function of the liver. In this case, the mother is weak, and cannot support the child. Many of these relationships can be linked to known physiological pathways now. Thus five element theory in traditional Chinese medicine is about five characteristics or five phases and their relationships in vivo instead of five substances in the environment. Strictly speaking, even though five characteristics correspond to five organs, the concept of characteristics is not exactly the concept of physical organs in anatomy.

The foregoing introduction to five element theory can often be found in the textbooks of traditional Chinese medicine. Even though five element theory is subjected to endless disputes and challenges, we believe that the core concept of five characteristics is correct. Under the context of quaternity spacetime, here we shall define five characteristics in rigorous geometry. As shown in Figure 11.6, five elements are represented by 2s, $2p_x$, $2p_y$, $2p_z$, and 2s electronic orbitals. Firstly, the 2s orbital at the top is a sphere of four-dimensional time with zero-dimensional space, representing metal. By rotatory operation, this orbital gradually condenses into a $2p_x$ orbital, which is a point of three-dimensional time with one-dimensional space. The condensation process corresponds to metal producing water. Water is drawn as a drop in the figure.

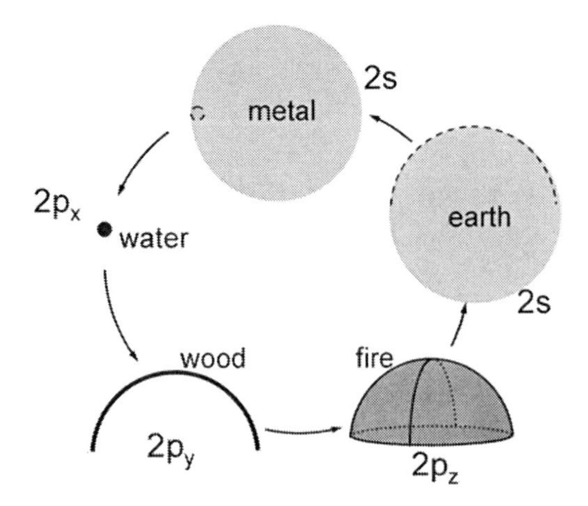


Figure 11.6. Geometrical and electronic definition of five characteristics.

**Table 11.2. Five elements with geometrical significance**

| Element | Metal | Water | Wood | Fire | Earth |
|---|---|---|---|---|---|
| Characteristic | Convergence | Fluid | Stretching | Warmness | Reception |
| Organ | Lungs | Kidneys | Liver | Heart | Spleen |
| Electron | 2s | $2p_x$ | $2p_y$ | $2p_z$ | 2s |
| Geometrical shape | Inner sphere | Point | Semicircle | Hemispherical surface | Outer sphere |
| Space dimension | 0 | 1 | 2 | 3 | 4 |
| Spherical quantity | Scalor | polor | metor | vitor | scalor |

Secondly, by rotatory operation, the $2p_x$ orbital transforms into a $2p_y$ orbital, which is a two-dimensional semicircle in shape. The stretching process corresponds to water producing wood. Wood is drawn as an arc. Thirdly, by rotatory operation, the $2p_y$ orbital transforms into a $2p_z$ orbital, which is hemispherical surface of one-dimensional time with three-dimensional space. The distribution of the electron is dense outside and empty inside, a characteristic of fire. Fourth, by rotatory operation, the $2p_z$ transforms into a 2s orbital, which is a full sphere of zero-dimensional time with four-dimensional space. The full sphere defines and supports a spherical layer, which is a characteristic of earth. For detail description of the shapes and transformation processes, please see Figures 2.14 and 2.12. Finally, from earth to metal, the sphere of zero-dimensional time with four-dimensional space changes its role into a sphere of four-dimensional time with zero-dimensional space. This process could be realized through transformation of 2s→ $2p_z$→ $2p_y$→ $2p_z$→ 2s (treated as a single step 2s→2s) in the opposite directions of quaternity axes. Table 11.2 lists five elements in geometrical shapes along with their characteristics.

We have established the correspondences between the cycle of five element generation and the cycle of electrons within neon shell in Figure 11.6. Under octet configuration, each electron is rotating counterclockwise as was shown in Figure 3.5. However, when there are vacancies in the atomic shell, some electrons may rotate back and forth (clockwise and counterclockwise) so as to fill the vacancies. To avoid colliding, the existence of a $2p_x$ electron in counterclockwise rotation may push a $2p_z$ electron to rotate counterclockwise as well because the transformation of $2p_x$→$2p_y$ cannot coexist with the transformation of $2p_z$→$2p_y$. Thus there are overcoming relationships between electrons in every other spherical quadrant in neon shell, similar to the cycle of five element overcoming.

## 11.5. Quaternity Interpretation of Eight Trigrams

In Figure 11.2c, we saw that two pairs of yin and yang produce four quadrants. As shown in Figure 11.7, three pairs of yin and yang would produce eight trigrams, forming an octagon. Relative to the observer sitting at the center, eight trigrams are located at eight orientations. Note that in the ancient Chinese literature, south was at the top and east was at the left, contrary to the orientations in the conventional maps. From the top trigram counting counterclockwise, they are Qian, Dui, Li, Zhen, Kun, Gen, Kan, and Xun, literally meaning sky, marsh, fire, thunder, ground, mountain, water, and wind, respectively. We may tentatively interpret the eight factors as heaven, biological activities, chemical reactions, electromagnetic phenomena, earth, solid, liquid, and gas in sequence.

Figure 11.7. Fuxi trigrams surrounding Taiji yin and yang fishes.

Furthermore, the overlay of two various trigrams produces 64 hexagrams, each having specific divine meanings. With six divided or whole lines stacked from bottom to top there are $2^6$ or 64 possible combinations, and hence 64 hexagrams. Based on the 64 hexagrams, "*Yi Jing*" consists of sixty-four short essays, enigmatically and symbolically expressed, on important themes, mostly of a moral, social, or political character.

People looking at the trigrams and hexagrams usually arrived at the conclusion that "*Yi Jing*" is the mother of binary numbers. In western nations, the binary system was first studied in detail by Leibnitz in 1678 and forms the basis for all computer and digital manipulations. Whether or not Leibnitz was inspired by the trigrams and hexagrams from "*Yi Jing*" in the binary number invention is moot. The significance of "*Yi Jing*" is beyond the binary system. Here we shall give Fuxi trigrams a fresh interpretation in terms of quaternity spacetime.

Each trigram consists of three lines: the top line, the middle line, and the bottom line. Each line is either a divided line or a whole line to represent yin and yang respectively. The relative positions and numbers of yin and yang lines within a trigram determine the meaning of that trigram. For example, Qian is a trigram with all yang; and Kun is a trigram with all yin. In the sense of yin and yang, Qian is active, dominating, and creative while Kun is receptive, yielding, and nurturing. The great attribute of heaven (Qian) and earth (Kun) is giving and maintaining life.

As was described in section 11.1, yin indicates space dimension and yang indicates time dimension. The middle line of a trigram always has extraordinary importance in determining the meaning of the trigram in "*Yi Jing*". If we regard the middle line as two lines (binary) and each of the top line and the bottom line as single, then Qian contains four time dimensions; Dui contains three time dimensions with one space dimension; Li contains two time dimensions with two space dimensions; Zhen contains one time dimension with three space dimensions; and Kun contains four space dimensions. The transformation from Qian to Dui is a rotatory operation, so are transformations from Dui to Li, from Li to Zhen, and from Zhen to Kun. The relationships between the trigrams are similar to those of $2s \rightarrow 2p_x \rightarrow 2p_y \rightarrow 2p_z \rightarrow 2s$ transformations. After maximum space has been reached, space and time switch roles in the opposite directions of quaternity axes during the counterclockwise rotations. In the transformations of Kun→ Gen → Kan → Xun → Qian, each step increases a time dimension while reducing a space dimension. Thus eight trigrams correspond to electron octet within neon shell. The arrangement of eight trigrams implies changes by rotatory operations.

Moreover, Taiji yin and yang fishes diagram was usually placed at the center of eight trigrams (Figure 11.7). Yin and yang fishes denote both 1s electron in helium shell; and eight trigrams denote electron octet in neon shell. From the theory of quaternity, the motion of both 1s electrons drives the motion of 2s2p electrons. They are coupled together by differential and integral chain rules. Taiji yin and yang fishes are encapsulated within eight trigrams. Even though the ancient Chinese didn't know anything about calculus and trigonometry, the unmistakable diagram indicates that they had mastered the gist of the natural law before the invention of eight trigrams. At least, I cannot imagine a better diagram than Figure 11.7 to express a dynamic four-dimensional structure such as a neon atom with ten electrons.

## 11.6. Summary

China has witnessed four thousand years of human civilization. The very headstream of this oriental civilization was documented in "*Yi Jing*", which gives us a valuable cultural inheritance. Through the above discussion, we have found that the ancient Chinese already developed a sound philosophy in medicine.

Yin and yang theory corresponds to space and time model in helium shell, and five element theory corresponds to space and time model in neon shell, both forming the foundation of traditional Chinese medicine. Fuxi eight trigrams delineate a four-dimensional spacetime structure visually; and four Chinese characters of Yuan, Heng, Li, and Zhen characterize four spherical quadrants literally.

Yin and yang theory is a general description of space and time relationship; and five element theory is more specific on the five characteristics or five phases of a four-dimensional spacetime. Five elements in traditional Chinese medicine (metal, water, wood, fire, and earth) correspond to scalor (time end), polor, metor, vitor, and scalor (space end) in quaternity spacetime.

Both theories are consistent with each other under the framework of quaternity spacetime. There is a large amount of research literature on "*Yi Jing*" and traditional Chinese medicine accumulated over long history. Of course, I cannot expect that all of the ancient knowledge reflects the truth. But I do realize that the foundation underlying traditional Chinese medicine is firm. This confirmation assures us that traditional Chinese medicine is a decent scientific discipline. It is an important step towards the serious integration of Eastern and Western sciences.

We have driven home the bright idea of spherical quantity in dynamic calculus through a multi-disciplinary approach touching on rather disparate fields in the domains of mathematics, physics, chemistry, biology, and philosophy.

Even though the goal of this book is extraordinarily ambitious, the law of nature is conservative. Careful readers surely find that the theme that has run through the entire book is unchanged. Spherical quantity in dynamic calculus is the complement of physical quantity in linear algebra. We have just brought the comprehensive theory in focus from a variety of visions.

In some cases, we exposed the unknown and made intellectual speculations or hypotheses, lending perceptual imagination in a coherent manner. We hope that this inferential composition was helpful in the understanding of the law of nature and would not dilute our great discovery of 2s2p orbital geometries.

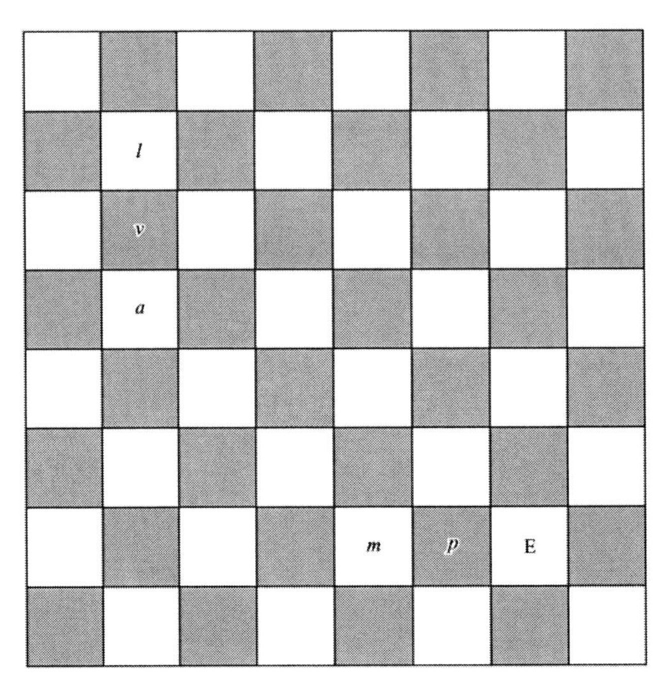


Motif 3. A periodic table of elements or physical quantities, a word puzzle with letters and vacancies, and a chessboard with law and order.

# APPENDIX I. REVIEW QUESTIONS

1.1. Please formulate wave equations for the cyclic motion of a free pendulum (Figure 1.4) and for the harmonic vibration of a ball under the action of springs (Figure 1.12). Are both equivalent mathematically in any way?

1.2. A circle may be regarded as a one-dimensional line such as a geodesic line around the earth (Figure Q1a), but a circle plus a center is regarded as a two-dimensional system (Figure Q1b, Q1c, and Q1d). Please explain this difference physically by looking at the following illustrations. (Hint: Think of a magnetic field line around electric current along a straight wire)

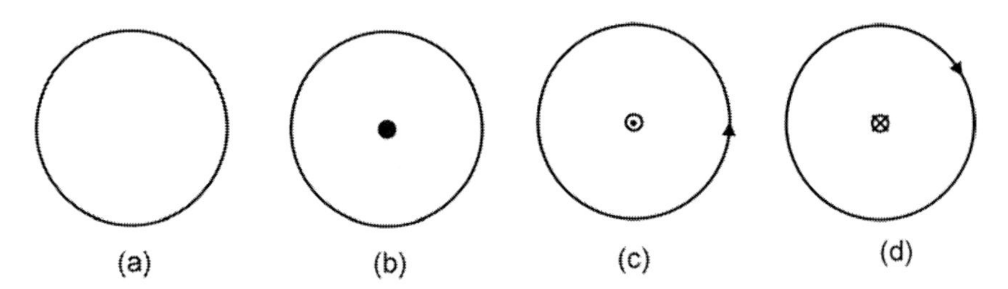


Figure Q1. Comparison of a circle without a center and three circles, each with a center.

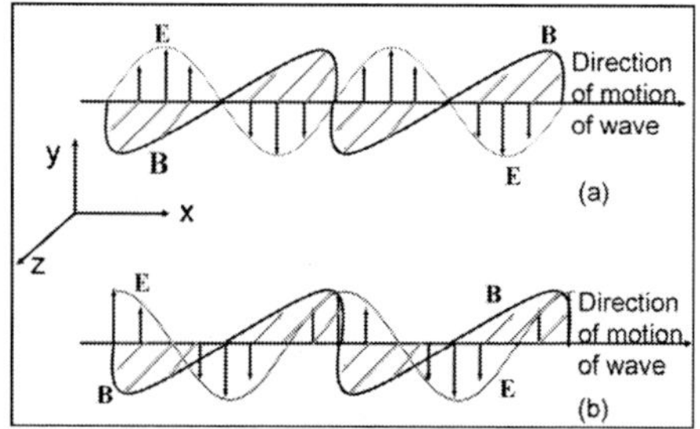


Figure Q2. Electric and magnetic field strengths in an electromagnetic wave. E and B are at right angles to each other. The entire pattern moves in a direction perpendicular to both E and B.

1.3. A sphere may be regarded as a one-dimensional radius, but a sphere plus a center is regarded as a two-dimensional system. Please explain this difference physically and biologically by looking at Figure 1.15. (Hint: A sphere indicates a 1s electron, but a sphere plus a center may refer to both 1s electrons oriented in radial and spherical surface directions, respectively. Biologically, a sphere may refer to an ovum, a haploid cell, but a sphere with a seed may refer to a fertilized ovum, a diploid zygote. )

1.4. Both Figure Q2a and Q2b show an electromagnetic wave. Which one is correct? (Hint: The wave function of the traveling wave in time is $C(\cos\alpha - i\sin\alpha)$, where the first and the second terms are orthogonal and separated by a phase of π/2. Many physics textbooks make a mistake!)

1.5. In parallel to equation

$$\cos x = 1 - \frac{x^2}{2!} + \frac{x^4}{4!} - \frac{x^6}{6!} + \ldots,$$

please express $\sin x$ in power series and discuss whether a smooth trigonometric function can be expressed by finite polynomial terms.

1.6. Please compare sine/cosine functions of the four roots to duality equation with +/- signs in four Cartesian quadrants (Figure 11.2). How is continuous two-dimensional spacetime related to Cartesian plane?

2.1. Given a circular ring on a plane whose outer diameter is three times as large as its inner diameter (Figure Q3), how much is its area by the unit of the area of the inner hole?

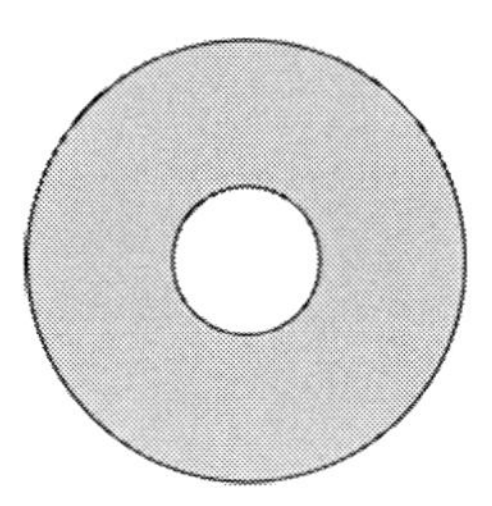

Figure Q3. A planar ring.

2.2. Flatland: In 1884, Edwin A. Abbott wrote a book titled "Flatland, a romance of many dimensions" describing the situation of two-dimensional creatures on a spherical surface. From the standpoint of the flatland inhabitants, the country is two-dimensional without the movements of rising up and sinking down. Straight lines, triangles, squares, pentagons, hexagons, and circles all appear as straight lines of the edges. But from the standpoint of the spherical center, those shapes appear as they are honestly. Quaternity theory regards the spherical surface as a three-dimensional geometrical shape since all shapes can move about the spherical surface, left and right, forward and backward, and up and down relative to the spherical central point. How does perspective shift alter geometrical shapes and dimensions in this case? Moreover, traditional geometry treats a point as an infinitesimal small point, a line as a very fine line, and a surface as one without thickness. This treatment deviates from real objects. In contrast, quaternity points have a certain size, and lines and surfaces possess proper thickness. Humans colonize the spherical surface of the earth. From the standpoint of humans, do we live in an environment like the flatland? Does the elimination of the thickness of geometrical elements on the ground (e.g., the tallness of mankind is negligible compared with the radius of the earth) change your answer? Does the imaginary flatland have any correspondence in the real world? If not, why not?

2.3. Dimension curling up mechanism: Trigonometric functions such as sine and cosine cast a radian angle into a real value; a compiler transforms a program of high-level programming language (such as C++ and Visual Basic programming languages) into machine language so that the program can be executed. How does the compiler insulate the dimension of machine codes from application programmers? Please compare compiler code insulation with trigonometric insulation of radian.

2.4. Please study the working principle of a four stroke engine and find out how to convert linear reciprocating motion into rotary motion by means of a crankshaft. Please study the mechanical function of a cam, a lever, and a spring and find out how to convert rotary motion into linear reciprocating motion. Please look at Figure Q4 and find out how the rotation of radian angle $\alpha$ around point A brings the motion of chord $\cos\alpha$ as point C orbits along the circle, like a crank driving the wheel. History tells us that the extent of exploitation of the wheels in any era is an unbiased indicator of a worker's productivity level of that era.

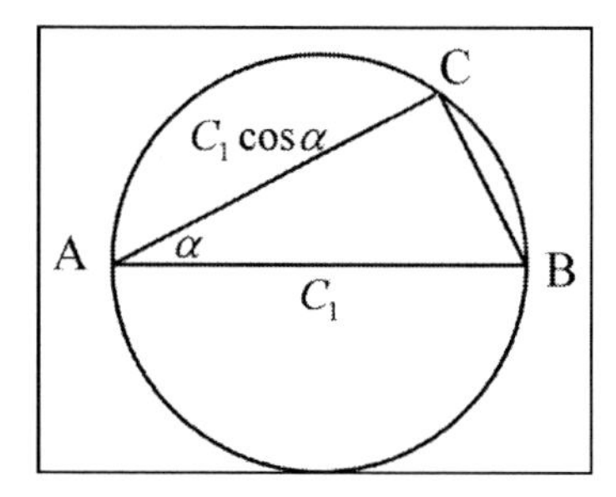


Figure Q4. The rotation of radian angle $\alpha$ determines the time component of an electron within helium shell.

2.5. Spacetime continuity: If one wave function is in the rotation pathway of another wave function, then they are continuous in spacetime. Please demonstrate that wave functions $\cos\Psi$ and $\cos\alpha$ are spacetime continuous if $\Psi = \cos\alpha$ by differential chain rule, but they are not continuous if $\Psi = 2\alpha$ instead. Spacetime continuity means that there is not any hole or break in a rotation pathway. For instance, the orbit of circular motion is continuous. Please demonstrate graphically that function $\cos(\Psi + \pi/4)$ is continuous to function $\cos\Psi$ via dynamic differentiation.

2.6. Mathematics is the abstraction of objective laws rather than the idealistic logic for applying indiscriminately and each mathematical operation has its special physical meaning. The plus sign, +, may be used in two different circumstances. It means addition when inserted between two physical quantities of similar dimensions; but it gives a complex number, or a vector when inserted between two physical quantities of orthogonal dimensions. In quaternity, a complex number is a wave function that undergoes rotatory operation so that the plus sign indicates a differential operation transforming the first term into the second term. The plus sign in dynamic calculus $\frac{d}{d\alpha}\cos\alpha = \cos(\alpha + \frac{\pi}{2})$ means a smooth functional shift from $\cos\alpha$ to $\cos(\alpha + \frac{\pi}{2})$ through the rotation of radian $\alpha$. Thus the plus sign indicates a dynamic differentiation. The plus sign in dynamic calculus signifies an action. For example, a man plus a man equals two men; a man plus a woman signifies marriage and sex, resulting in three people or more in the end; and a teacher plus a student means education. Could you give more vivid examples on both conventional and quaternity usages of the plus sign?

2.7. Unity but diversity: a pair of $2p_x$ electrons correspond to two points displaced from a center symmetrically; a pair of $2p_y$ electrons correspond to two unbound semicircular arcs connected by the two points into a circle; and a pair of $2p_z$ electrons correspond to two unbound hemispherical surfaces connected by the circle into a spherical surface; and a pair of 2s electrons correspond to the central point and an open solid sphere bounded by the spherical surface (Figure 2.14). These four symmetric pairs constitute four basic structures in neon shell and have interesting analogs in biological anatomy: a pair of eyes are symmetric and synchronized; a pair of ears are separated but interconnected; a nose has two symmetric nostrils; and the mouth and the anus are channeled by the alimentary canal. They function properly in response to four orthogonal vehicles of light, sound, smell, and flow, respectively. Is the analogy far-fetched or relevant to the general expression of the natural law? Could you list more symmetric and interrelated pairs of organs in a human body?

2.8. A magic number: 1.618*1.618≈2.618; 2.618*6/5≈Π. Π is the ratio of the circumference to the diameter of a circle. In dimension diagram, the diameter or the chord is represented by a spherical quantity whereas the circumference or the arc may be represented by a radian angle. A diameter is a diameter when viewed from the inner shell; but a diameter is a semicircle when viewed from the outer shell. Thus Π is the ratio of a spherical quantity viewed from the outer shell to that viewed from the inner shell. It is also interesting that Π symbolizes a door and was assumed to be the ratio of the height to the width of a door in ancient Greek buildings, say, 22/7. Can you remember the value of Π to ten significant digits?

3.1. As infinitesimal calculus and complex number drive mathematics towards idealism farther and farther, quaternity definition of dynamic calculus and dimensional interpretation of complex number bring it back to reality. Please describe the differences and connections of physical quantity and spherical quantity (Figure 1.1). Why can they express harmonic oscillation in a similar mathematical form?

3.2. Bidirectional light hypothesis: Light is an electromagnetic wave. From the property of a traveling wave and Maxwell's equations, it is easy to derive

$$\varepsilon_0 \mu_0 c^2 = 1$$

where $\varepsilon_0$ and $\mu_0$ are the electrical permittivity and magnetic permeability of free space respectively, and $c$ is the speed of light. Hence the speed of light can be written as

$$c = \pm \frac{1}{\sqrt{\varepsilon_0 \mu_0}}.$$

Traditionally, we adopt the positive value and discard the negative one without second thoughts. However, since each mathematical formula corresponds to certain physical reality, light might be composed of two elements traveling in opposite directions if we keep the negative as well as the positive values. Where light reaches, darkness disappears immediately.

So is it possible that darkness or the black is traveling in the opposite direction of light? Or is the negative value only a mathematical miscalculation?

4.1. There are two fictitious methods for characterizing the property of a man. The first one is to investigate his social relationships by asking who his parents are, who his wife is, who his children are, and what his job is. The second one is to detect his physical response upon an electric shock, say, how much the probability of his jumping over a preset hurdle is at the moment of the dangerous shock. Which approach, do you think, is more pertinent and relevant to the purpose of knowing the man? How quaternity theory and quantum mechanics are different in characterizing an electron?

4.2. Imagine the different pictures of a forest viewed from the perspective of an eagle in the sky and from the perspective of an ant residing in one of the trees!

4.3. What might be the different descriptions of today's event from the standpoint of yesterday and the standpoint of tomorrow? Do both descriptions deviate from today's reality as we are experiencing now? How does your answer help in evaluating the perspective and objectivity of historical records?

4.4. When David was a kid, his father expected him to become a CEO. David was indeed the CEO of a software company in the later part of his career, of which his son is very proud. What might be the different images of David from the expectation of his father and from the impression of his son? Do both images have any kinds of symmetry? (Hint: Is there any symmetry or complementarity between the images of a coffee mug viewed from the left and from the right directions?)

4.5. Standpoint shift: The earth is only one of many planets in the solar system, yet it is the only one that is known to bear life in the universe. From quarks, atoms, molecules, cells, lives, moons, planets, stars, to galaxies, human race is at the center of the sequence and is the most delicate entity of space and time convergence from the microcosm and the macrocosm. Human race is at the center of the universe in that we always examine the environment from the human perspective; the earth that we colonize is the center of the celestial sphere; and the sun and the moon revolve around the earth as we can see. Can you list some descriptions where human-centric conception is assumed, geocentric conception is assumed, and heliocentric conception is assumed, respectively? (Hint: Paradoxically, modern science is human-centric, but it suggests heliocentric viewpoint in astronomy).

5.1. By the principle of quaternity, space and time are complementary. Thus man is complementary to woman; forward is in contrary to backward; and a brave man is in contrast to a coward. How to define the significance of complementarity mathematically in set theory?

5.2. Rotatory operation in reality may refer to the functional or material connection between two things such as a chisel and wood, a plough and soil, a man and food, a bird and worms, a pen and ink, cheese and milk. Rotatory operation makes spacetime connections. With your wide-opened eyes, examine how things on the earth are spacetime continuous harmonically instead of a hodgepodge.

5.3. Dream: If a rotation pathway represents the rule of physiological activity for human, then human physiological activity must at least include a pair of semicircles in diurnal cycle, say, doing consciously in daytime versus dreaming passively at night. What dreaming in sleep tries to accomplish is to recover the body to its original state in yesterday morning under the direction of the natural law and through a complementary biochemical channel in sleep to that of daytime activity. Please compare the pair of doing and dreaming with both electrons in helium shell that are symmetry in spacetime (Figure Q5), i.e., the time component of one electron is equivalent to the space component of another, and vice versa.

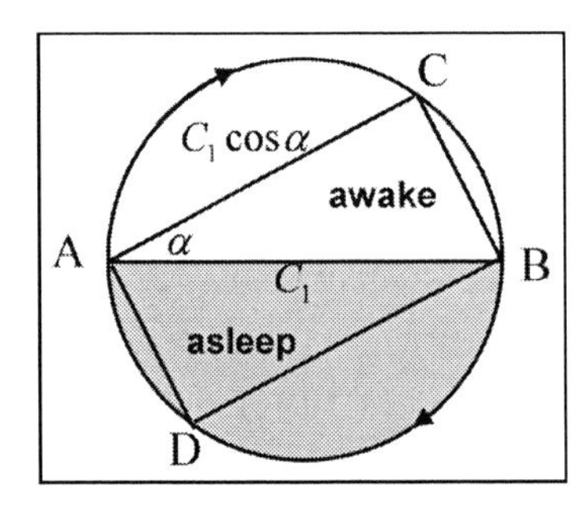


Figure Q5. The complementary relationship between doing and dreaming analogous to that of both electrons in helium shell.

6.1. By the principle of rotatory operation, two electrons within helium shell satisfy harmonic oscillation equation $\Omega = \Omega^{02}$, and eight electrons within neon shell obey harmonic oscillation equation $\Phi = \Phi^{08}$ or $\Phi = u^4\Phi^{04}$ where the times of rotatory operations (and hence electron numbers) must be an even number for the wave equations to have proper solutions because a cosine function returns to the same cosine form only after being differentiated twice, and so does a sine function. Consider a possible solution of harmonic oscillation equation $\Phi = \Phi^{06}$ and its implication in chemistry? (Hint: Six-electron structure is not a stable configuration for an atom, but it might be a perfect pattern for a molecule such as benzene as shown in Figure Q6).

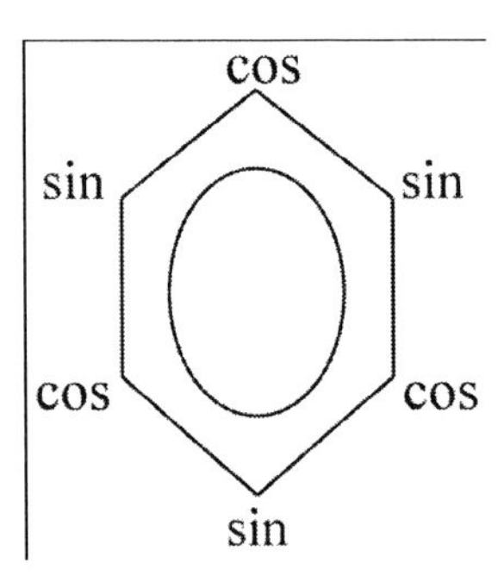


Figure Q6. The structure and waveforms of a benzene molecule.

6.2. Collect all kinds of snails and examine whether their spiral curves are directed in the same chiral orientation (Figure Q7).

7.1. Please visit a rope-building factory to learn various braiding patterns of ropes. What are the basic properties of a rope? How ropes are used in a boat and how to tie various kinds of knots for those usages? Moreover, it is well believed that humans recorded things and events by tying knots in ropes long before the advent of characters. Please list various usages of ropes in business and imagine how the world would be without ropes.

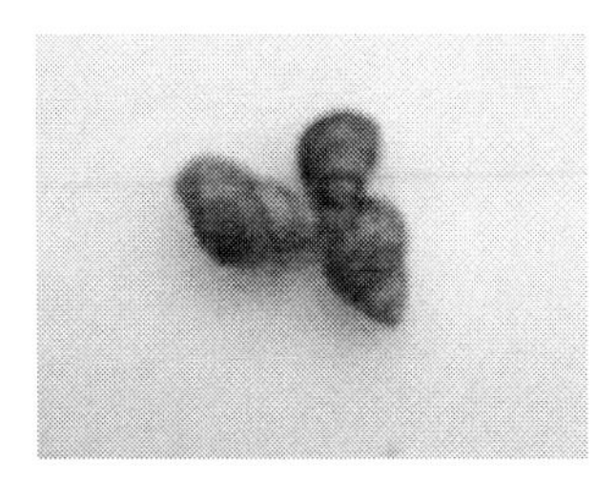

Figure Q7. Three field snails.

7.2. Please calculate the length of a helix around a cylinder with radius R and height H (Figure Q8).

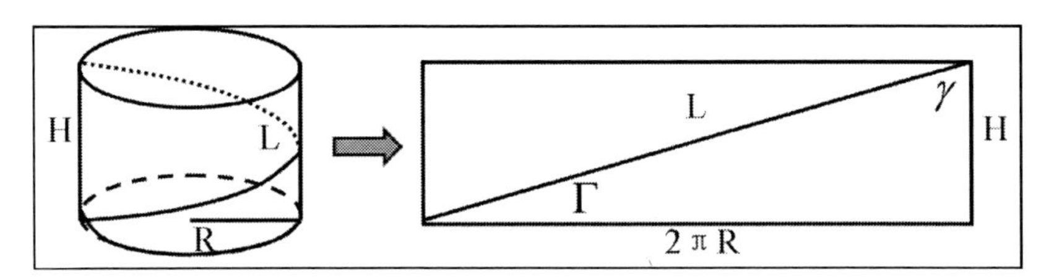


Figure Q8. The relationship between a helix around a cylinder and the hypotenuse of a right triangle.

7.3. Please write down the definition of curvature and torsion of a three-dimensional curve in Cartesian coordinates and calculate how much they are for a strand of B-DNA double helix.

7.4. Please calculate the length of curve of $y = \cos x, x \in [0, \pi/2]$. (Answer: 1.91).

8.1. The variable of a certain biological property observes trigonometric function with its probability density as

$$f(x) = A \sin x ,$$

where $A$ is a constant. Suppose $x$ value is confined to the scope of $[0, \pi]$, based on the nature of a density function, please calculate the value of $A$, expectation $E(x)$, and variance $D(x)$. Please draw the graph of the trigonometric function and the graph of a normal distribution function with equal expectation and variance on the same worksheet and explain their difference. (Answer: $A$=1/2, $E(x)=\pi/2$ , $D(x)=\pi^2/4-2$)

8.2. The 80/20 rule: Italian economist Vilfredo Pareto proposed the 80/20 rule for wealth distribution after discovering that, in Italy, 80 percent of the country's wealth was held by 20 percent of the country's population. Thus it was inferred that for the typical product category, eighty percent of the products sold will be consumed by twenty percent of the customers. The 80/20 rule was later adopted for measuring progress. For example, after spending about 20% of the time, many programmers claim that they have completed 80% the work. Later reality strikes from the back. Upon having spent another 80% of the time, the same programmers come to realize that they have resolved only 20% of the remaining problem.

a) Suppose in a more egalitarian community, wealth distribution follows a probability density function

$$f(x) = \frac{1}{2}\sin x, x \in [0, \pi],$$

where $x$ refers to relative wealth value. How much percentage of the wealthiest people owns half of total community wealth? (Answer: 33%)

b) Suppose a learning progress with time consumption follows the curve of sine

$$f(x) = \sin x, x \in [0, \pi / 2]$$

where $f(x)$ reaches 100% after one spends $\pi/2$ unit time interval, how many percent of time must be spent to finish the first and last 20% workload? (Answer: 12% and 41%)

8.3. A biological variable has a density probability of

$$y = \frac{a}{2}\cos a(x - \mu), x \in [\mu - \pi/2a, \mu - \pi / 2a],$$

where $a$ is a constant and $\mu$ is the expectation of $x$. If the variance of $x$ is $\sigma^2$, then how much is $a$ in terms of $\sigma$ ? (Answer: $a = \frac{1}{\sigma}\sqrt{\pi^2/4 - 2}$ )

8.4. Are there any structural similarities between an apple and the earth?

8.5. A seed grows into a tree via rotatory operation $\int(-\partial/\partial t)dl$ in the sense that the height of the tree increases as time elapses. Thus the seed represents dense time and the tree is the unfolding space, both being continuous and symmetric in quaternity. The tree bears another seed through a mechanism other than the reversal of growth, thereby completing a full cycle. Could you give more examples of continuity, symmetry, and cycle like the life cycle of the tree?

8.6. Please compare the blood circulation in a human body with electronic orbitals within a neon atom. Note that the heart pulse from right atrium to right ventricle may be characterized by wave function $(\cos\alpha - \omega\sin\alpha)$ while the heart pulse from left atrium to left ventricle may be characterized by wave function $(\cos\beta + r\sin\beta)$. Four chambers of a heart correspond to four terms in equation (1.9) for 1s electrons. The systemic circulation corresponds to $2s2p_x2p_y2p_z$ electron quartet in counterclockwise rotations while the pulmonary circulation corresponds to another quartet in the negative direction of quaternity axes. Do cardiograms help in establishing the analogy?

8.7. How many loops are shown in Figure 8.1? Which loop represents properly the whole universe? If you walk along the loop, can you experience ever larger cycles all the time without an end? Which part of the loops represents traveling light?

8.8. Please open an old-fashioned mechanical watch and study the working principle of cog wheels in mesh. How do cog wheels in mesh ensure the accuracy of time counting? Can you mention some other devices where cog wheels in mesh play a crucial role?

9.1. The earth revolves around the sun making a year in each circle. During the same period, a longan tree in front of my house experiences four seasons. Does the longan tree undergo a cycle in any sense similar to the circular motion of the earth? Are both cyclic phenomena tightly coupled together in any mechanism? During the same period, a flock of geese migrate from Canada to the United States in the fall and return to the north in the spring when the weather is getting warmer. How to characterize their migrating cycle in comparison with the earth's orbit and the longan's experience? (Hint: Circle is cycle and vice verse)

9.2. Angular momentum: The earth has orbital angular momentum attributable to its annual revolution around the sun, and spin angular momentum to its daily rotation about the north-south axis. Suppose the earth does not rotate at all, there would be only one day in a year resulting from revolution, and the revolving speed and the circumference of the earth's elliptical orbit would determine the length of the day. Can the earth's angular momentum be converted into spin momentum? How are they coupled together? (Hint: Look at Figure Q9 for a visual aid on the relationship between a circle and a twisted band and tell how the length of the circumference influences the torsion of the twisted band?)

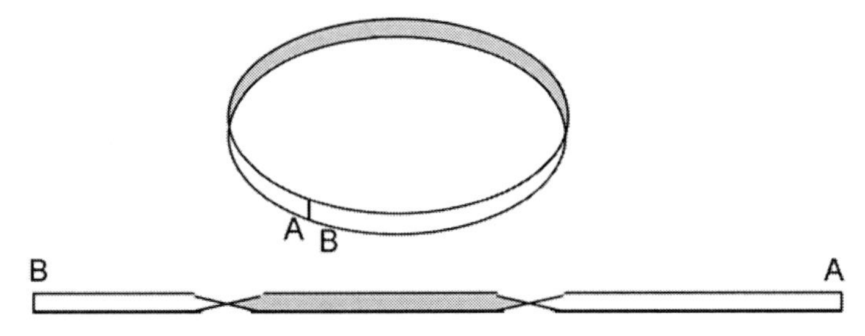


Figure Q9. The topological transformation of a circular band into a 360° twisted linear band by cutting at A-B and translating both ends in the opposite directions.

10.1. Assuming the pattern of nature repeats in various levels to a certain extent as exemplified by the periodic table of elements, where electron configuration somewhat repeats in every row, please compare the structure of a human society with the anatomy of a human body.

10.2. Dimension algebra: Ignoring the positive and negative signs of wave functions, the spacetime dimensions of eight electrons in neon shell are in the sequence of

$$1 + \dot{\Psi} + u(1 + \dot{\Psi}) + u^2(1 + \dot{\Psi}) + u^3(1 + \dot{\Psi}) +$$
$$u^4(1 + \dot{\Psi}) + u^5(1 + \dot{\Psi}) + u^6(1 + \dot{\Psi}) + u^7(1 + \dot{\Psi}),$$

which equals the product of four factors $(1+\dot{\Psi})(1+u)(1+u^2)(1+u^4)$. How does dynamic calculus modify the sign of each dimension in the power series? (Hint: see equations 2.18 and 2.25)

10.3. Ideal gas obeys the law $PV = nRT$, where $P$ denotes pressure, $V$ volume, $T$ absolute temperature, $n$ number of moles of gas molecules, and $R$ universal gas constant.

a) Regarding each human individual as a free molecule, we might consider a human society to be ideal gas where $P$ denotes working pressure, $V$ surviving roomage or job opportunity, $T$ knowledge and technology (or the capability of an average man), $n$ number of people, and $R$ human activity constant. In a society, is working pressure lower when more job opportunities are available? Does technology intensify the working pressure of workers? Does technology create more opportunities for employment? Does technology progress mean higher degree of mixing between people of different origins? If yes, then please evaluate seriously the applicability of ideal gas law to a human society with the conceptual substitution.

b) Regarding each star as a free molecule, can the whole universe or a galaxy be treated as star gas that obeys ideal gas law?

c) Regarding electron cloud as ideal gas, does electronic behavior obey ideal gas law? Can $nR$ be treated as the entropy of the electron so that $PV = ST$ ?

10.4. Light is an electromagnetic wave. If light travels by electromagnetic induction as shown in Figure Q10(a), then as photons move forward at an average speed of $c = 3.0 \times 10^8$ m/s, how much is the speed of the corresponding electromagnetic wave traveling along semicircle AWB as shown in Figure Q10(b)? (Answer: $0.5\pi c$ )

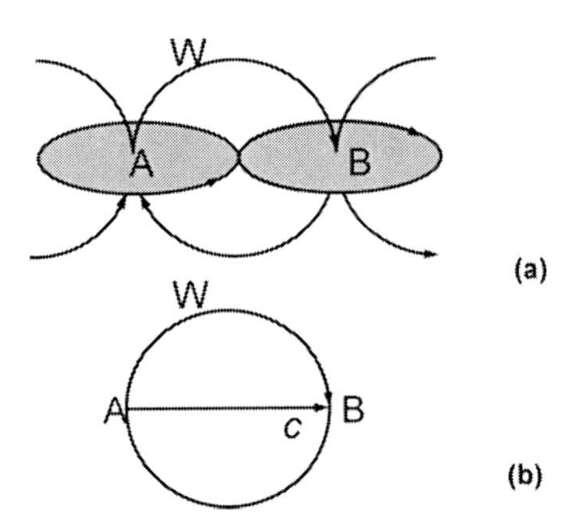


Figure Q10. Schematic diagram of light traveling through electromagnetic induction.

11.1. According to Chinese folklore, if a toxic herb grows in a region, then it is likely to find an antidotal herb against it in the nearby area. Does the ancient empirical medicine agree with the principle of space and time complementarity?

11.2. Chinese ancestors had created eight trigrams (Figure 11.7) by observing celestial bodies in the sky and studying entities on the ground before the advent of characters five thousand years ago. Please read the trigrams carefully and assign each trigram a binary number in a logical way. Furthermore, how to relate the eight trigrams to electronic octet in neon shell? Please refer to the vast Chinese literature of "*Yi Jing*" (or "*I Ching*") and try to explain the philosophy of quaternity, keeping new terminology to a minimum.

11.3. Quantitative to qualitative change: History shows that social and economic orders are in a process of change. Marx's dialectical materialism maintains that what causes change is simply alteration in the quantity of things, which eventually leads to something qualitative new. For example, as I increase the temperature of water, it not only becomes warmer, but finally reaches the point at which this quantitative change changes it from a liquid into a vapor. Reversing the process, by gradually decreasing the temperature of water, I finally freeze it from a liquid to solid ice.

a) Look at Figure 1.6 and examine quantitative to qualitative change in cosine function algebraically. Does the value of $\cos x$ changes sign when variable $x$ gradually increases from 0 to over $\pi/2$? Does the slope of the curve of changes sign when $x$ gradually increases from $\pi/2$ to over $\pi$?

b) Look at Figure 2.12 and examine quantitative to qualitative change in orbital type. Does $2p_x$ orbital changes into $2p_y$ orbital when radian variable $\Psi$ increases from 0 to over $\pi/2$? Does $2p_y$ orbital changes into $2p_z$ orbital when $\Psi$ increases from $\pi/2$ to over $\pi$? Please describe these changes in geometry.

11.4. Please study the process of space dimension increment of an electron in consecutive rotations (Figure 2.13) and tell why the future of a youngster is generally bright but the course of a career is usually sinusoidal.

11.5. Please compare the trend of function $y = \arctan x$ with that of a logistic curve function in the biological application.

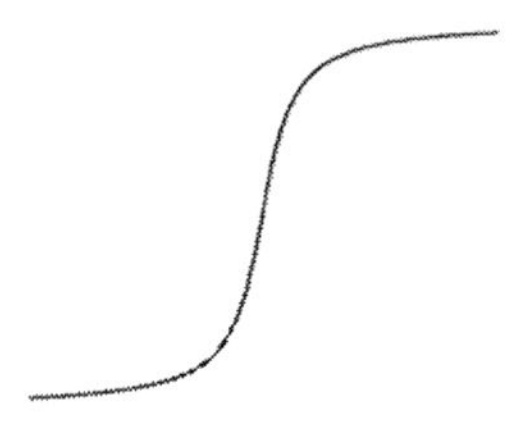

Figure Q11. The Sigmoidal curve of $y = \arctan x$ as a logistic curve in biology.

11.6. He who would behold trigonometry in the greatest glory could not find its use better than in the implementation of dynamic calculus. In helium shell, both 1s electrons transform between each other:

$$\cos\alpha\cos\beta \Leftrightarrow -v\sin\alpha\sin\beta,$$

$$K = \frac{-v\sin\alpha\sin\beta}{\cos\alpha\cos\beta} = -v.$$

The value of $K$ is independent of phases $\alpha$ and $\beta$, which are complementary at any moment. Please compare the property of $K$ with that of a chemical equilibrium constant.

# APPENDIX II. GLOSSARY OF KEY CONCEPTS

The invention of a fundamental theory from scratch invariably requires a large new terminology. In order to avoid misunderstanding, it is important to be aware of the fresh, non-classical meanings of the following words and phrases that we have deliberately created in this book.

**Anisotropic orbitals** Electronic orbitals unequally distributed along the X, Y, and Z directions geometrically. Specifically, a $2p_x$ is a linear segment along X-axis; a $2p_y$ is a planar sector on X-Y plane; a $2p_z$ is a hemispherical surface with a vertex on Z-axis; and a 2s orbital is a solid sphere.

**Asymmetric carbon** A carbon atom bonded to four different substituent groups giving its optical activity.

**Atomic orbital** An unoccupied spacetime dimension in an atomic sphere. When occupied by an electron, it is called an electronic orbital.

**Atomic shell octet configuration** The configuration of eight atomic shells including hydrogen, helium, neon, argon, krypton, xenon, radon, and Uuo shells in a Uuo atom analogous to electron octet in neon shell. Each atomic shell corresponds to an inert atom except hydrogen.

**Central force** The force that guides the surrounding bodies along circular or elliptical orbits; or the force that attracts electrons in various harmonic oscillations in the atomic sphere.

**Centromere** The state point that connects with four chords (two time and two space components) and moves from one end to the other along a semicircular arc in dimension diagram during the rotation of a spherical quantity.

**Chirality** The structural characteristic of a molecule that makes it impossible to superimpose it on its mirror image.

**Circular complex function** A complex function comprising a complete set of either time or space dimensions within an inert atomic shell. Every two adjacent terms of a circular complex function bear differential or integral relationship.

**Clockwise rotation** A differential operation with respect to a space dimension along with an integral operation over a time dimension upon a spherical quantity in the meantime.

**Complementarity of orbitals** The relationship of two electronic orbitals sharing a full spatial sphere but mutually exclusive at any moment during the whole period of their transformations, such as the relationship between $2p_x$ and $3p_z$ electrons and that between 1s and $3d^{1s}$ electrons.

**Complementarity of space and time** A spacetime principle in which the number of space dimensions plus that of time dimensions equals four in the four-dimensional electronic orbitals. Reducing a time dimension inevitably increases a space dimension during the spacetime trade-off, and vice versa.

**Complex function** A wave function that spans two space or time dimensions with a dimension factor as complex function identifier.

**Complex number notation, complex number identifier** A dimension factor that functions as the imaginary unit *i* in a complex function.

**Continuous variable** A variable that at least theoretically can assume an infinite number of values between any two fixed points in its valid range.

**Coordination operation** A dot operation between both time and space circular complex functions to produce a complete set of spherical quantities in an inert atomic shell. It is different from multiplication. When both circular complex functions are placed parallel, only adjacent terms between both functions are associated in coordination operation.

**Course of dynamic calculus** The semicircular arc in a dimension diagram, along which a dynamic calculus is performed upon a spherical quantity. This is the course of an electron in quaternity coordinates, which has diverse manifestations in geometric paths.

**Density gradient** of an electron An electronic orbital in dynamic transformation from one state to another rendering density gradient of the orbital in time series.

**Differential chain rule** A differential rule that involves wave functions in two spherical layers geared together: $-\frac{d\Theta}{dt} = -\frac{d\Theta}{d\Psi}\cdot\frac{d\Psi}{dt}$, where $-\frac{d\Theta}{dt}$ initiates wave functions for electron octet in neon shell while $-\frac{d\Psi}{dt}$ indicates wave functions in helium shell.

**Dimension** A diameter in dimension diagram for an electron to traverse via its subtended semicircular arc during dynamic differential or integral transformation.

**Dimension diagram** A diagram showing a time or a space dimension as a diameter of a semicircle and two orthogonal time or space components as chords subtended by the semicircle, both chords and the diameter always forming a right triangle so that Pythagorean theorem prevails in dynamic differential or integral operation.

**Dimension encapsulation, dimension curling up** The mechanism by which dimensions of an inner sphere are hidden in a gateway function from an outer spherical layer. Given $-\frac{d\Theta}{dt} = -\frac{d\Theta}{d\Psi}\cdot\frac{d\Psi}{dt}$, the trigonometric expression of $-\frac{d\Theta}{dt} = -A_1\cos\Psi$ constitutes a dimension encapsulation of 1s electrons.

**Dimension factor** A gateway function serving as a dimension indicator and complex number identifier when placed before a trigonometric function. But it may be regarded as a differential or integral operator when associated with a spherical quantity directly.

**Discrete variable** A variable that has only certain fixed numerical values, with no intermediate values possible in between, such as the dimension of a physical quantity.

**Dualism** The view that the world consists of or is explicable as two fundamental entities, such as mind and body. Under spherical view, mind and body are represented by time and space quantities respectively, which are intertwined through rotatory operation.

**Duality equation** A partial differential equation describing the spherical quantity of electrons in two dimensions in helium shell: $\frac{\partial^2 \Omega}{\partial t^2} = v^2 \frac{\partial^2 \Omega}{\partial l^2}$.

**Dynamic calculus** Differential or integral operation on a spherical quantity implemented by smooth rotation of a radian angle in sine or cosine function for $\pi/2$ displacement, conceptually different from infinitesimal calculus. Dynamic calculus stands for a physical process rather than a mathematical limit.

**Electron cloud** The volatile medium of an electron in the atomic sphere. An electron within an atom is normally smeared out over a large region of space even though it is always a particle when actually detected.

**Electronic orbital** An electron in harmonic oscillation within an inert atom such as helium and neon that can be characterized by a spherical quantity in dynamic calculus.

**Electronic rope** Two or more electrons in a spherical layer twisted circularly by rotatory operations. For example, if we regard rotatory operator $\int(-\partial/\partial t)dl$ upon an electron as a kind of twisting tension, then electron transformations of $\Omega_0^0 = \Omega_2$ and $\Omega_2^0 = \Omega_0$ constitute the art of making a rope with $\Omega_0$ and $\Omega_2$ as two participating strands.

**Electronic wave** Although an electron is traditionally regarded as a particle, it is a wave of electron cloud characterized by a wave function in the atomic sphere.

**Endosymbiosis** According to endosymbiotic hypothesis by Lynn Margulis, self-replicating cellular organelles such as mitochondria and chloroplasts were originally free-living organisms that entered into a symbiotic relationship with nucleated anaerobic cells. Mitochondria are believed to descend from aerobic bacteria, while chloroplasts derive from cells that were similar to modern cyanobacteria.

**Entity** An object or a living organism in its entirety such as a photon, an electron, an atom, a DNA molecule, a cell, a bacterium, an organ, a plant, an animal, the earth, the solar system, and the universe. Harmonic oscillation is the theme of the behavior of an entity.

**Euclidean space** A linear vector space that satisfies certain properties of the scalar product operation on pairs of tangent vectors.

**Feynman diagram** A diagram illustrating the indirect interaction of electrons through emitting and receiving a virtual photon in between.

**Five element theory** An ancient Chinese theory regarding five elements (metal, water, wood, fire, and earth) as the basic constituents of matter. In traditional Chinese medicine, five characteristics or five phases are derived from five elements to explain the interactions and relationships between biological phenomena.

**Gateway function** A complex number identifier serves as an intermediate variable in calculus chain rule and as a connector between spherical quantities of an inner sphere and an outer sphere. For example, $\dot{\Psi}$ in wave function $A_1 A_2(\cos\Psi\cos\psi - \dot{\Psi}\sin\Psi\cos\psi)$ is a gateway function.

**Gateways combination operator** A subscript symbol "$_{\otimes}$" placed between two gateway function variables such as $\overline{\Gamma}_i \,_{\otimes}\, \phi_j$ to denote valid f-orbitals where $\phi_j$ attaches to wave function $\overline{\Gamma}_i$.

**Geodesic curve** The shortest curve between two points on a spherical surface. The whole curve is a circle dividing the sphere into two equal hemispheres.

**Geometric elements** The basic elements of geometry such as points, lines, planes, and solids.

**Gyroscope** A device consisting of a spinning mass, typically a disk or wheel, mounted on a base so that its axis can turn freely in one or more directions and thereby maintain its orientation regardless of any movement of the base. Under spherical view, we replace each directions of movement with an oscillatory cycle or a ring. A human may be regarded as a multiple-ring gyroscope, each ring representing a physical, physiological, or psychological cycle in a certain manner.

**Harmonic oscillation, simple harmonic oscillation** A two-dimensional uniform circular motion or its diverse projections in one, three, and four space dimensions.

**Helium shell** The spacetime sphere containing both 1s electrons.

**Integral chain rule** The counterpart of differential chain rule in dynamic calculus that involves integral operations in two spherical layers. For example, $\int A_2 \cos\psi dl = \frac{dl}{d\psi}\int A_2 \cos\psi d\psi = -\frac{1}{\psi'}(A_2 \sin\psi) = A_2 \sin\psi \cdot \int C_2 \cos\beta dl$ where $\int A_2 \cos\psi dl$ is an integral operation over a space dimension in neon shell and $\int C_2 \cos\beta dl$ is an integral operation over a space dimension in helium shell triggered by $-\frac{1}{\psi'}$ after the fullness of $\int A_2 \cos\psi d\psi$ dimension, both forming a chain.

***LC* circuit** The oscillatory circuit composed of a capacitor and an inductor disregarding electric resistance along the wires.

**Meiosis** The division of spherical quantities in an inert atomic shell into two circular complex functions. It is somewhat analogous to the meiosis of cells because time and space variables are separated during the process and hence the number of variables is reduced by half as a result.

**Membrane function** The wave function constructed by encapsulating a spherical layer, functioning as a membrane of the spherical layer as if plasma membrane would enclose the cytoplasm in a cell.

**Metor** A spherical quantity in the second spherical quadrant.

**Natural law, the law of nature** The rules and properties of dynamic calculus that govern spherical quantities, in stark contrast to physical laws based on physical quantities.

**Neon shell** The spacetime sphere containing electron octet of $2s^2 2p^6$.

**Newtonian time** Absolute, true and mathematical time, of itself, and from its own nature, flows equably, without relation to anything external. A linear or uniform time in contrast to sinusoidal time in spherical view.

**Orbital hybridization** A traditional but wrong hypothesis that when carbon is part of an organic structure, its 2s and 2p orbitals may mix together to produce orbitals of different property. There are three possible types of orbital hybridization, namely $sp^3$, $sp^2$, and sp hybridization.

**Orbital synchronization** The way electronic orbitals of various kinds such as $2p_x$, $2p_y$, $2p_z$, and 2s within neon shell transform by rotatory operations at the same phase at any moment.

**Orthogonal** In the atomic spacetime, spherical quantities $\Phi_0$ and $\Phi_1$ are orthogonal if they are at exact dimension interval or have whole differential or/and integral relationships. A spherical quantity may transform into its orthogonal spherical quantity.

**Partial rotatory operation, partial rotation** The partial effect of a rotatory operation relative to an observer when the rotatory operation does not proceed swiftly from the viewpoint of the observer.

**Path of vector calculus** A line segment, a semicircular arc, a hemispherical surface, or a solid sphere along which a vector calculus is carried out in one, two, three, or four space dimensions respectively. This is the geometric path of an electron in Euclidean space. There are paths of vector differential calculus as well as paths of vector integral calculus.

**Periodic table of physical quantities** A periodic table listing all physical quantities according to their space and time dimensions in parallel to Mendeleev's periodic table of chemical elements.

**Physical laws, the laws of physics** The rules and properties of one or more physical quantities; or the relationships that govern two or more physical quantities.

**Physical quantity** A quantity that assumes a fixed dimension and a continuous varying magnitude.

**Polor** A spherical quantity in the first spherical quadrant.

**Quaternity axes** Four dimensional factors $1$, $u$, $u^2$, and $u^3$ that serve as spacetime axes in quaterntiy coordinate system.

**Quaternity coordinates** A schematic diagram that adopts four dimensional factors as four axes ($1$, $u$, $u^2$, and $u^3$) to illustrate the intricate dimensional relationships between various spherical quantities in dynamic calculus instead of facilitating metric measurements between positions in Cartesian coordinates.

**Quaternity equation** A partial differential equation describing spherical quantities of electrons in four dimensions in neon shell: $\frac{\partial^4 \Phi}{\partial t^4} = u^4 \frac{\partial^4 \Phi}{\partial l^4}$.

**Quaternity periodic table of elements** A periodic table of chemical elements tabulated according to quaternity spacetime under the columns of s-orbitals, $p_x$-orbitals, $p_y$-orbitals, $p_z$-orbitals, and s-orbitals.

**Quaternity perspective** The perspective of observing an entity from outside its activity sphere and at a proper distance and time interval that match the size and oscillating rhythm of the entity.

**Quaternity spacetime** Four-dimensional spacetime such as neon shell characterized by polor, metor, vitor, and scalor, respectively.

**Quaternity space** The notion of four space dimensions within a spherical layer.

**Quaternity, quaternity theory, the theory of quaternity** A four-dimensional spacetime theory that defines the structure of an inert atom by spherical quantities in dynamic calculus.

**Rotatory operation, rotation, counterclockwise rotation** A differential operation with respect to a time dimension along with an integral operation over a space dimension upon a spherical quantity in the meantime.

**Rotatory vector** A position vector on X-Y plane that rotates about the origin O for $\pi/2$ to represent the smooth transformation of a spherical quantity from X dimension to Y dimension.

**Scalability** The feasibility of repeatedly using four spherical quantities (polor, metor, vitor, and scalor) in sequence to characterize a multilayered sphere such as a krypton atom.

**Scalor** A spherical quantity in the fourth spherical quadrant.

**Selective electron transfer** Electron transfer through an asymmetric carbon along a selective pathway from a covalent bond to another due to the chirality of the carbon atom.

**Semicircle** A semicircular arc subtending a diameter that connotes the course of an electron in harmonic oscillation or the course of a spherical quantity in dynamic calculus.

**Set** A collection of objects (or elements) or a container of objects. In this book, a set refers to a part of spacetime continuum of certain dimensions, or a spherical quantity.

**Space** True physical existence that transforms sinusoidally in a certain dimension with relation to time.

**Spacetime continuity** The property of the atomic spacetime where space and time components of electrons are transforming smoothly and continuously from one state to another without any interruption.

**Spacetime continuum** A continuous spacetime sphere as a whole relatively independent of its environment.

**Sphere** Traditionally, a sphere is a three-dimensional surface, all points of which are equidistant from a fixed point. But in the atomic spacetime, a sphere always refers to a solid spherical object, not a surface. Helium sphere is two-dimensional in the atomic spacetime, and neon sphere is four-dimensional in the atomic spacetime.

**Spherical layer** An outer sphere minus a concentric inner sphere or the medium between two concentric spherical surfaces.

**Spherical quadrants** The four regions into which a sphere is divided by four spacetime axes ($1$, $u$, $u^2$, and $u^3$), designated first, second, third, and fourth, counting counterclockwise from the region bounded by $1$ and $u$ axes.

**Spherical quantity** A quantity that assumes a fixed magnitude and a continuous varying dimension in complement to physical quantity. True physical existence that transforms by dynamic calculus instead of linear algebra. (Please also refer to time and space definitions).

**Spherical view** The scientific approach of adopting spherical quantities in dynamic calculus to characterize electrons within atoms; the point of view that entities transform by dynamic calculus of spherical quantities.

**Symmetry of space and time** The equivalence of a time and a space dimensions, with respect to which differential operations are performed upon a spherical quantity $\Omega$ observing $-\frac{\partial \Omega}{\partial t} = v\frac{\partial \Omega}{\partial l}$.

**The atomic spacetime** The spacetime within an atomic sphere where time and space components of electrons obey dynamic calculus instead of linear algebra. For example, helium shell is a two-dimensional atomic spacetime, and neon shell is a four-dimensional atomic spacetime.

**The atomic structure** The organization and relationship of electronic orbitals within an atomic shell based on spherical quantities in dynamic calculus.

**The natural pattern** The order of entities created and governed by the law of nature.

**Time** True physical existence that transforms sinusoidally in a certain dimension with relation to space.

**Uniform circular motion** The kinematic movement of an object in a circle at a constant speed. The projection of uniform circular motion onto a one-dimension profile is harmonic oscillation.

**Virtual photon** The hypothetical photon through which electrons interact with each other within an atomic sphere. It physically implements the motion of an electron by dynamic calculus.

**Vitor** A spherical quantity in the third spherical quadrant.

**Wave function** A physical quantity satisfies a certain wave equation for describing the motion of a particle. Or a spherical quantity expressed in sine and/or cosine functions.

**Yin and yang theory** A fundamental theory underlying traditional Chinese medicine. Yin and yang are two complementary, interdependent, and opposite properties or entities within a greater whole. Yin refers to feminine, passive, and negative principle in nature; and yang refers to masculine, active, and positive principle in nature. They are constantly transforming into each other.

**$\pi$-bond** In quaternity theory, a $\pi$-bond is the partial overlap of two $2p_y$ flat sectors in one atom with two $2p_y$ flat sectors in another atom, or is the partial overlap of two $2p_z$ hemispheres in one atom with two $2p_z$ hemispheres in another atom on the edge. A $\pi$-bond is a directionally oriented and unsaturated covalent bond in contrary to a $\sigma$-bond.

# APPENDIX III. THE LAW OF NATURE IN MATHEMATICAL FORMULAE

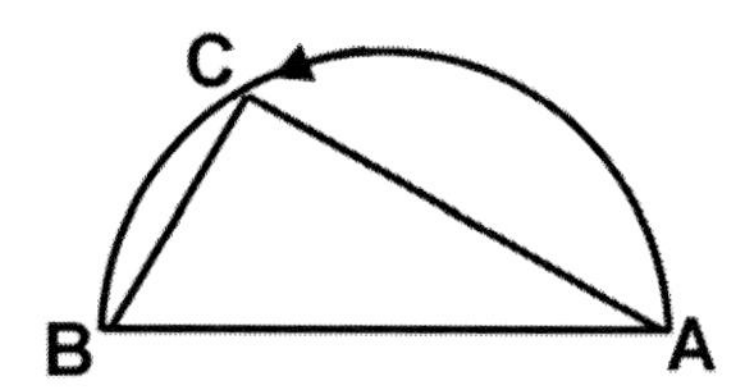


We inherit the ancient Chinese worldview that the earth is square land under domed sky, but give it a fresh interpretation under dimension diagram. As point C traces the arc, the displacements of AC and BC are always measured by chords. Diameter AB is the dimension that C traverses via the semicircle. The law of nature is spherical quantity in dynamic calculus, which invariably describes an entity tracing the course of the semicircle, but manifests in diverse mathematical forms. Here we list some of them that have been shown in the text. Please read the contexts and see how they are fundamentally equivalent as the paradigms of dynamic calculus. The law of nature must be simple to allow diversity and single to achieve unity. In this way, it rules the natural world robustly and unifies the biological kingdoms harmonically.

Harmonic oscillation equation

$$\frac{d^2\Psi}{dt^2} = -\omega^2\Psi \tag{1.1}$$

$$\frac{\partial^2\psi}{\partial l^2} = -\frac{1}{r^2}\psi \tag{1.4}$$

Duality equation versus Schrödinger's equation

$$\frac{\partial^2\Omega}{\partial t^2} = v^2\frac{\partial^2\Omega}{\partial l^2} \tag{1.13}$$

$$i\hbar\frac{\partial\xi}{\partial t} = -\frac{\hbar^2}{2m}\frac{\partial^2\xi}{\partial x^2} \tag{1.38}$$

Faraday's law, a Maxwell's equation

$$\nabla \times E = -\frac{\partial B}{\partial t} \tag{1.21}$$

$$-\frac{\partial\Omega_0}{\partial t} = \frac{\partial\Omega_2}{\partial l} \tag{1.19}$$

$$-\int\Omega_0 dl = \int\Omega_2 dt \tag{1.20}$$

Dynamic calculus of circular functions

$$\frac{d}{d\alpha}\cos\alpha = \cos(\alpha + \frac{\pi}{2}) \tag{1.54}$$

$$\frac{d}{d\beta}\sin\beta = \sin(\beta + \frac{\pi}{2}) \tag{1.55}$$

$$\int\cos\alpha\ d\alpha = \cos(\alpha - \frac{\pi}{2}) \tag{1.56}$$

$$\int\sin\beta\ d\beta = \sin(\beta - \frac{\pi}{2}) \tag{1.57}$$

The symmetry of space and time

$$-\frac{\partial\Omega_0}{\partial t} = v\frac{\partial\Omega_0}{\partial l}, \tag{1.63}$$

$$u\frac{\partial\Phi_1}{\partial l_3} = -\frac{\partial\Phi_1}{\partial t_1} \tag{3.37}$$

Probability density of an electron in Euclidean space

$$P = -\ln(\cos\beta) \tag{1.75}$$

Wave functions in terms of discrete linear polynomial functions

$$\cos x = 1 - \frac{x^2}{2!} + \frac{x^4}{4!} - \frac{x^6}{6!} + \ldots$$

(Q1.5)

$$\sin x = x - \frac{x^3}{3!} + \frac{x^5}{5!} - \frac{x^7}{7!} + \ldots$$

(Q1.5)

$$e^x = 1 + x + \frac{x^2}{2!} + \frac{x^3}{3!} + \frac{x^4}{4!} + \ldots \qquad (5.57)$$

Rotatory operation (rotation is marked by superscript $^{0}$)

$$\Phi^0 \equiv \int \left( -\frac{\partial \Phi}{\partial t} \right) dl \qquad (2.3)$$

$$\frac{\partial \Phi^0}{\partial l} = -\frac{\partial \Phi}{\partial t} \qquad (2.4)$$

$$\varphi_{00}^{-0} = -\frac{dA_2 \cos\psi}{dl} \int A_1 \cos \Psi dt \qquad (5.1)$$

Complementary radian angles

$$\Psi + \psi = \pi/2 \qquad (2.11)$$

$$\cos\Theta = \sin\theta \qquad (5.59)$$

Integral chain rule

$$\int A_2 \cos\psi \; dl = -\frac{1}{\psi'} \int A_2 \cos\psi \; d\psi$$

$$= -\frac{1}{\psi'} (A_2 \sin\psi) \qquad (2.13, 2.16)$$

$$= A_2 \sin\psi \cdot C_2 \int \cos\beta dl$$

Quaternity equation

$$\frac{\partial^4 \Phi}{\partial t^4} = u^4 \frac{\partial^4 \Phi}{\partial l^4} \qquad (2.7)$$

Cyclic rotations of electron octet (rotations are marked by arrows)

$$\Phi_0 \mapsto \Phi_1 \mapsto \Phi_2 \mapsto \Phi_3 \mapsto \Phi_4 \mapsto \Phi_5 \mapsto \Phi_6 \mapsto \Phi_7 \mapsto \Phi_0 \tag{2.29}$$

Dynamic calculus of spherical quantity in trigonometry

$$-\frac{d\Psi}{dt} = -C_1 \cos\alpha \tag{2.30}$$

Dimension encapsulation

$$-\frac{\partial \Phi_0}{\partial t} = -A_1 \cos\Psi \tag{2.33}$$

Differential chain rule

$$\begin{aligned} -\frac{\partial \Phi_0}{\partial t} &= -A_1 \cos\Psi, \Psi \mapsto \Psi + \frac{\pi}{2} \\ &= -A_1 \dot{\Psi} \sin\Psi \\ &= -A_1 \sin\Psi \cdot (-C_1 \cos\alpha). \end{aligned} \tag{2.34}$$

The relation of probability density and flux

$$\frac{\partial P(x,t)}{\partial t} = -\frac{\partial J(x,t)}{\partial x}$$

(2.38)

Spherical quadrants in vector calculus

$$\begin{cases} \int \Phi_0 dl_3 = \int_R \Phi_{0x} dx, \\ \int \Phi_1 dl_2 = \int_C \Phi_{1x} dx + \Phi_{1y} dy, \\ \int \Phi_2 dl_1 = \iint_H \Phi_{2x} dydz + \Phi_{2y} dzdx + \Phi_{2z} dxdy, \\ \int \Phi_3 dl_0 = \iiint_V (\frac{\partial \Phi_{3x}}{\partial x} + \frac{\partial \Phi_{3y}}{\partial y} + \frac{\partial \Phi_{3z}}{\partial z}) dxdydz. \end{cases} \tag{3.38}$$

$$\begin{cases} \frac{\partial \Phi_1}{\partial l_3} = \frac{\partial \Phi_{1x}}{\partial x}, \\ \frac{\partial \Phi_2}{\partial l_2} = \frac{\partial \Phi_{2y}}{\partial x} - \frac{\partial \Phi_{2x}}{\partial y}, \\ \frac{\partial \Phi_3}{\partial l_1} = (\frac{\partial \Phi_{3z}}{\partial y} - \frac{\partial \Phi_{3y}}{\partial z})\mathbf{i} + (\frac{\partial \Phi_{3x}}{\partial z} - \frac{\partial \Phi_{3z}}{\partial x})\mathbf{j} + (\frac{\partial \Phi_{3y}}{\partial x} - \frac{\partial \Phi_{3x}}{\partial y})\mathbf{k}, \\ \frac{\partial \Phi_4}{\partial l_0} = \frac{\partial \Phi_4}{\partial x}\mathbf{i} + \frac{\partial \Phi_4}{\partial y}\mathbf{j} + \frac{\partial \Phi_4}{\partial z}\mathbf{k}. \end{cases} \tag{3.39}$$

Orbital synchronizations

Green's theorem: $u\int \Phi_0 dl_3 = \int \Phi_1 dl_2$ (3.32)

Stokes's theorem: $u\int \Phi_1 dl_2 = \int \Phi_2 dl_1$ (3.34)

Gauss's theorem: $u\int \Phi_2 dl_1 = \int \Phi_3 dl_0$ (3.36)

Partial clockwise rotation effect

$$A_p = P\sqrt{1-\left(\frac{d}{c}\right)^2} \qquad (4.6)$$

Dimension factors as operators

$$\omega\Omega = -\frac{\partial\Omega}{\partial t} \qquad (5.13)$$

$$r\Omega = \int \Omega dl \qquad (5.15)$$

$$\eta_i = q^i \eta_0, i = 1,2,...,7 \qquad (5.103)$$

Electronic rope

$$\Omega_0{}^0 = \Omega_2\,;\ \Omega_2{}^0 = \Omega_0 \qquad (5.30)$$

$$-\frac{\partial p}{\partial t} = \frac{\partial E}{\partial l} \qquad (5.31)$$

Circular complex functions as exponential functions

$$\Psi = C_1 e^{-\omega\alpha} \qquad (5.48)$$

$$\psi = C_2 e^{r\beta} \qquad (5.49)$$

$$\Theta = A_1 e^{-\dot{\Psi}\Psi}, \theta = A_2 e^{-\frac{1}{\psi'}\psi} \qquad (5.54)$$

Orbital complement in set theory

$$\Omega_0 \cup \overline{\Omega}_0 = \Omega,\ \Omega_0 \cap \overline{\Omega}_0 = \emptyset \qquad (5.77)$$

$$\Omega_2 \cup \overline{\Omega}_2 = \Omega,\ \Omega_2 \cap \overline{\Omega}_2 = \emptyset \tag{5.78}$$

Stepwise *LC* oscillatory circuit of DNA

$$L\frac{d^2 I}{dt^2} + R\frac{dI}{dt} + \frac{I}{C} = 0 \tag{7.6}$$

Human sex identity

$$M = \cos\Psi\cos\psi - \dot{\Psi}\sin\Psi\cos\psi\ , \tag{8.2}$$

$$W = \cos\psi\cos\Psi - \psi'\sin\psi\cos\Psi\ , \tag{8.3}$$

The probability density of biological variables

$$f(x) = A\sin(ax + b) \tag{8.5}$$

The Poisson distribution corollary

$$\frac{\partial N}{\partial x} = -N - \frac{\partial M}{\partial t} \tag{8.9}$$

The eccentricity of an ellipse

$$\cos\theta = \sin\delta \tag{9.1}$$

Pythagorean theorem

$$(C_1\cos\alpha)^2 + (-C_1\omega\sin\alpha)^2 = C_1^2 \tag{9.2}$$

$$(C_2\cos\beta)^2 + \left|(C_2 r\sin\beta)^2\right| = C_2^2 \tag{9.3}$$

$$\cos\alpha = \sin\beta \tag{9.4}$$

Central force

$$F = L\frac{C_1 C_2}{r^2}(1 - \frac{2^2 L^2}{2! r^2} + \frac{2^4 L^4}{4! r^4} - \frac{2^6 L^6}{6! r^6} + \ldots) \tag{9.13}$$

# INDEX

## A



## B



## C

## D

## E



## F

## R



## S

## T



## U



## V

## W



## X



## Y



## Z